Chemistry of Non-stoichiometric Compounds

Chemistry of Non-stoichiometric Compounds

Koji Kosuge

Professor of Solid State Chemistry, Kyoto University

OXFORD NEW YORK TOKYO
OXFORD UNIVERSITY PRESS
1994

Oxford University Press, Walton Street, Oxford OX2 6DP

Oxford New York Toronto
Delhi Bombay Calcutta Madras Karachi
Kuala Lumpur Singapore Hong Kong Tokyo
Nairobi Dar es Salaam Cape Town
Melbourne Auckland Madrid
and associated companies in
Berlin Ibadan

Oxford is a trade mark of Oxford University Press

Published in the United States
by Oxford University Press Inc., New York

This English edition has been translated from the original Japanese publication
Chemistry of Non-stoichiometric Compounds

A catalogue record for this book is available from the British Library

Library of Congress Cataloging in Publication Data
Kosuge, Kōji, 1937–
[Futeihi kagōbutsu no kagaku. English]
Chemistry of non-stoichiometric compounds / K. Kosuge.
Includes bibliographical references and index.
1. Crystals—Defects. 2. Inorganic compounds. I. Title.
QD921.K676613 1993 548′.8—dc20 93-21864

ISBN 0-19-855555-5

Typeset by Integral Typesetting, Gt. Yarmouth, Norfolk
Printed in Great Britain by Biddles Ltd, Guildford and Kings Lynn

PREFACE TO THE ENGLISH EDITION

I am very pleased that the English edition of *Chemistry of non-stoichiometric compounds* has been published.

After publication of the Japanese edition, a group of compounds, the superconducting oxides with higher values of T_c (which are closely related to non-stoichiometric compounds) has been discovered.

In the process of translation into English, I tried to introduce these compounds as examples of non-stoichiometric compounds, adding a new section. I had to decide, however, to stop my trial, because many papers on these compounds have been published from the viewpoint of the relation between the non-stoichiometry and physical properties, i.e. it is too early to adopt these compounds in a textbook.

Consequently I have simply made a few small revisions and corrected some errors that have been noticed.

I should like to express my thanks to Oxford University Press and also to the Baifukan Press for producing a book of such pleasing appearance.

Last, but by no means least, thanks are due to Miss Chiyoe Sakata, who has patiently typed the various drafts and corrections, and also to my wife, who helped practically in preparing the manuscript.

Kyoto K.K.
September 1993

PREFACE

The term 'stoichiometry' is often used and is well understood in Chemistry, and the law of definite proportions and the law of multiple proportion are well-known examples deduced from the stoichiometric relation. The existence of non-stoichiometric compounds cannot be explained by a simple interpretation of the above mentioned laws, however, it is no exaggeration that all inorganic compounds exhibit non-stoichiometry.

In recent years inorganic compounds have increasingly been put to practical use, mainly in electric, magnetic, and optical devices. An understanding of the chemical and physical properties of inorganic compounds is indispensable to the progress of materials science and ceramics. Because all inorganic compounds show non-stoichiometry, it is important to understand the nature of non-stoichiometry and the relation between defect structures (i.e. non-stoichiometry) and the properties that they show.

The author believes that a course on non-stoichiometric compounds is an important part of Inorganic Chemistry or Solid State Chemistry and this is what motivated him to write this book.

Although this book has primarily been written as a textbook for university students, it is hoped that it may be useful to research workers engaged in the development of new materials. Chapters 1 and 2 are based on lectures given to undergraduate and graduate students in Kyoto University as part of a course on Inorganic Chemistry; Chapter 3 was written specifically for the book.

Chapter 1 deals with classical non-stoichiometric compounds. By classical, the author means that the basic concept of the phase stability has been well established from a thermodynamical point of view, and does not mean that research in this field has been fully completed. In these compounds the origin of non-stoichiometry is 'point defects'. In the first half of the chapter, the fundamental relation between point defects and non-stoichiometry is described in detail, based on (statistical) thermodynamics, and in the second half various examples, referred to the original papers, are shown.

Chapter 2 describes non-stoichiometric compounds derived from 'extended defects'. Research in this field showed significant growth towards the end of the 1960s (although important papers by Professor A. Magnéli and Professor A. D. Wadsley were published in the 1950s), initiated by the proposition of the 'shear structure' model by Professor J. S. Anderson and

Professor B. G. Hyde. Since then the resolution of electron microscopy has been much developed, which has enabled observation of a variety of extended defects and accelerated research in this field. This chapter also serves as an introductory text to modern 'Structural Inorganic Chemistry'.

In Chapter 3, four examples of non-stoichiometric compounds used as practical materials are described from a chemical point of view. The sections on ionic conducting materials and hydrogen-absorbing alloys concentrate on how to utilize the characteristic properties of these compounds, in relation to their non-stoichiometry. In the section on magnetic and electrical materials, methods of sample preparation, focusing on the control of non-stoichiometry, and the relation between non-stoichiometry and the properties of the compounds are presented.

The compounds discussed are mainly metal oxides, metal sulfides, and metal hydrides. Non-stoichiometry of alloys, which can be interpreted using the electron theory, is not treated. The author hopes that readers will inform him if they find any part of this book misleading or erroneous.

Acknowledgement

I would like to gratefully acknowledge the late Professor S. Kachi, who introduced me to Solid State Chemistry, and taught and encouraged me over twenty-five years. I would like also to acknowledge the helpful discussions I have had with Professor Y. Bando on the basic problems in Solid State Chemistry. I started to write this book in 1980, when I stayed as a research fellow at the Australian National University. It is a pleasure to express my thanks to Professor B. G. Hyde for all that he did to make my visit a most stimulating and memorable experience, and also for his seminar on modern crystal chemistry, which was very helpful in writing Chapter 2. It is also a pleasure to acknowledge here in part the wide assistance I have received. I am indebted to Dr T. Tanaka, Dr H. Okinaka, and Mr T. Iwasaki for collecting basic material for Chapter 3. To my co-workers Professors Y. Oka, Y. Ueda, and N. Nakayama, I am indebted for many helpful suggestions and comments. Special thanks are due to Professor Y. Ueda, who kindly read the manuscript and suggested improvements in many aspects.

Lastly, I wish to thank the staff of Baifukan, especially Mr T. Nohara, for co-operation in bringing the book to press.

Kyoto K.K.
1985

CONTENTS

1

NON-STOICHIOMETRIC COMPOUNDS DERIVED FROM POINT DEFECTS

1.1 Introduction

In this chapter, we discuss 'classical' non-stoichiometry derived from various kinds of point defects. To derive the phase rule, which is indispensable for the understanding of non-stoichiometry, the key points of thermodynamics are reviewed, and then the relationship between the phase rule, Gibbs' free energy, and non-stoichiometry is discussed. The concentrations of point defects in thermal equilibrium for many types of defect structure are calculated by simple statistical thermodynamics. In Section 1.4 examples of non-stoichiometric compounds are shown referred to published papers.

The technical term 'non-stoichiometric compounds' has been used for a long time, in contradiction to the term 'stoichiometric compounds'. The existence of non-stoichiometric compounds, which have also been called Bertholides compounds, cannot be explained from the law of definite proportion in its simplest meaning. Proust insisted that only stoichiometric compounds (also named Daltonide compounds) existed, whereas Bertholet maintained the existence of not only stoichiometric compounds but also non-stoichiometric compounds. This is a very famous argument in the history of chemistry. In the early years of the twentieth century, Kurnakov investigated the physical and chemical properties of intermetallic compounds in detail and found that the maximum or minimum in melting point, electrical resistivity, and also in the ordering temperature of lattices does not necessarily appear at the stoichiometric composition. An important discovery of Dingman was that stoichiometric $FeO_{1.00}$ is non-existent under ordinary conditions. (At present, we can synthesize stoichiometric $FeO_{1.00}$ under high pressure.[1])

Non-stoichiometry, which originates from various kinds of lattice defect, can be derived from the phase rule. As an introduction, let us consider a trial experiment to understand non-stoichiometry (this experiment is, in principle, analogous to the one described in Section 1.4.8). Figure 1.1 shows a reaction vessel equipped with a vacuum pump, pressure gauge for oxygen gas, pressure controller for oxygen gas, thermometer, and chemical balance. The temperature of the vessel is controlled by an outer-furnace and the vessel has a special window for *in-situ* X-ray diffraction. A quantity of metal powder

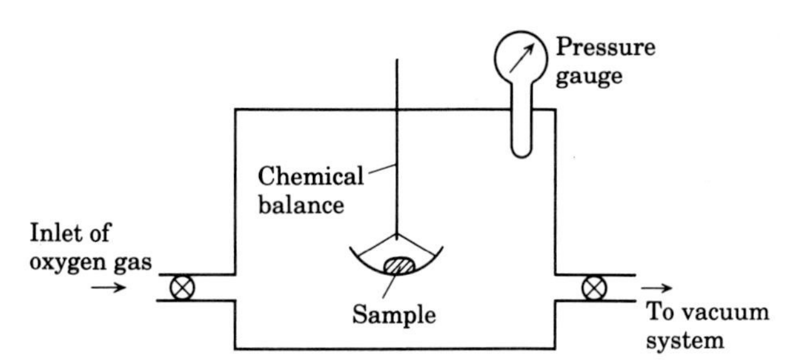

FIG. 1.1 Schematic drawing of apparatus for the determination of phase diagrams for non-stoichiometric compounds.

is placed on the chemical balance, and then the vessel is evacuated at room temperature. After the introduction of a known quantity of oxygen gas into the vessel, the system is heated to a temperature T_1. By examination of the time dependence of oxygen pressure and weight gain, chemical equilibrium can be identified. After the system has attained chemical equilibrium, the same procedure is repeated by the introduction of oxygen gas. In Fig. 1.2, the composition at equilibrium is plotted against the oxygen pressure at T_1. Observed weight gain is entirely due to the oxidation of the metal, accordingly, the chemical composition MO_x can be calculated.

After a series of experiments has been performed as a function of temperature, we can obtain the relation between the composition, equilibrium oxygen pressure, and temperature as shown in Fig. 1.2 (the curves for different temperature never cross on the P_{O_2}-composition plane). In this case, only phase (I) and (II) are assumed to be existent. Phase (I) is known as a

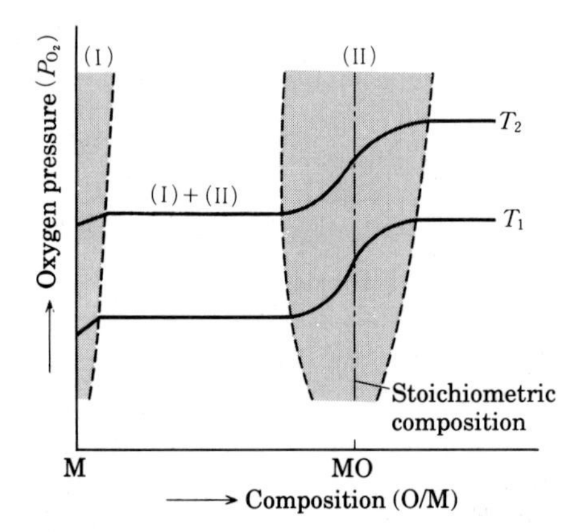

FIG. 1.2 Oxygen partial pressure (P_{O_2}) versus composition (O/M) curves at different temperatures ($T_2 > T_1$).

primary solid solution having the same structure of M as M(O), and phase (II) is a non-stoichiometric compound $M_{1-\delta}O$.

From this experiment, the following important results are obtained:

1. In the one phase region, the composition and equilibrium oxygen (partial) pressure change in one-to-one correspondence at constant temperature.

2. In the two phase region, the oxygen pressure is constant or independent of the gross composition at constant temperature.

3. The higher the temperature, the higher the oxygen pressure.

These result can always be observed for a binary system such as M–O. From the first result, the non-stoichiometric compound $M_{1-\delta}O$ can be synthesized reproducibly by controlling the oxygen pressure at a fixed temperature, by utilizing the oxygen pressure versus composition curves similar to those of Fig. 1.2. Also, a compound having the same value of non-stoichiometry, δ, can be prepared at different temperatures by choosing the oxygen pressure. It is clear that the stoichiometric compound ($\delta = 0$) can be prepared only under specific conditions (oxygen pressure and temperature).

Next, let us consider a similar experiment under constant oxygen (partial) pressure by use of a pressure controller. An amount of metal powder is put on the chemical balance, the system is evacuated, and then the oxygen pressure is controlled to be $P_{O_2}^1$ by a pressure controller. The system is heated to a temperature T_1. By oxidation of the metal, the weight of solid begins to increase which leads to a decrease of the oxygen pressure. The oxygen pressure is recovered to $P_{O_2}^1$ by the pressure controller, and finally the weight becomes constant, which indicates that the system has attained chemical equilibrium. The temperature of the system is then lowered and the experiment is repeated under a constant oxygen pressure of $P_{O_2}^1$. A series of experiments changing P_{O_2} gives us the curves of the composition versus temperature as a function of oxygen pressure P_{O_2}, as shown in Fig. 1.3. A curve of the composition–temperature in this figure corresponds to a cross-section of Fig. 1.2 at a constant oxygen pressure. In this experiment, the one phase region can be clearly distinguished from the two phase region.

In a primary solid solution M(O) (phase (I)) as appears in this system, the oxygen atoms take up positions between the lattice sites of metals, i.e. the interstitial positions. The chemical composition of phase (II) is usually expressed as $M_{1-\delta}O$. In the case of $\delta \neq 0$, the crystal has various kinds of defects. In Fig. 1.4, schematic structures corresponding to various values of δ are depicted by simple models (a more detailed description is given in Section 1.3). There are no defects for the crystal $\delta = 0$, i.e. the stoichiometric

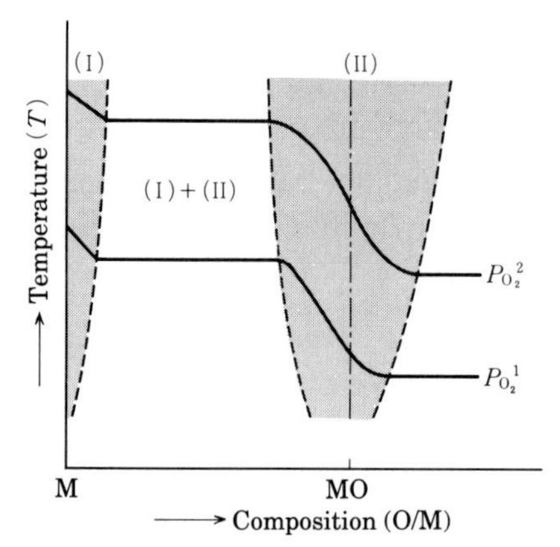

FIG. 1.3 Temperature (T) versus composition (O/M) curves at different oxygen partial pressure ($P_{O_2}^2 > P_{O_2}^1$).

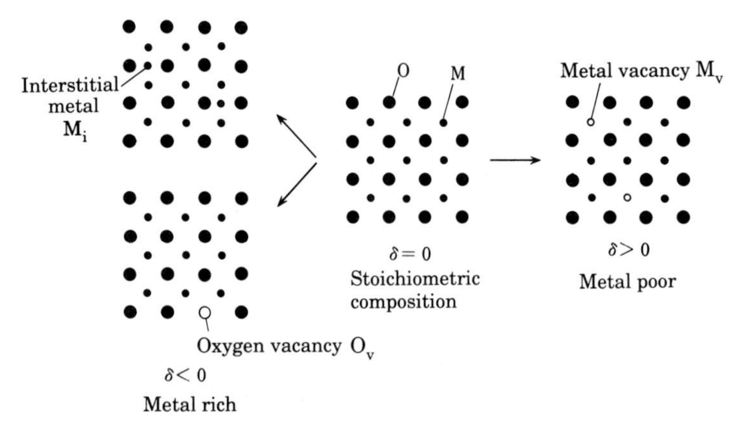

FIG. 1.4 Lattice defects in the $M_{1-\delta}O$ system.

composition.* For the case $\delta > 0$, the crystal has metal vacancies, M_v, which are distributed randomly in the metal sites (there is a possibility of interstitial oxygen, O_i). On the other hand, for the case $\delta < 0$, the crystal has excess metal atoms which are situated at interstitial positions M_i. In this case, oxygen vacancies, O_v, can usually be considered to be non-existent.

Thus, compounds with $\delta \neq 0$ show disorder of the crystal structure, and it can be easily imagined that the physical and chemical properties which

* This expression is not strictly correct, only at absolute zero is this state realized. With an increase of temperature, however, the free energy of the system decreases with increasing entropy. Therefore at higher temperatures, the crystal develops lattice defects in both the metal and oxygen sites, known as Schottky defects (see Section 1.3.2).

are structure sensitive may be dependent on the non-stoichiometry. In fact, the electrical resistivity of the semiconductor and the chemical diffusion of the compound, for example, are significantly influenced by the lattice defects, i.e. non-stoichiometry, as mentioned later.

1.2 The phase rule and non-stoichiometry

1.2.1 Partial molal quantities—review of thermodynamics

The relation between the non-stoichiometry and the equilibrium oxygen pressure mentioned in Section 1.1 can be deduced from the phase rule. For the purpose of the derivation of the phase rule, we shall review fundamental thermodynamics. Gibbs' free energy G is defined by the relation

$$G = H - TS \tag{1.1}$$

where H and S are the enthalpy and the entropy of the system, respectively, and T is the absolute temperature. The enthalpy H is defined as

$$H = E + PV \tag{1.2}$$

where E, P, and V are the internal energy, pressure, and volume of the system, respectively. The complete differential $\mathrm{d}G$ of G for the system doing work against pressure only is

$$\mathrm{d}G = V\,\mathrm{d}P - S\,\mathrm{d}T \tag{1.3}$$

which is known as the combined statement of the first and second laws of thermodynamics. Applying this equation to 1 mole of an ideal gas ($PV = RT$) at constant temperature, we have

$$\mathrm{d}G = RT\,\mathrm{d}\ln P \tag{1.4}$$

This equation is valid only for an ideal gas, and by introducing the function fugacity, f, the equation is rewritten as

$$\mathrm{d}G = RT\,\mathrm{d}\ln f \tag{1.5}$$

which can be applied to real gases. By integration we get

$$G + I = RT\ln f \tag{1.6}$$

The integration constant I is chosen so that the fugacity approaches the pressure as the pressure approaches zero. It is clear that for an ideal gas the fugacity is numerically equal to the pressure at all pressures.

Equations (1.4), (1.5), and (1.6) can be applied only to single component gas systems. Next, we shall derive the corresponding equations for the mixed gas systems. In a phase under consideration there are many gases designated by $1, 2, 3, 4, \ldots, i, \ldots$ with $n_1, n_2, n_3, n_4, \ldots, n_i, \ldots$ moles (the mixed gas system is thermodynamically considered to be one phase except for special cases). The partial pressure of gas 'i', p_i, is defined as the product of the mole fraction of that species and the total pressure (P), namely

$$p_i = \frac{n_i}{\sum_j n_j} P \tag{1.7}$$

The free energy of the system (G) increases when 1 mole of the gas 'i' is added to a very large quantity of the mixed gas phase composed of $1, 2, 3, 4, \ldots, i, \ldots$ gases. The increment of G is generally called a partial molal free energy, designated as $\bar{G}_i$ (a general description of partial molal quantity will be given in a later part of this section).

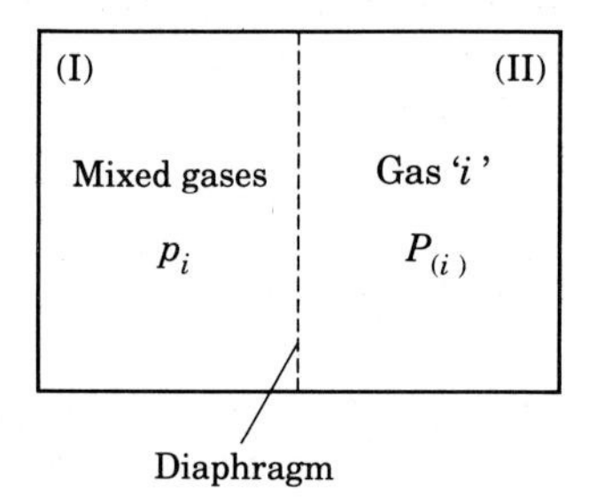

FIG. 1.5 Chamber separated by a diaphragm which is permeable to only gas 'i'.

As shown in Fig. 1.5, if two parts (named room (I) and room (II)) of a chamber containing a mixed gas are separated by a diaphragm which is permeable to component 'i' of the mixed gases but not to the others, then an equilibrium may be established, at which the partial pressure of component 'i' is the same in the two parts. If initially room (II) is evacuated, the total pressure of (II) at equilibrium is equal to the partial pressure of 'i' gas in the room (I), i.e. $P_{(i)}^{(\mathrm{II})} = p_i^{(\mathrm{I})}$. At the chemical equilbrium, the transfer of a small amount of component 'i' across the diaphragm is accompanied by no change in the free energy of the system, which leads to $\mathrm{d}G_{(i)} = \mathrm{d}\bar{G}_i$. Applying eqn (1.4) to room (II), $\mathrm{d}G_{(i)} = RT\,\mathrm{d}\ln P_{(i)}$. Thus, the following relation can be derived, assuming that the constituent gases behave ideally,

$$d\bar{G}_i = RT \, d \ln p_i \tag{1.8}$$

which corresponds to eqn (1.4) for the mixed gas system. For a non-ideal gas mixture, we get

$$d\bar{G}_i = RT \, d \ln f_i \tag{1.9}$$

where f_i is the fugacity for component 'i' in room (I). It follows immediately from eqn (1.9) that

$$\bar{G}_i + I = RT \ln f_i \tag{1.10}$$

The relation between the partial molal free energy and the fugacity for the mixed gas system has been obtained.

Before the derivation of a general expression for the partial molal quantities, a simple example of partial molal quantity is treated. Consider liquids metal A and metal B, which form a complete solution of composition A_xB_{1-x}. Let the volume of 1 g atom of liquid A and liquid B be v_A and v_B, respectively. The volume of the solution, $V(x)$, the mole fraction of which is x: $(1 - x)$, generally satisfies the following relation,

$$V(x) \neq xv_A + (1 - x)v_B \tag{1.11}$$

which means that the volume of the non-ideal solution does not show an additive property (an ideal solution can be defined by $V(x) = xv_A + (1 - x)v_B$). This is clearly due to the difference between the interactions A–A, A–B, and B–B in the solution. Partial molal quantities are defined so as to have the additive property formally.

The definition of the partial molal volume for the above-mentioned solution is as follows. When 1 g atom of liquid A is added to a very large quantity of the solution A_xB_{1-x}, the volume of the solution increases by ΔV_A. The quantity ΔV_A is called the partial molal volume of liquid A, denoted as $\bar{V}_A(x)$. The partial molal volume of liquid B, $\bar{V}_B(1 - x)$, can also be defined in a similar way. It is to be noted that the $\bar{V}_A(x)$ and $\bar{V}_B(1 - x)$ thus defined are functions of x, though the v_A and v_B in eqn (1.11) are not. If dn_A moles of A and dn_B moles of B are added to a large quantity of the solution A_xB_{1-x}, the increment of the volume dV' is given by

$$dV' = \bar{V}_A(x) \, dn_A + \bar{V}_B(1 - x) \, dn_B \tag{1.12}$$

Strictly speaking, eqn (1.12) is correct at constant temperature and pressure. Because the partial molal volume is defined as having an additive property, as mentioned above, we have

$$V(x) = x\bar{V}_A(x) + (1 - x)\bar{V}_B(1 - x)$$
$$= \frac{n_A}{n_A + n_B}\bar{V}_A(x) + \frac{n_B}{n_A + n_B}\bar{V}_B(1 - x) \quad (1.13)$$

where $V(x)$ is the volume of 1 mole of the solution A_xB_{1-x}. If $V'(x)$ is defined as

$$V'(x) = (n_A + n_B)V(x) \quad (1.14)$$

eqn (1.13) becomes

$$V'(x) = n_A\bar{V}_A + n_B\bar{V}_B \quad (1.15)$$

where $\bar{V}_A(x)$ and $\bar{V}_B(1 - x)$ is abbreviated to $\bar{V}_A$ and $\bar{V}_B$, respectively. On differentiation at constant temperature and pressure, eqn (1.15) yields

$$\mathrm{d}V'(x) = \bar{V}_A\,\mathrm{d}n_A + \bar{V}_B\,\mathrm{d}n_B + n_A\,\mathrm{d}\bar{V}_A + n_B\,\mathrm{d}\bar{V}_B \quad (1.15')$$

From the thermodynamical point of view, it is evident that V' in eqn (1.12) and $V'(x)$ in eqn (1.15) are the same variable. By comparing eqn (1.12) with eqn (1.15), it follows that

$$n_A\,\mathrm{d}\bar{V}_A + n_B\,\mathrm{d}\bar{V}_B = 0 \quad \text{or} \quad x\,\mathrm{d}\bar{V}_A + (1 - x)\,\mathrm{d}\bar{V}_B = 0 \quad (1.16)$$

Equation (1.16) is one example of the Gibbs–Duhem equation, which is one of the most useful formulae in thermodynamics. Thus, the partial molal quantity defined as the quantity satisfying the additive property is easily understandable.

We shall now derive the general equation for the partial molal quantity. Consider a system of a solution containing n_1 moles of species 1, n_2 moles of species 2, n_3 moles of species 3, . . . , n_i moles of species i. The system is divided into two parts by a wall, by this operation, it becomes clear that there are two types of thermodynamic variables: one is unchanged by the operation and the other is changed. The former are called 'intensive variables' which do not depend on the quantity of the substance: P (pressure) and T (temperature) are intensive variables. On the other hand, G (Gibbs' free energy), H (enthalpy), E (internal energy), S (entropy), and V (volume), which depend on the quantity of the substance, are called 'extensive variables'. The partial molal quantity is obviously related to the extensive variables. Here we shall use symbol W' to represent any one of the extensive variables, where the extensive quantity so primed refers to any arbitrary amount of the substance rather than to 1 mole (as in eqns (1.12) and (1.15)). Accordingly the extensive variable per mole, W, is expressed for the above mentioned example as

$$W = \frac{W'}{\sum_j n_j} \tag{1.17}$$

This equation corresponds to eqn (1.14) for the case of W' being V' (volume). The quantity W' is a function of $P, T, n_1, n_2, n_3, \ldots, n_i, \ldots$ which can be written in the form

$$W' = W'(P, T, n_1, n_2, n_3, \ldots) \tag{1.18}$$

On differentiation at constant pressure and temperature, eqn (1.18) yields

$$\mathrm{d}W' = \sum_j \left(\frac{\partial W'}{\partial n_j}\right)_{P, T, n_1, n_2, \ldots, n_{j-1}, n_{j+1}, \ldots} \mathrm{d}n_j \tag{1.19}$$

If $\mathrm{d}n_j$ moles of liquid metal j are added to a large quantity of the solution, the W' increases by $\mathrm{d}W'$ at constant pressure and temperature. Considering the above mentioned example for the partial molal volume, the partial molal quantity $\bar{W}_j$ may be defined as

$$\bar{W}_j = \left(\frac{\partial W'}{\partial n_j}\right)_{P, T, n_1, n_2, \ldots, n_{j-1}, n_{j+1}, \ldots} \tag{1.20}$$

Equation (1.19) can be rewritten by virtue of eqn (1.20)

$$\mathrm{d}W' = \sum_j \bar{W}_j \, \mathrm{d}n_j \tag{1.21}$$

which is comparable with eqn (1.12). Because the partial molal quantity $\bar{W}_j$ has an additive property, it follows immediately that

$$W = \sum_j \frac{n_j}{\sum_i n_i} \bar{W}_j$$

$$= \left(\frac{1}{\sum_i n_i}\right) \times \sum_j n_j \bar{W}_j \tag{1.22}$$

By comparing eqn (1.17) with eqn (1.22), we get

$$W' = \sum_j n_j \bar{W}_j \tag{1.23}$$

On differentiation at constant pressure and temperature, we obtain

$$\mathrm{d}W' = \sum_j n_j \,\mathrm{d}\bar{W}_j + \sum_j \bar{W}_j \,\mathrm{d}n_j \tag{1.24}$$

By comparison with eqn (1.21), it follows

$$\sum_j n_j \,\mathrm{d}\bar{W}_j = 0 \tag{1.25}$$

which is the general form of the Gibbs–Duhem equation.

Now let us return to the binary solution A–B and try to obtain the partial molal quantities $\bar{W}_j$ from the molal quantities W. For the binary system, it follows from eqn (1.22) that

$$W = N_A \bar{W}_A + N_B \bar{W}_B, \qquad N_A = \frac{n_A}{n_A + n_B}, \qquad N_B = \frac{n_B}{n_A + n_B} \tag{1.22$'$}$$

Differentiating eqn (1.22′) at constant pressure and temperature and combining eqn (1.25), $N_A \,\mathrm{d}\bar{W}_A + N_B \,\mathrm{d}\bar{W}_B = 0$, with the result, we have

$$\mathrm{d}W = \bar{W}_A \,\mathrm{d}N_A + \bar{W}_B \,\mathrm{d}N_B \tag{1.22$''$}$$

Multiplying through by $N_A/\mathrm{d}N_B$, adding this to eqn (1.22′) and then transposing all terms,

$$\bar{W}_B = W + N_A \frac{\mathrm{d}W}{\mathrm{d}N_B}$$

$$= W + (1 - N_B) \frac{\mathrm{d}W}{\mathrm{d}N_B} \tag{1.26}$$

noting that $\mathrm{d}N_A = -\mathrm{d}N_B$. An analogous equation holds for $\bar{W}_A$,

$$\bar{W}_A = W + (1 - N_A) \frac{\mathrm{d}W}{\mathrm{d}N_A} \tag{1.27}$$

By means of eqns (1.26) and (1.27), $\bar{W}_A$ and $\bar{W}_B$ can be graphically obtained from a knowledge of the composition dependence of W as shown in Fig. 1.6. Selecting any arbitrary point Z, corresponding to a definite mole fraction N_B, on the W versus composition curve and drawing the tangent XY to the curve at point Z, it can be easily shown with the aid of eqns (1.26) and (1.27) that the ordinates of the tangent at $N_B = 0$ and at $N_B = 1$ are equal to the partial molal quantities $\bar{W}_A$ and $\bar{W}_B$, respectively. This convenient intercept method is very important for the understanding of the partial molal quantity, especially in the case of W being G relating to the chemical equilibrium.

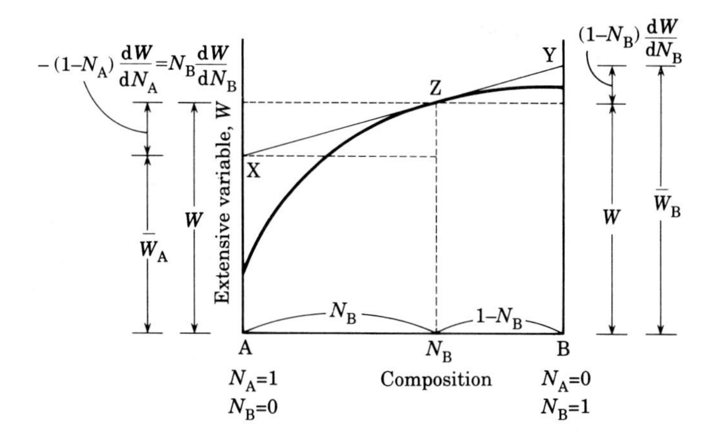

FIG. 1.6 Extensive variable (W) versus composition at fixed temperature. By drawing tangent XY to the curve at arbitrary composition N_B, the partial molal quantities $\bar{W}_A$ and $\bar{W}_B$ can be obtained.

We shall now consider the activity a or a_i, which is a very important quantity for the consideration of the chemical equilibrium. The activity is defined as the ratio of the fugacity of the substance in its present state to its fugacity in its standard state. Thus

$$a = \frac{f}{f^0}, \qquad a_i = \frac{f_i}{f_i^0} \tag{1.28}$$

where f^0 and f_i^0 are the fugacity in the standard state. (Thermodynamic variables in the standard state are marked by superscript zero.) The standard state of a substance is chosen as the pure solid or liquid form of the substance at 1 atm and at the temperature under consideration or as the gas at 1 atm at the temperature under consideration. It is noted that in many cases the activity of a substance in the standard state is unity and that the standard state changes with the experimental variation of temperature but not with the experimental variation of pressure. Combining eqn (1.28) with eqns (1.5) and (1.9), we get

$$\left.\begin{aligned} dG &= RT\, d\ln a \\ d\bar{G}_i &= RT\, d\ln a_i \end{aligned}\right\} \tag{1.29}$$

because f_i and f_i^0 are constant at constant temperature. Integrating eqn (1.29) from the standard state ($a^0 = 1, a_i^0 = 1$) to any arbitrary state at constant temperature, we obtain

$$\left.\begin{aligned} G - G^0 &= RT \ln a \\ \bar{G}_i - G_i^0 &= RT \ln a_i \end{aligned}\right\} \tag{1.30}$$

As the partial molal (Gibbs) free energy $\bar{G}_i$ is usually called the chemical potential, designated by μ_i, it follows immediately that

$$\mu_i = \mu_i^0 + RT \ln a_i \tag{1.31}$$

Now let us consider the following chemical reaction at constant pressure and temperature

$$l\text{L} + m\text{M} + \cdots \rightleftharpoons q\text{Q} + r\text{R} + \cdots \tag{1.32}$$

where the capital letters denote the chemical elements or chemical species and the small letters the number of gram atoms or moles. The free energy change of this chemical reaction can be expressed as

$$\Delta G = (q\mu_\text{Q} + r\mu_\text{R} + \cdots) - (l\mu_\text{L} + m\mu_\text{M} + \cdots) \tag{1.33}$$

In the standard state, eqn (1.33) becomes

$$\Delta G^0 = (q\mu_\text{Q}^0 + r\mu_\text{R}^0 + \cdots) - (l\mu_\text{L}^0 + m\mu_\text{M}^0 + \cdots) \tag{1.34}$$

Subtracting eqn (1.33) from (1.34)

$$\begin{aligned} \Delta G^0 - \Delta G = {} & q(\mu_\text{Q}^0 - \mu_\text{Q}) + r(\mu_\text{R}^0 - \mu_\text{R}) + \cdots - l(\mu_\text{L}^0 - \mu_\text{L}) \\ & - m(\mu_\text{M}^0 - \mu_\text{M}) - \cdots \end{aligned} \tag{1.35}$$

On substitution from eqn (1.31), we get

$$\Delta G^0 - \Delta G = -RT \ln \frac{a_\text{Q}^q a_\text{R}^r \cdots}{a_\text{L}^l a_\text{M}^m \cdots} \tag{1.36}$$

$$= -RT \ln K \tag{1.37}$$

where K is the equilibrium constant which is expressed as

$$K = \frac{a_\text{Q}^q a_\text{R}^r \cdots}{a_\text{L}^l a_\text{M}^m \cdots} \tag{1.38}$$

Equation (1.38), which is called, in a most strict sense, the law of mass action, is very useful for the consideration of chemical equilibrium. In the chemical equilibrium state $\Delta G = 0$ in eqn (1.36), we therefore get

$$\Delta G^0 = -RT \ln K \tag{1.39}$$

This equation shows that the equilibrium constant K depends only on temperature, because ΔG^0 is a function of temperature only.

1.2.2 *The phase rule and its application*

We shall derive the phase rule by use of the knowledge of thermodynamics described in Section 1.2.1. Consider a system composed of species 1, 2, 3, . . . , $i, \ldots, c$ with $n_1, n_2, n_3, \ldots, n_c$ moles, respectively.* The free energy G' of the system at constant pressure and temperature can be expressed as (see eqn (1.21))

$$\mathrm{d}G' = \sum_{j=1}^{c} \bar{G}_j \, \mathrm{d}n_j \tag{1.40}$$

or by use of the chemical potential

$$\mathrm{d}G' = \sum_{j}^{c} \mu_j \, \mathrm{d}n_j \tag{1.41}$$

The system under consideration consists of phases 1, 2, 3, . . . , p ('p' means the number of coexisted phases in the system under consideration), which are in an equilibrium state. Thus it follows that

$$\mathrm{d}G'_{(1)} = \sum_{j}^{c} \mu_j^{(1)} \, \mathrm{d}n_j^{(1)}, \mathrm{d}G'_{(2)} = \sum_{j}^{c} \mu_j^{(2)} \, \mathrm{d}n_j^{(2)}, \ldots, \mathrm{d}G'_{(p)} = \sum_{j}^{c} \mu_j^{(p)} \, \mathrm{d}n_j^{(p)} \tag{1.42}$$

Suppose that a process occurred by which $\mathrm{d}n_i^{(1)}$ moles of species i (component) were taken from phase 1 and added to phase 2. Then the free energy change, $\Delta G'$, in G' becomes

$$\begin{aligned}\Delta G' &= \mathrm{d}G'_{(1)} + \mathrm{d}G'_{(2)} \\ &= \mu_i^{(1)} \, \mathrm{d}n_i^{(1)} + \mu_i^{(2)} \, \mathrm{d}n_i^{(2)} \\ &= (\mu_i^{(1)} - \mu_i^{(2)}) \, \mathrm{d}n_i^{(1)} \quad (\because \mathrm{d}n_i^{(2)} = -\mathrm{d}n_i^{(1)})\end{aligned} \tag{1.43}$$

This process can be regarded as a chemical reaction in a system at equilibrium, which guarantees the condition $\Delta G' = 0$. Then we have

$$\mu_i^{(1)} = \mu_i^{(2)} \tag{1.44}$$

In a similar manner it can be shown that similar relations apply to all components in all phases. These relations are expressed by a set of equations

* Hereafter, 'c' means the number of components of the system for the case of the phase rule. The number of components is not necessarily equal to the number of kinds of constituent elements of the system. For example, in the chemical equilibrium $CaCO_3 \rightleftharpoons CaO + CO_2$ the number of components is two (CaO and CO_2) rather than three, because the composition of all phases can be expressed in terms of CaO and CO_2.

such as:

$$\left.\begin{aligned} \mu_1^{(1)} &= \mu_1^{(2)} = \cdots = \mu_1^{(p)} \\ \mu_2^{(1)} &= \mu_2^{(2)} = \cdots = \mu_2^{(p)} \\ &\vdots \\ \mu_c^{(1)} &= \mu_c^{(2)} = \cdots = \mu_c^{(p)} \end{aligned}\right\} \tag{1.45}$$

The state of a system is specified if we specify the amounts of each component (in mole fraction) in each phase, the pressure, and the temperature. If all mole fractions but one are specified in each phase, the remainder can be calculated, because the sum of the mole fractions equals unity. Thus there are $(c - 1)$ variables for the specification of the composition in each phase. The total number of variables of the system is, therefore, equal to $p \times (c - 1) + 2$, including the freedoms of pressure and temperature. The number of degrees of freedom of the system (denoted F) is equal to the number of possible variables (it is possible to take the chemical potentials as variables, instead of the mole fractions) required to specify the system, $p \times (c - 1) + 2$, minus the number of variables not being independently changed or the number of restrictions, $(p - 1) \times c$ (see eqn (1.45)). Therefore

$$\begin{aligned} F &= p(c - 1) + 2 - (p - 1)c \\ &= c + 2 - p \end{aligned} \tag{1.46}$$

This relation is known as the phase rule of Gibbs or simply the phase rule.

As a simple application of the phase rule, let us consider the M–O system described in Section 1.1, in which the number of components, c, is equal to 2 (M and O). The degree of freedom F is equal to $4 - p$. In the single phase region of the solid, i.e. phase (I) or (II), $F = 2$ because the solid phase always coexists with the gas phase (O_2) under the experimental conditions, i.e. $p = 2$. The experiment shown in Fig. 1.2 was carried out at constant temperature, and so the number of degrees of freedom, or the variance, is 1, indicating that we may arbitrarily fix either the composition (the mole ratio of M/O) or the pressure (P_{O_2}).* For example, if we arbitrarily fix the composition, the pressure is thereby uniquely determined: in the solid one phase region the composition changes with the oxygen pressure at constant temperature.

In the two phase region of the solids, i.e. phase (I) + (II), F equals 1, because the solid phases always coexist with the oxygen gas phase, i.e. $p = 3$.

* In the case of a closed system such as that described in Section 1.1, the pressure P of the system is equal to the oxygen pressure, the change of which corresponds to the change of the chemical potential of oxygen in the solid. We can control the oxygen activity of the gas phase by use of mixed gases, described in Section 1.4. In this case the pressure of the system does not equal the partial pressure of oxygen.

Table 1.1

Application of the phase rule to the binary (M–O) and ternary oxide system (M_1–M_2–O) in a closed system[a]

	M–O system	M_1–M_2–O system
Component c	2	3
Freedom F	$4 - p$	$5 - p$
	Freedom F	
Condensed one phase[b] system ($p = 2$)	2	3
Condensed two phase system ($p = 3$)	1	2
Condensed three phase system ($p = 4$)	0	1
Condensed four phase system ($p = 5$)	—	0

[a] In a closed system, the condensed phase always coexists with the oxygen gas phase, where oxygen gas pressure, P_{O_2}, equals the pressure of the system.
[b] Condensed phase: solid and liquid phase.

Under the condition of constant temperature, there is no freedom at equilibrium, which means that both the pressure (P_{O_2}) and the compositions of the coexisting solid phases are definitely determined. This is the reason why the composition* versus the oxygen pressure curves in the two phase region shows a straight line parallel to the abscissa as shown in Fig. 1.2. The result of Fig. 1.3 can be explained in a similar way. Thus, eqn (1.46) is very important in understanding the phase equilibrium. Table 1.1 summarizes the relationship between the composition, the pressure (P_{O_2}), and the temperature for the binary M–O and ternary M_1–M_2–O systems for the case of the closed system, which is derived from the phase rule.

1.2.3 *Gibbs free energy and non-stoichiometry*

The relationship between the composition, temperature, and pressure derived from the phase rule can also be obtained by considering the effect of free energy change on the composition. Suppose that the binary M–O system, similar to Figs 1.2 and 1.3, has three phases of solids {M(O), MO, and MO_3} and the free energy G (per mole) versus composition curves for each phase at temperature T_1 are given as shown in Fig. 1.7 (in this figure, x denotes the mole (atom) fraction of oxygen atoms O, for example, $x = 0.5$ corresponds to the compound MO). From eqns (1.26) and (1.27) and Fig. 1.6, the chemical

* In the two phase region of solids, the change of the overall composition corresponds to the change of the ratio of two solid phases, and the composition of each phase is fixed.

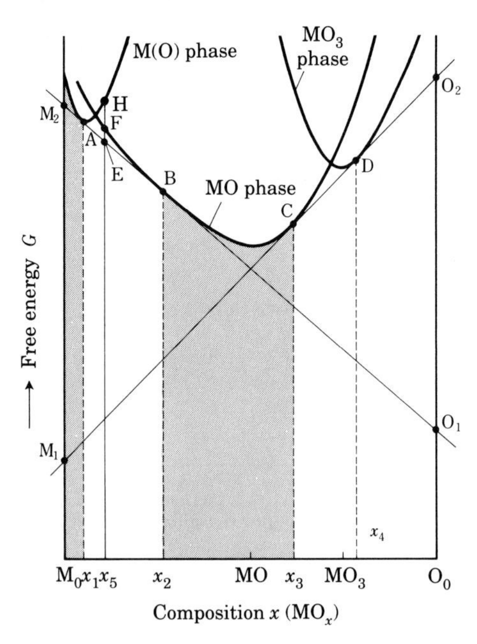

FIG. 1.7 Free energy (G) versus composition (MO_x) for the phases M(O), MO, and MO_3 at fixed temperature.

potential of M and O at any composition can be obtained as the ordinates of the tangent to the curve. For the $\{M(O) + MO\}$ phase region, the following conditions, from eqn (1.45), must be satisfied

$$\left.\begin{aligned} \mu_{M}^{M(O)} &= \mu_{M}^{MO} = \mu_{M}\ \text{gas} \\ \mu_{O}^{M(O)} &= \mu_{O}^{MO} = \tfrac{1}{2}\mu_{O_2}\ \text{gas} \end{aligned}\right\} \tag{1.47}$$

These conditions can be satisfied by drawing the common tangent to the G curves of M(O) and MO. As shown in Fig. 1.7, the chemical potentials of M and O for the M(O) phase with the composition x_1 are equal to those for the MO phase with the composition x_2, and the values correspond to $\overline{M_0M_2}$ and $\overline{O_0O_1}$, respectively. If the experimental conditions are similar to those described in Section 1.1, the solid phases must coexist with the gas phase. It may be adequate for the gas phase to be pure O_2, because the vapour pressure of other species is very low in this case. The chemical potential of O for the gas phase is equal to $\overline{O_0O_1}$, which corresponds to the oxygen pressure. Thus we can understand the coexistence of the M(O) phase with x_1 and the MO phase with x_2 from the free energy change of composition.

At the composition x_5 between x_1 and x_2, let the mixed phase $\{M(O) + MO\}$ have the mole ratio of $\alpha_{x_1}/\alpha_{x_2}$. Then we have

$$\alpha_{x_1} + \alpha_{x_2} = 1 \tag{1.48}$$

The number of oxygen atoms in the mixed phase is equal to the sum of that in the M(O) and MO phases:

$$x_1\alpha_{x_1} + x_2\alpha_{x_2} = x_5 \tag{1.49}$$

From eqns (1.48) and (1.49)

$$\left.\begin{aligned} \alpha_{x_1} &= \frac{x_2 - x_5}{x_2 - x_1} \\ \alpha_{x_2} &= \frac{x_5 - x_1}{x_2 - x_1} \end{aligned}\right\} \tag{1.50}$$

This relation is called the lever law or lever rule (Fig. 1.8), which shows

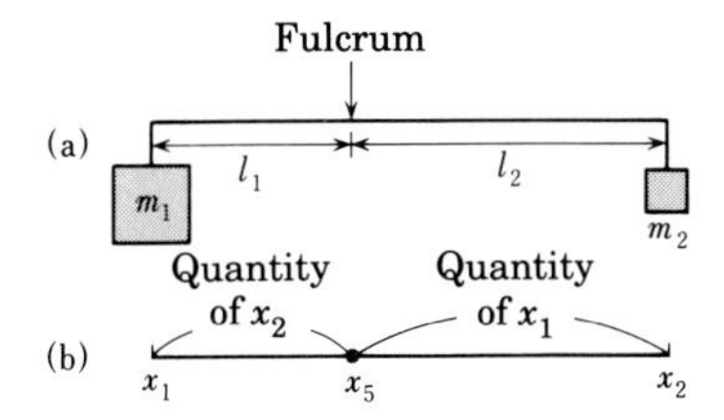

FIG. 1.8 'Lever rule' in phase diagrams. In Fig. 1.7, the composition x_5 shows two phase mixtures, the compositions of which are x_1 (M(O) phase) and x_2 (MO phase). The mixing ratio of the two phases ($\alpha_{x_1}/\alpha_{x_2}$) equals $(x_2 - x_5)/(x_5 - x_1)$, (b), which is analogous to the lever rule in classical mechanics, (a).

the ratio of mole fraction (in this case, $\alpha_{x_1}/\alpha_{x_2} = (x_2 - x_5)/(x_5 - x_1)$) in the mixed phase. (It is also possible by a similar calculation to get the ratio of mass fractions in the mixed phase.) The free energy at the gross composition x_5 (G_{x_5}), which is composed of the M(O) and MO phases, can be obtained by use of eqn (1.50) as

$$\begin{aligned} G_{x_5} &= \alpha_{x_1}G_{x_1} + \alpha_{x_2}G_{x_2} \\ &= \frac{x_2 - x_5}{x_2 - x_1}G_A + \frac{x_5 - x_1}{x_2 - x_1}G_B \\ &= G_B + \frac{x_2 - x_5}{x_2 - x_1}(G_A - G_B) \end{aligned} \tag{1.51}$$

which indicates that $G_{x_5} = \overline{x_5 E}$. This value of the free energy is smaller than that of the M(O) and MO phases having the composition MO_{x_5}, shown by points H and F on the G versus x curves of each phase in Fig. 1.7. It is to be noted that this result holds in the case of any x value being between x_1 and x_2. Thus, it is concluded that if a common tangent can be drawn to the free energy versus composition curves for adjacent phases, the free energy for the two phase mixture is lower than that of the pure phases under consideration, between the compositions corresponding to these two points of tangent. In Fig. 1.7, the slope M_1O_2 is drawn tangent to the G curves of the MO and MO_3 phases. The MO and MO_3 phases coexist in the gross composition range $x_3 < x < x_4$, and therefore the MO phase exists as a single phase in the wide composition range $x_2 < x < x_3$, i.e. a non-stoichiometric compound $M_{1-\delta}O$, where $1 - 1/x_2 < \delta < 1 - 1/x_3$ in this case. If the MO phase coexists with the gas (O_2) phase, the chemical potential of oxygen in the O_2 gas phase, also in the solid phase, changes continuously from $\overline{O_0 O_1}$ to $\overline{O_0 O_2}$.

1.3 Statistical thermodynamics of point defects and non-stoichiometry

1.3.1 Point defects of elements

As mentioned above, the non-stoichiometric compounds originate from the existence of point defects in crystals. Let us consider a crystal consisting of mono-atoms. In ideal crystals of elements, atoms occupy the lattice points regularly. In real crystals, on the other hand, various kinds of point defects can exist in thermodynamic equilibrium. First, we shall consider 'vacancies', which are empty regular lattice points. Consider a crystal composed of one element which has N atoms sited on regular lattice points and N_v vacancies, i.e. number of lattice points = $(N + N_v)$, the increment of the free energy from the ideal crystal can be calculated as

$$\begin{aligned}\Delta G &= \Delta H - T\,\Delta S \\ &= N_v \varepsilon_v - kT \ln {}_{N+N_v}C_{N_v} \qquad (1.52)\end{aligned}$$

where ε_v is the energy to create one vacancy. A vacancy is produced by removing one atom from a regular lattice point and then placing it at the surface in a regular lattice point. The increase in the enthalpy term corresponds to the creation energy of the N_v vacancies. The total entropy change is determined by the number of possible ways in which the N_v vacancies may be distributed among the $(N + N_v)$ sites. Under the condition

$(\partial \Delta G/\partial N_v) = 0$, we get a minimum of the free energy ΔG when

$$\frac{N_v}{N} = \exp -\frac{\varepsilon_v}{kT} \tag{1.53}$$

assuming $N \gg N_v$. If ε_v is 1 eV, which is reasonable for usual metals, and T is 1000 K, we find $N_v/N = 10^{-5}$. This indicates that 0.001 per cent of lattice points are vacant at thermal equilibrium.

Another type of lattice defect for elements is 'interstitial' atoms, in which an atom is transferred from a regular lattice point to an interstitial position, normally unoccupied by an atom. Consider a crystal which has N atoms sited on regular lattice points and $N_{v\to i}$* atoms sited on interstitial lattice points (the number of interstitial lattice points is N_i, which is fixed by the crystal structure under consideration), by a similar calculation, the free energy increment from the ideal crystal is expressed as

$$\Delta G = N_{v\to i}\varepsilon_i - kT \ln {}_{N+N_{v\to i}}C_{N_{v\to i}} \cdot {}_{N_i}C_{N_{v\to i}} \tag{1.54}$$

where ε_i is the energy to produce one interstitial atom. The total entropy change is determined by the number of possible ways in which the $N_{v\to i}$ vacancies are distributed among the $(N + N_{v\to i})$ lattice points and the $N_{v\to i}$ interstitials atoms are distributed among the N_i lattice points. The free energy is a minimum when the following condition is satisfied:

$$N_{v\to i} = \sqrt{NN_i} \exp -\frac{\varepsilon_i}{2kT} \tag{1.55}$$

assuming $N, N_i \gg N_{v\to i}$. For the case $N \simeq N_i$, it follows that

$$\frac{N_{v\to i}}{N} = \exp -\frac{\varepsilon_i}{2kT} \tag{1.56}$$

For the usual metals, interstitial atoms are very rare, because the energy ε_i is very large; for example, ε_i is estimated to be 3 eV for Cu, i.e. $N_i/N = 10^{-8}$ at 1000 K.

1.3.2 Point defects of compounds

From consideration of the point defects of elements mentioned in Section 1.3.1, the nature of the point defects of compounds MX, where M is metal and X is O, S, N, H (volatile elements) etc., can be easily understood. Possible point defects of the compounds MX are as follows:

---

* This notation means that the number of vacancies is equal to the number of interstitial atoms. Therefore it is assumed that an interstitial atom is produced by removing an atom from a regular lattice site, placing it at the surface in a regular lattice site, and then transferring it to an interstitial lattice point.

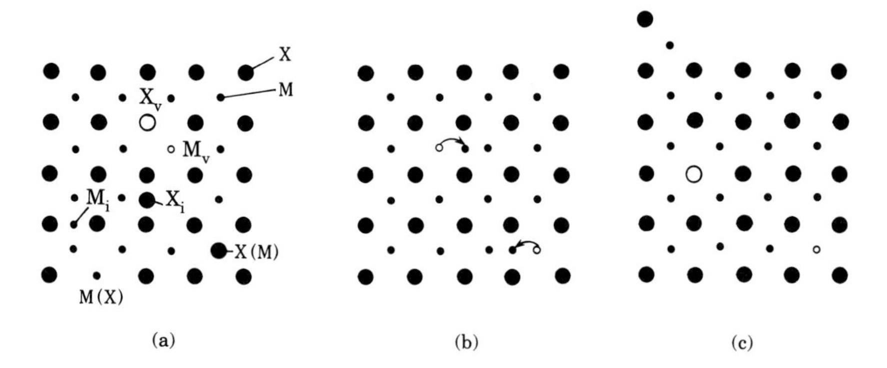

FIG. 1.9 Point defects of compounds: (a) Various point defects, (b) Frenkel type defects, (c) Schottky type defects.

M_v or X_v: metal vacancy or anion vacancy

M_i or X_i: interstitial metal atom or interstitial anion

M(X) or X(M): metal on anion site or anion on metal site (substitution)

These symbols, which will be used hereafter in this book, are after Kröger and Vink[2] (see Fig. 1.9). By combinations of these defects, a variety of defect structure types is produced. If both the combination and the concentration of point defects are adequate, the composition of a compound is stoichiometric, but this is a special case, and compounds with point defects are usually non-stoichiometric compounds. Typical examples of defect structure types are:

1. *Frenkel type* in which metal atoms on regular sites move to interstitial sites, leaving metal vacancies, i.e. ($M_v = M_i$) (Fig. 1.9(b)). Anti-Frenkel type defects, in which anion atoms on regular sites move to interstitial sites, are also possible, but are rarely observed because the ionic radii of anions are usually larger than those of the metals under consideration. Frenkel type is stoichiometric.

2. *Schottky type* in which equal numbers of M_v and X_v are created (Fig. 1.9(c)), in this case, the composition is also stoichiometric, and anti-Schottky type, in which equal numbers of metal atoms and anions on respective regular sites move to respective interstitial lattice points leaving vacancies M_v and X_v.

In ionic crystals showing semiconducting or insulating properties, other defect structures such as F centres, V centres, and Koch–Wagner type are well known, but descriptions of these defect structures are not included here.

The number of interstitial atoms N_F in the Frenkel type and the number of vacancies N_S in the Schottky type at thermal equilibrium can be obtained, following a similar calculation to that for the concentration of point defects of elements mentioned in Section 1.3.1, as

$$N_F = \sqrt{NN_i} \exp -\frac{\varepsilon_F}{2kT} \tag{1.57}$$

$$N_S = N \exp -\frac{\varepsilon_S}{2kT} \tag{1.58}$$

where ε_F and ε_S are the energy to create one defect of the Frenkel and Schottky type, respectively. As mentioned above, both Frenkel and Schottky defects only give stoichiometric compounds, and therefore we cannot discuss non-stoichiometry on the basis of these defects. Imbalance between the numbers of metal defects ($M_v + M_i + M(X)$) and anion defects ($X_v + X_i + X(M)$) is necessary for non-stoichiometry. In the following, we shall consider very simple examples of non-stoichiometric compounds.

1.3.3 Small deviation from stoichiometry. I. Metal vacancies

Let us consider the compounds which show a small deviation from the stoichiometric composition and whose non-stoichiometry is derived from metal vacancies. The free energy of these compounds, which take the composition MX in the ideal or non-defect state, can be calculated by the method proposed by Libowitz.[3,4] To readers who are well acquainted with the Fowler–Guggenheim style of statistical thermodynamics,[5] the method here adopted may not be quite satisfactory; however, the Libowitz method is understandable even to beginners who know only elementary thermodynamics and statistical mechanics. It goes without saying that the result calculated by the Libowitz method is essentially coincident with that calculated by the Fowler–Guggenheim method.

Consider a crystal in which M atoms randomly occupy N_M lattice points in N lattice points and anions occupy N_X lattice points in N lattice points. The free energy of the crystal may be written

$$G = N\mu'_{MX} + (N - N_M)\varepsilon_v^M + (N - N_X)\varepsilon_v^X - kT \ln {}_N C_{N_M} \cdot {}_N C_{N_X} \tag{1.59}$$

where μ'_{MX} is the formation energy of one molecule of MX (stoichiometric), and ε_v^M and ε_v^X are the formation energy of one metal and one anion vacancy, respectively. The first term of eqn (1.59) is the formation energy of the stoichiometric MX compound, the second and the third are the vacancy formation energy for metal and anion, respectively, from the stoichiometric compound, and the last term corresponds to the entropy. A deviation from stoichiometry (δ) is defined as

$$\delta = \frac{N_X - N_M}{N_X} \tag{1.60}$$

At chemical equilibrium, $(\partial G/\partial N)_{P,T,N_M,N_X} = 0$, hence it follows that

$$\mu'_{MX} + \varepsilon_v^M + \varepsilon_v^X - kT\left[\ln \frac{N}{N - N_M} \cdot \frac{N}{N - N_X}\right] = 0 \tag{1.61}$$

(The equation $(\partial G/\partial N)_{P,T,N_M,N_X} = 0$ gives the number of lattice points at chemical equilibrium under the condition of constant composition. Vacancies derived from this are called intrinsic defects.)

From the definition of the chemical potential, μ_X and μ_M are given:

$$\frac{\partial G}{\partial N_X} = \mu_X = -\varepsilon_v^X - kT \ln \frac{N - N_X}{N_X} \tag{1.62}$$

$$\frac{\partial G}{\partial N_M} = \mu_M = -\varepsilon_v^M - kT \ln \frac{N - N_M}{N_M} \tag{1.63}$$

On elimination of ε_v^X from eqns (1.61) and (1.62), we obtain

$$\mu_X = \mu'_{MX} + \varepsilon_v^M + kT \ln \delta \tag{1.64}$$

assuming $N = N_X$ (this assumption may be reasonable for the crystals treated in this chapter, i.e. there are no intrinsic defects). By using the relation $\mu_X = kT \ln a_X$, eqn (1.64) can be rewritten as

$$\ln a_X = \frac{\mu'_{MX} + \varepsilon_v^M}{kT} + \ln \delta \tag{1.65}$$

where a_X is the activity of anions. If the temperature dependence of a_X and δ are measured, the value of $(\mu'_{MX} + \varepsilon_v^M)$ can be obtained by plotting $\ln(a_X/\delta)$ against $1/T$. (In the usual case, the value of μ'_{MX} can be obtained by other experiments and therefore ε_v^M values are obtained from this graph.) Figure 1.10 shows the numerical calculation of eqn (1.65) in the range $0 < \delta < 0.01$,* assuming $-\mu'_{MX} = 2\varepsilon_v^M = 20$ kcal mol^{-1}. At constant temperature, a_X increases with increasing δ, and at constant composition a_X increases with increasing temperature. The free energy G can be reduced to

$$G = N\{\mu'_{MX} + \varepsilon_v^M\delta + kT[(1 - \delta)\ln(1 - \delta) + \delta \ln \delta]\} \tag{1.66}$$

where

$$N_X = N \tag{1.67}$$

* It is noted that by increasing δ to some extent the interaction energy between vacancies plays an important role in non-stoichiometric compounds, as mentioned below.

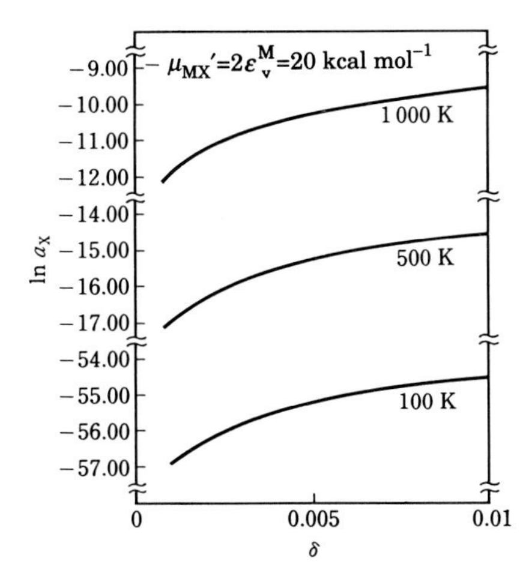

FIG. 1.10 Computed result of eqn (1.65).

This example, which may be the most simple model of non-stoichiometric compounds, is important for understanding the following examples.

1.3.4 Small deviation from stoichiometry. II. Imbalanced Frenkel defects

As noted above, the existence of anion vacancies is considered to be very rare except in special compounds such as $VO_{1+\delta}$ (see Table 1.7 below). Consider a crystal $M_{1-\delta}X$ which contains both metal vacancies and interstitial metal atoms in low concentration, i.e. M occupies N_M lattice points in N lattice sites of metal, X occupies N_X in N lattice sites (generally, $N_X \neq N$, but in this calculation we assume $N_X = N$), and, moreover, interstitial M occupies N_i^M in $N\alpha$,* where α is a constant which is fixed by crystal structure. If the conditions $N \gg (N - N_M)$, $(N - N_X)$, N_i^M are satisfied, it is not necessary to take the interaction energy between defects, as mentioned below, into consideration. The free energy of the crystal may be written as

$$G = N\mu'_{MX} + (N - N_M)\varepsilon_v^M + N_i^M\varepsilon_i^M + (N - N_X)\varepsilon_v^X - kT[\ln {}_N C_{(N-N_M)} \cdot {}_{N\alpha}C_{N_i^M} \cdot {}_N C_{N-N_X}] \quad (1.68)$$

where μ'_{MX} is the formation energy of one mole of MX, and ε_v^M, ε_i^M, and ε_v^X are the formation energies of a metal vacancy, an interstitial metal atom,

* In the case $(N - N_M) = N_i^M$, the defect is a Frenkel type, but we are now intending to calculate the free energy for the case $(N - N_M) \neq N_i^M$, we call this type of defect imbalanced Frenkel type.

and an anion vacancy, respectively. Deviation from stoichiometry, δ, is

$$\delta = \frac{N_X - (N_M + N_i^M)}{N_X} \tag{1.69}$$

because the total number of metal atoms equals $(N_M + N_i^M)$. At $\delta = 0$, i.e. $N_X = N_M + N_i^M$, the compound is stoichiometric (Frenkel defect). Hence, at chemical equilibrium, $(\partial G/\partial N)_{P,T} = 0$, it follows, after some rearrangement

$$\mu'_{MX} + \varepsilon_v^M + \varepsilon_v^X - kT\left[\ln \frac{N^2}{(N - N_M)(N - N_X)} + \alpha \ln \frac{N\alpha}{N\alpha - N_i^M}\right] = 0 \tag{1.70}$$

The chemical potentials of metal atoms, both in the regular and in the interstitial sites, and that of the anions in the regular sites are obtained from eqn (1.68) as

$$\mu_M = \frac{\partial G}{\partial N_M} = -\varepsilon_v^M + kT \ln \frac{N_M}{N - N_M} \tag{1.71}$$

$$\mu_i^M = \frac{\partial G}{\partial N_i^M} = \varepsilon_i^M + kT \ln \frac{N_i^M}{N\alpha - N_i^M} \tag{1.72}$$

$$\mu_X = \frac{\partial G}{\partial N_X} = -\varepsilon_v^X + kT \ln \frac{N_X}{N - N_X} \tag{1.73}$$

At chemical equilibrium, $\mu_M = \mu_i^M$, thus we have

$$\varepsilon_i^M + \varepsilon_v^M = kT \ln \frac{N_M}{N - N_M} \cdot \frac{N\alpha - N_i^M}{N_i^M} \tag{1.74}$$

Here let us consider the usual Frenkel type defect, i.e. $N_i^M = (N - N_M) \equiv N_F$. Assuming $N \gg N_F$, eqn (1.74) can be rewritten,

$$\begin{aligned} \varepsilon_i^M + \varepsilon_v^M &= kT \ln \frac{N - N_F}{N_F} \cdot \frac{N\alpha - N_F}{N_F} \\ &\simeq kT \ln \frac{N^2\alpha}{N_F^2} \end{aligned} \tag{1.75}$$

and then

$$N_F = N\sqrt{\alpha} \exp -\frac{\varepsilon_v^M + \varepsilon_i^M}{2kT} \tag{1.76}$$

Comparing this with eqn (1.57), it immediately follows that $N\alpha = N_i$, $\varepsilon_v^M + \varepsilon_i^M = \varepsilon_F$. An intrinsic disorder parameter c_i, as a measure of the disorder of metal atoms at the stoichiometric composition, is defined as (by use of eqn (1.76))

$$c_i = \frac{N_F}{N} = \sqrt{\alpha} \exp -\frac{\varepsilon_v^M + \varepsilon_i^M}{2kT} \tag{1.77}$$

Now let us come back to the consideration of the general case. Eliminating ε_v^X from eqns (1.70) and (1.73), we obtain

$$\mu_X = (\mu'_{MX} + \varepsilon_v^M) + kT \ln\left(\delta + \frac{N_i^M}{N}\right) \tag{1.78}$$

assuming $N_X = N$, $N \gg N_i^M$. Replacing $(\mu'_{MX} + \varepsilon_v^M)$ with $kT \ln Q$, we get from eqn (1.78)

$$\frac{a_X}{Q} = \delta + \frac{N_i^M}{N} \tag{1.79}$$

If $a_X(0)$ denotes the activity for $\delta = 0$, we obtain the relation $a_X(0) = Qc_i$ from eqn (1.79). Hence

$$\frac{a_X(\delta)}{a_X(0)} = \frac{\delta + \dfrac{N_i^M}{N}}{c_i} \tag{1.80}$$

From eqns (1.74) and (1.77), we have

$$-2kT \ln \frac{c_i}{\sqrt{\alpha}} = kT \ln \frac{\alpha N^2}{(N - N_M)\{(N - N_M) - N\delta\}}$$

assuming $N_X = N$, $N \gg N_i^M$. Then it follows that

$$\frac{\alpha}{c_i^2} = \frac{\alpha N^2}{(N - N_M)\{(N - N_M) - N\delta\}} \equiv \rho \tag{1.81}$$

Solving for $(N - N_M)$ yields

$$N - N_M = \frac{\rho N\delta \pm \sqrt{\rho^2 N^2 \delta^2 + 4\rho N^2 \alpha}}{2\rho}$$

$$= \frac{\rho N\delta + \sqrt{\rho^2 N^2 \delta^2 + 4\rho N^2 \alpha}}{2\rho} \tag{1.82}$$

It is noted that $N - N_M > 0$. Combining this with eqn (1.69), we get

$$\frac{N_i^M}{N} = \frac{\sqrt{\delta^2 + 4c_i^2}}{2} - \tfrac{1}{2}\delta \tag{1.83}$$

Substituting this equation for N_i^M from eqn (1.80), we have

$$\frac{P_{X_2}(\delta)}{P_{X_2}(0)} = 1 + \frac{\delta^2 + \delta\sqrt{\delta^2 + 4c_i^2}}{2c_i^2} \tag{1.84}$$

by virtue of $a_X \propto (P_{X_2})^{1/2}$. (It is assumed that the chemical species of the gas phase equilibrated with the solid phase is X_2. For this case, the following relations hold: $a_X^{\text{solid}} = \frac{1}{2}a_{X_2}^{\text{gas}} \propto (P_{X_2})^{1/2}$.) Equation (1.84) shows the relationship between δ, P_{X_2}, and c_i, under the condition of low concentration of metal defect.

In Fig. 1.11, the ratio of the pressure of the gas ($P_{X_2}(\delta)$) in equilibrium with the non-stoichiometric solid to that ($P_{X_2}(0)$) in equilibrium with the stoichiometric solid ($\delta = 0$) is shown as a function of various values of c_i, computed from eqn (1.84). This figure shows that compounds which have high levels of intrinsic defects can deviate from stoichiometry by small changes of relative gas pressure, and conversely a deviation from stoichiometry for compounds which have low levels of intrinsic defects necessitates large changes in the pressure. If it is assumed that the formation energy of a Frenkel defect ε_F $(=\varepsilon_v^M + \varepsilon_i^M)$ depends little on temperature, the value of c_i increases with increasing temperature, which results in a tendency to more deviation at higher temperatures.

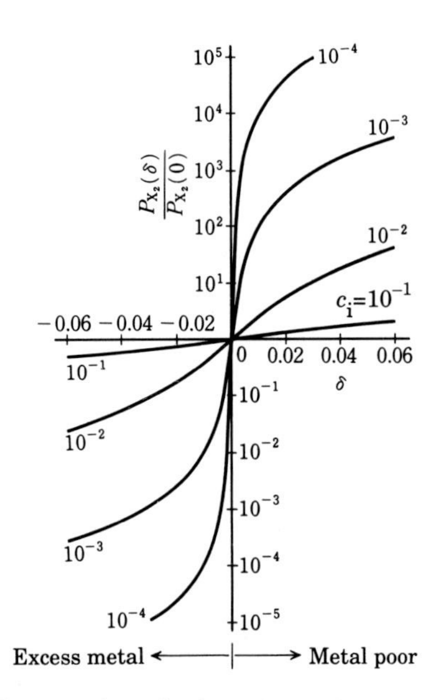

FIG. 1.11 Computed result of eqn (1.84). Oxygen partial pressure ($P_{X_2}(\delta)$) is plotted against non-stoichiometry δ at fixed c_i values.

1.3.5 Large deviation from stoichiometry. I. Random arrangement of vacancies

Let us consider a crystal similar to that discussed in Sections 1.3.3 and 1.3.4, which, in this case, shows a larger deviation from stoichiometry. It is appropriate to assume that there are no interstitial atoms in this case, because the Frenkel type defect has a tendency to decrease deviation. Consider a crystal in which M occupies N_M sites in N lattice points and X occupies N_X sites in N lattice points. It is necessary to take the vacancy–vacancy interaction energy into consideration, because the concentration of vacancies is higher. The method of calculation of free energy (enthalpy) related to ε_{vv}^{M} is shown in Fig. 1.12. The total free energy of the crystal may be written

$$G = N\mu'_{MX} + (N - N_M)\varepsilon_v^M + (N - N_X)\varepsilon_v^X + \tfrac{1}{2}Z_M \frac{(N - N_M)^2}{N} \varepsilon_{vv}^M$$

$$- kT[\ln {}_N C_{N_M} \cdot {}_N C_{N_X}] \qquad (1.85)$$

We now consider the role of the vacancy–vacancy interaction energy in the total free energy. If $\varepsilon_{vv}^{M} > 0$, this energy contributes to the increase of the free energy. The free energy gains by vacancies being isolated from one another (no vacancy–vacancy interaction), which seems that the interaction between vacancies is repulsive. The crystal has a tendency to form a vacancy ordered lattice (see Section 1.3.6). In contrast, if $\varepsilon_{vv}^{M} < 0$, this energy contributes to

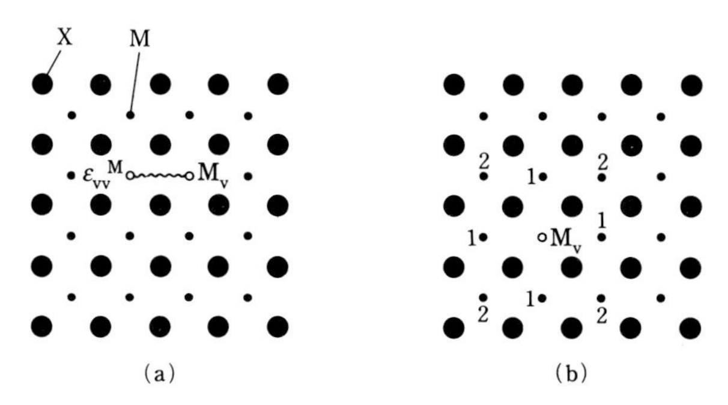

FIG. 1.12 Interaction energy, ε_{vv}^{M}, between metal vacancies and its calculation. (a) ε_{vv}^{M} denotes the interaction energy (enthalpy) between the first nearest vacancies. (b) A metal vacancy has Z_M metal sites as first nearest neighbours (labelled '1'). The probability of being a vacancy in a metal site equals $(N - N_M)/N$. The interaction energy between a metal vacancy and its first nearest neighbour vacancies is $Z_M[(N - N_M)/N]\varepsilon_{vv}^{M}$. The total interaction energy (enthalpy) for a crystal, therefore, becomes $\frac{1}{2}Z_M[(N - N_M)^2/N]\varepsilon_{vv}^{M}$.

the decrease of the free energy. The free energy gains by vacancies being near one another, which seems that the interaction between vacancies is attractive. In this case, vacancies have a tendency to cluster.

Imposing the chemical equilibrium condition $(\partial G/\partial N)_{P,T} = 0$ on eqn (1.85), we have

$$\mu'_{\mathrm{MX}} + \varepsilon_{\mathrm{v}}^{\mathrm{M}} + \varepsilon_{\mathrm{v}}^{\mathrm{X}} + \frac{Z_{\mathrm{M}}\varepsilon_{\mathrm{vv}}^{\mathrm{M}}}{2}\left\{1 - \left(\frac{N_{\mathrm{M}}}{N}\right)^2\right\} - kT\left[\ln\frac{N}{N - N_{\mathrm{M}}}\cdot\frac{N}{N - N_{\mathrm{X}}}\right] = 0 \tag{1.86}$$

The chemical potential of M and X is

$$\mu_{\mathrm{M}} = \left(\frac{\partial G}{\partial N_{\mathrm{M}}}\right) = -\varepsilon_{\mathrm{v}}^{\mathrm{M}} - Z_{\mathrm{M}}\varepsilon_{\mathrm{vv}}^{\mathrm{M}}\frac{N - N_{\mathrm{M}}}{N} - kT\ln\frac{N - N_{\mathrm{M}}}{N_{\mathrm{M}}} \tag{1.87}$$

$$\mu_{\mathrm{X}} = \left(\frac{\partial G}{\partial N_{\mathrm{X}}}\right) = -\varepsilon_{\mathrm{v}}^{\mathrm{X}} - kT\ln\frac{N - N_{\mathrm{X}}}{N_{\mathrm{X}}} \qquad)1.88)$$

If we eliminate $\varepsilon_{\mathrm{v}}^{\mathrm{X}}$ from eqns (1.86) and (1.87) and assume $N_{\mathrm{X}} = N$, we obtain

$$\mu_{\mathrm{X}} = \mu'_{\mathrm{MX}} + \varepsilon_{\mathrm{v}}^{\mathrm{M}} + \frac{Z_{\mathrm{M}}\varepsilon_{\mathrm{vv}}^{\mathrm{M}}}{2}\left[1 - \left(\frac{N_{\mathrm{M}}}{N}\right)^2\right] - kT\ln\frac{N}{N - N_{\mathrm{M}}} \tag{1.89}$$

By using the relations $\delta = (N - N_{\mathrm{M}})/N$ and $\mu_{\mathrm{X}} = kT\ln a_{\mathrm{X}}$ and putting $A = Z_{\mathrm{M}}\varepsilon_{\mathrm{vv}}^{\mathrm{M}}/2$, it follows that

$$\ln a_{\mathrm{X}} = \frac{\mu'_{\mathrm{MX}} + \varepsilon_{\mathrm{v}}^{\mathrm{M}}}{kT} + \frac{A}{kT}(2\delta - \delta^2) + \ln\delta \equiv f(\delta) \tag{1.90}$$

From eqn (1.90), we have

$$\frac{\partial f(\delta)}{\partial\delta} = f'(\delta) = (1 - \delta)\{(1/[\delta(1 - \delta)] + 2A/kT\}$$

The sign of $f'(\delta)$ is determined by the sign of the term { }, because the term $(1 - \delta)$ is always positive. The sign of $f'(\delta)$ is as follows:

Condition	Range of δ	Sign of $f'(\delta)$
$-2A/kT < 4$	$0 < \delta < 1$	+
$-2A/kT > 4$	$0 < \delta < l_1$	+
	$l_1 < \delta < l_2$	−
	$l_2 < \delta < 1$	+

where $l_1, l_2 = \frac{1}{2}[1 \mp \sqrt{1 + 2kT/A}]$.

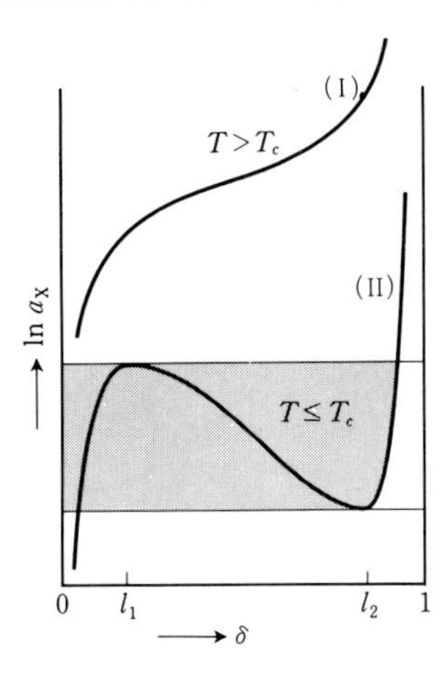

FIG. 1.13 $\ln a_X$ versus δ curves (eqn (1.90)).

A critical temperature T_c is defined by $-2A/kT_c = 4$. Hence we get

$$T_c = -\frac{A}{2k} = -\frac{Z_M \varepsilon_{vv}^M}{4k} \tag{1.91}$$

For the existence of T_c, $\varepsilon_{vv}^M < 0$ is a necessary condition, i.e. the vacancy–vacancy interaction energy is attractive. In Fig. 1.13 the δ dependency of $f(\delta)$ is schematically shown. If $T > T_c$, $f(\delta)$ increases monotonously with increasing δ. If $T < T_c$, on the other hand, there seems to be three solid phases coexisting in the shaded region of a_X, which is against the phase rule. Let us solve eqns (1.85) and (1.90) numerically making the following assumptions for simplicity, to make the situation clear.

$$\left.\begin{aligned} &\varepsilon_v^M = 2p\varepsilon_{vv}^M \qquad (p < 0)\\ &\mu'_{MX} + \varepsilon_v^M = \varepsilon_{vv}^M\\ &Z_M = 4\\ &\frac{T}{T_c} = \alpha \end{aligned}\right\} \tag{1.92}$$

The sign of the parameter p has to be minus, because we are intending to solve the equations in the condition $\varepsilon_{vv}^M < 0$. The parameter α corresponds to a reduced temperature.

From eqn (1.90), we get

$$\ln a_X = -\frac{1}{\alpha} + \left(-\frac{2}{\alpha}\right)(2\delta - \delta^2) + \ln \delta \tag{1.93}$$

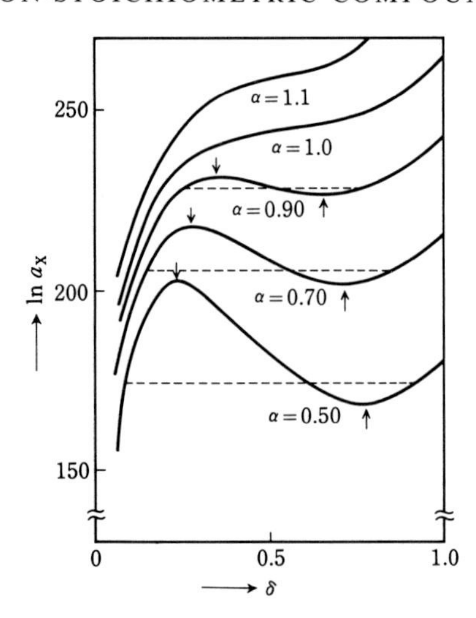

FIG. 1.14 Computed results of eqn (1.93) (the relation between $\ln a_X$, δ, and α) adopting the assumptions in eqn (1.92). Arrows on the curves show the minimum or maximum points.

From eqn (1.85), we obtain

$$G(\delta, \alpha) \equiv -\frac{G - N\mu'_{MX}}{N\varepsilon^{M}_{vv}} = -[2p\delta + 2\delta^2 - \alpha\{(1-\delta)\ln(1-\delta) + \delta \ln \delta\}] \tag{1.94}$$

if $N_X = N$ (no anion vacancy). It is apparent that the function $G(\delta, \alpha)$ can be used as a substitute for the free energy G. As mentioned above, there is no problem for the case $\varepsilon^M_{vv} > 0$, because the a_X versus δ curve shows similar behaviour to curve (I) depicted in Fig. 1.13, therefore, we discuss the case $\varepsilon^M_{vv} < 0$ ($p = -1$) here. Numerical calculations for eqns (1.93) and (1.94) are shown in Figs 1.14 and 1.15, respectively. In the temperature region $\alpha < 1$ ($T < T_c$), $\ln a_X$ versus δ curves have a maximum at lower δ and a minimum at higher δ, which just corresponds to curve (II) in Fig. 1.13. The value of $\ln a_X$ increases with increasing temperature. On the other hand, $G(\delta, \alpha)$ versus δ curves have two minima at δ_1 and δ_2 ($\delta_1 < \delta_2$) and one maximum in the temperature region $\alpha < 1$, and the value of $G(\delta, \alpha)$ decreases with increasing temperature. The function $G(\delta, \alpha)$ happens to be symmetrical about $\delta = \frac{1}{2}$ for $p = -1$, and therefore the value of $G(\delta_1, \alpha)$ equals that of $G(\delta_2, \alpha)$, where $\frac{1}{2} - \delta_1 = -\frac{1}{2} + \delta_2$. Thus the line drawn tangent to the curve $G(\delta, \alpha)$ at δ_1 overlaps that at δ_2, that is, there is a common tangent at δ_1 and δ_2. (In this case, it happens that there is a common tangent at δ_1 and δ_2, where the function $G(\delta, \alpha)$ has minima. In general, there is no correlation between the

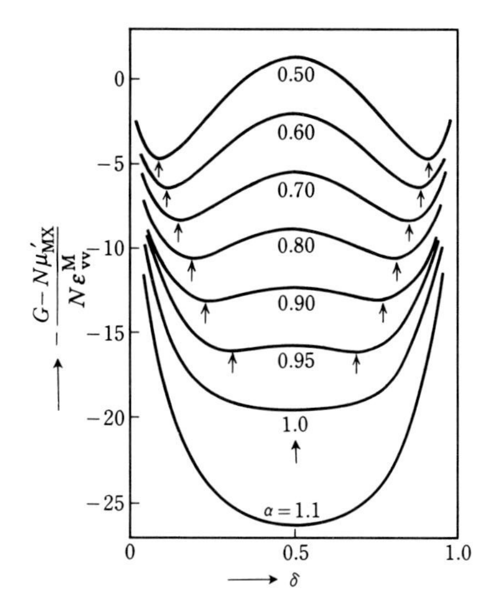

FIG. 1.15 Computed results of eqn (1.94) (the relation between G, δ, and α). Arrows show the minimum points of free energy.

δ values of the common tangent and the minimum of $G(\delta, \alpha)$.) This means that in the composition range $\delta_1 < \delta < \delta_2$ two phases having the composition δ_1 and δ_2 coexist. It can be deduced from the phase rule for the binary system M–X that the activity a_X is constant in the two phase region ($\delta_1 < \delta < \delta_2$) at constant temperature and pressure.

Figure 1.16 shows the relation between the free energy $G(\delta, \alpha)$ and the activity at temperature $\alpha = 0.85$, both of which are normalized at $\delta = 0.5$. $G(\delta, 0.85)$ has two minima at $\delta_1 = 0.185$ and $\delta_2 = 0.815$ and the line PQ is a common tangent to this curve. Therefore two phases coexist in the composition range $0.185 < \delta < 0.815$. In the $\ln a_X$ versus δ curve, the curve ABCDEFG is from numerical calculation of eqn (1.93), but the observable change has to be ABDFG, i.e. a_X is constant (B → D → F) in the two phase region.

To examine the phase relation of this system deduced from eqn (1.94), let us consider the condition $\partial G(\delta, \alpha)/\partial \delta = 0$:

$$\ln \frac{\delta}{1-\delta} = 2\alpha^{-1}(2\delta - 1) \equiv \mathrm{Y} \tag{1.95}$$

The values of δ which satisfy eqn (1.95) correspond to those giving a maximum or minimum of $G(\delta, \alpha)$. The equation can be solved by a graphical method as shown in Fig. 1.17(a). The solution is obtained as the intersection of the two curves, $\ln \delta/1 - \delta$ and $2\alpha^{-1}(2\delta - 1)$. It is to be noted that the

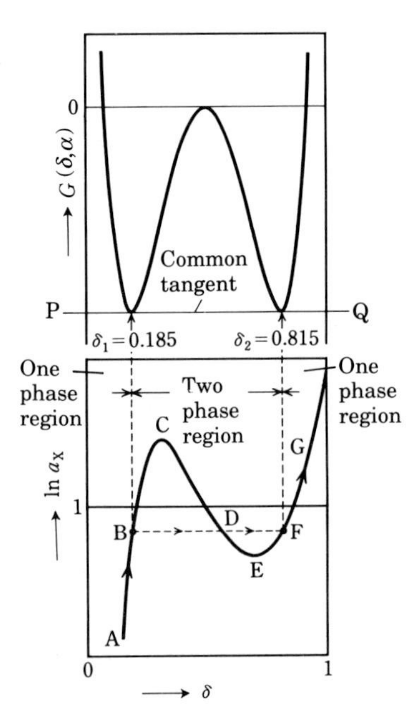

FIG. 1.16 Schematic drawing of the relationship between $G(\delta, \alpha)$, $\ln a_X$, and δ for $\alpha = 0.85$. The line PQ is a common tangent to the curve $G(\delta, \alpha)$, which results in the two phase mixture in the composition region $0.185 < \delta < 0.815$. Accordingly, $\ln a_X$ changes with δ along A → B → D → F → G.

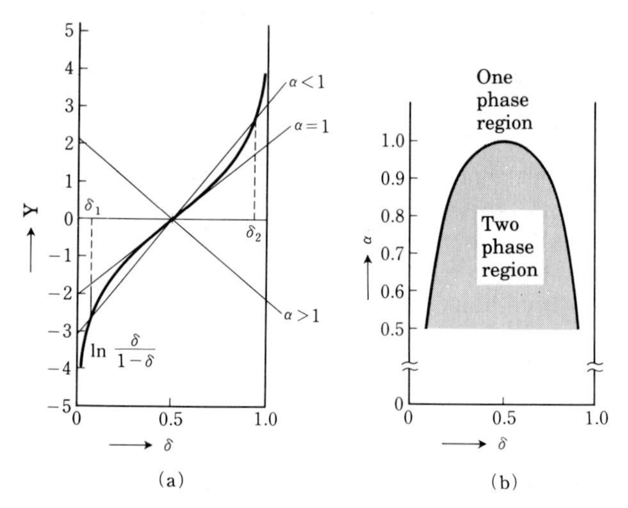

FIG. 1.17 (a) A graphic solution of eqn (1.95), (b) the obtained phase diagram (phase separation type).

tangent of the curve $2\alpha^{-1}(2\delta - 1)$ continuously changes with the change of temperature. If $\alpha > 1$, the two curves intersect only at $\delta = 0.5$, corresponding to a minimum of $G(\delta, \alpha)$. If $\alpha < 1$, the two curves intersect at δ_1, δ_2, and 0.5, where $\delta_1 = 1 - \delta_2$. The former two values correspond to minima and the latter one to a maximum of $G(\delta, \alpha)$. As is clearly shown, $G(\delta_1, \alpha) = G(\delta_2, \alpha)$, that is, there is a common tangent at $G(\delta_1, \alpha)$ and $G(\delta_2, \alpha)$. Therefore the two phases coexist in the composition range $\delta_1 < \delta < \delta_2$, where the activity is constant. The temperature dependence of δ_1 and δ_2 is shown in Fig. 1.17(b). In the hatched region two phases coexist.

In Fig. 1.14, the dotted lines for each curve show the activity of the coexisting phases at chemical equilibrium. Similarly in Fig. 1.16 the dotted line BDF shows the activity of the coexisting phases ($\delta = 0.185$ and 0.815). The coexisting phases, which have the same structure, differ in the concentration of vacancies. This phenomenon is generally called phase separation or spinodal decomposition (it is observed not only in the solid phases but also in the liquid phases), and originates from the sign of the interaction energy $\varepsilon_{\mathrm{vv}}^{\mathrm{M}}$ being minus in this case.

1.3.6 Large deviation from stoichiometry. II. Ordered structure

In Section 1.3.5, a simple model structure was discussed on the assumption that vacancies are randomly distributed, although the interaction energy between vacancies is taken into consideration. This model is plausible at higher temperatures because the entropy term contributes much to the total free energy at higher temperatures, that is, the random state must stabilize the system under consideration. On lowering the temperature the contribution of the entropy term to the free energy must decrease, which results in the clustering and ordering of vacancies. An example of the former case is described as spinodal decomposition, see Section 1.3.5. Here let us consider the latter case.

Consider a crystal of composition $\mathrm{M_{0.5}X}$, the metal vacancies are regularly arranged among the lattice sites at lower temperatures, shown in Fig. 1.18 as a basic model of a vacancy-ordered structure with a two-dimensional lattice (in this figure, the anion atoms are omitted for clarity). This structure is realized if the composition of the crystal is $\mathrm{M_{0.5}X}$, and metal atoms M fully occupy the B-sites and metal vacancies fully occupy the A-sites, this only occurs at absolute zero temperature (perfect order). The occupation probabilities, p_{A} and p_{B}, denote the ratio of number of metal atoms on the A-sites (n_{A}) to the number of lattice points of the A-sites ($\frac{1}{2}N$) and the ratio of number of metal atoms on the B-sites (n_{B}) to the number of lattice points of the B-sites ($\frac{1}{2}N$), respectively. Thus p_{A} and p_{B} can be expressed as

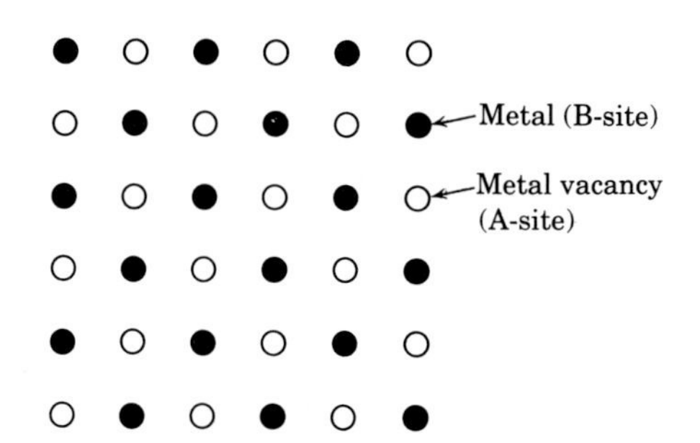

FIG. 1.18 Vacancy-ordered structure of $M_{0.5}X$. Anions X are omitted.

$$\left. \begin{aligned} p_A &= \frac{n_A}{\frac{1}{2}N} \\ p_B &= \frac{n_B}{\frac{1}{2}N} \end{aligned} \right\} \tag{1.96}$$

The crystal having the composition $M_{0.5}X$, which is the stoichiometric composition of the vacancy-ordered structure, shows disordering such as p_A and $p_B \to \frac{1}{2}$, with increasing temperature. With decreasing number of vacancies in the composition $M_{1-\delta}X$, p_A increases towards 1 at absolute zero temperature. (In this example δ ranges from 0 to 0.5 in the expression $M_{1-\delta}X$.)

From the definition of p_A and p_B, we obtain the following relationship;

$$\begin{aligned} &\text{vacancy-ordered structure: } p_B > p_A \\ &\text{vacancy-disordered structure: } p_A = p_B \end{aligned} \tag{1.97}$$

Thus we can observe the order–disorder transition by changing temperature at constant δ and also changing δ at constant temperature.

Using a method similar to that described above, the total free energy G can be expressed as follows:

$$\begin{aligned} G = N\mu'_{MX} + \{N - (n_A + n_B)\}\varepsilon_v^M + (N - N_X)\varepsilon_v^X + H_{vv}^M \\ - kT \ln {}_{N/2}C_{n_A} \cdot {}_{N/2}C_{n_B} \cdot {}_N C_{N_X} \end{aligned} \tag{1.98}$$

where H_{vv}^M is the interaction energy between vacancies. (In principle the interaction energy between metal and vacancy ε_{mv}^M has to be taken into consideration. It was found,[6] however, that such a term does not affect the essence of the discussion.) This energy, which depends only on the distance

Table 1.2
Structure of vacancy-ordered $M_{0.5}X$ (see Fig. 1.18)

Metal site	Number of metal sites	Nearest neighbour metal site (number)		
		First	Second	Third
A	$\frac{1}{2}N$	B (4)	A (4)	A (4)
B	$\frac{1}{2}N$	A (4)	B (4)	B (4)

between vacancies, can be given by

$$\begin{aligned} H_{vv}^{M} &= \tfrac{1}{2}(1-p_A)\tfrac{1}{2}N\{4(1-p_B)\varepsilon_{vv}^{M_1} + 4(1-p_A)\varepsilon_{vv}^{M_2} + 4(1-p_A)\varepsilon_{vv}^{M_3}\} \\ &\quad + \tfrac{1}{2}(1-p_B)\tfrac{1}{2}N\{4(1-p_A)\varepsilon_{vv}^{M_1} + 4(1-p_B)\varepsilon_{vv}^{M_2} + 4(1-p_B)\varepsilon_{vv}^{M_3}\} \\ &= N\{(1-p_A)^2(\varepsilon_{vv}^{M_2} + \varepsilon_{vv}^{M_3}) + 2(1-p_A)(1-p_B)\varepsilon_{vv}^{M_1} \\ &\quad + (1-p_B)^2(\varepsilon_{vv}^{M_2} + \varepsilon_{vv}^{M_3})\} \end{aligned} \tag{1.99}$$

where $\varepsilon_{vv}^{M_1}$, $\varepsilon_{vv}^{M_2}$, and $\varepsilon_{vv}^{M_3}$ are the interaction energy between the first, second, and third nearest neighbour vacancies (see Fig. 1.18 and Table 1.2). From eqns (1.98) and (1.99), after rearrangement, we have

$$\begin{aligned} G &= N\mu'_{MX} + N\{1 - \tfrac{1}{2}(p_A + p_B)\}\varepsilon_v^M + (N - N_X)\varepsilon_v^X \\ &\quad + N\{[(1-p_A)^2 + (1-p_B)^2](\varepsilon_{vv}^{M_2} + \varepsilon_{vv}^{M_3}) + 2(1-p_A)(1-p_B)\varepsilon_{vv}^{M_1}\} \\ &\quad + \frac{N}{2}kT\left\{(1-p_A)\ln(1-p_A) + p_A\ln p_A + (1-p_B)\ln(1-p_B)\right. \\ &\qquad \left. + p_B\ln p_B + 2\ln\frac{N-N_X}{N} + 2\frac{N_X}{N}\ln\frac{N_X}{N-N_X}\right\} \end{aligned} \tag{1.100}$$

From the condition $(\partial G/\partial N)_{P,T} = 0$, it follows that

$$\begin{aligned} &\mu'_{MX} + \varepsilon_v^M + \varepsilon_v^X + \{(1-p_A^2)(\varepsilon_{vv}^{M_2} + \varepsilon_{vv}^{M_3}) + (1-p_B^2)(\varepsilon_{vv}^{M_2} + \varepsilon_{vv}^{M_3}) + 2(1-p_Ap_B)\varepsilon_{vv}^{M_1}\} \\ &\qquad + \tfrac{1}{2}kT\left\{\ln(1-p_A) + \ln(1-p_B) + 2\ln\frac{N-N_X}{N}\right\} = 0 \end{aligned} \tag{1.101}$$

noting that p_A and p_B are functions of N as given in eqn (1.96). The chemical potentials of X (μ_X) and M (μ_M^A and μ_M^B) are given as follows

$$\mu_X = \frac{\partial G}{\partial N_X} = -\varepsilon_v^X + kT\ln\frac{N_X}{N-N_X} \tag{1.102}$$

$$\mu_{\mathrm{M}}^{\mathrm{A}} = \frac{\partial G}{\partial n_{\mathrm{A}}} = -\varepsilon_{\mathrm{v}}^{\mathrm{M}} + 4[(p_{\mathrm{A}} - 1)(\varepsilon_{\mathrm{vv}}^{\mathrm{M}_2} + \varepsilon_{\mathrm{vv}}^{\mathrm{M}_3}) + (p_{\mathrm{B}} - 1)\varepsilon_{\mathrm{vv}}^{\mathrm{M}_1}] - kT \ln \frac{1 - p_{\mathrm{A}}}{p_{\mathrm{A}}} \tag{1.103}$$

$$\mu_{\mathrm{M}}^{\mathrm{B}} = \frac{\partial G}{\partial n_{\mathrm{B}}} = -\varepsilon_{\mathrm{v}}^{\mathrm{M}} + 4[(p_{\mathrm{B}} - 1)(\varepsilon_{\mathrm{vv}}^{\mathrm{M}_2} + \varepsilon_{\mathrm{vv}}^{\mathrm{M}_3}) + (p_{\mathrm{A}} - 1)\varepsilon_{\mathrm{vv}}^{\mathrm{M}_1}] - kT \ln \frac{1 - p_{\mathrm{B}}}{p_{\mathrm{B}}} \tag{1.104}$$

At chemical equilibrium the chemical potential of M at an A site $\mu_{\mathrm{M}}^{\mathrm{A}}$ must equal that of M at a B site $\mu_{\mathrm{M}}^{\mathrm{B}}$. Hence we obtain

$$4[\varepsilon_{\mathrm{vv}}^{\mathrm{M}_1} - (\varepsilon_{\mathrm{vv}}^{\mathrm{M}_2} + \varepsilon_{\mathrm{vv}}^{\mathrm{M}_2})](p_{\mathrm{B}} - p_{\mathrm{A}}) = kT \ln \frac{(1 - p_{\mathrm{A}})p_{\mathrm{B}}}{p_{\mathrm{A}}(1 - p_{\mathrm{B}})} \tag{1.105}$$

Let us define the parameter α, corresponding to temperature, as

$$\alpha = \frac{kT}{4\{\varepsilon_{\mathrm{vv}}^{\mathrm{M}_1} - (\varepsilon_{\mathrm{vv}}^{\mathrm{M}_2} + \varepsilon_{\mathrm{vv}}^{\mathrm{M}_3})\}} \tag{1.106}$$

Then we have

$$\alpha = \frac{p_{\mathrm{B}} - p_{\mathrm{A}}}{\ln \dfrac{(1 - p_{\mathrm{A}})p_{\mathrm{B}}}{p_{\mathrm{A}}(1 - p_{\mathrm{B}})}} \tag{1.107}$$

which gives the temperature dependence of the occupation probabilities. The ratio of M to X is

$$\frac{\mathrm{M}}{\mathrm{X}} = \frac{n_{\mathrm{A}} + n_{\mathrm{B}}}{N} = \tfrac{1}{2}(p_{\mathrm{A}} + p_{\mathrm{B}}) \tag{1.108}$$

assuming $N_{\mathrm{X}} = N$. Hence the deviation from stoichiometry is

$$\delta = \frac{N - (n_{\mathrm{A}} + n_{\mathrm{B}})}{N} \tag{1.109}$$

Combined with eqn (1.108), we get

$$(p_{\mathrm{A}} + p_{\mathrm{B}}) = 2(1 - \delta) \tag{1.110}$$

The relationship between p_{A}, p_{B}, and δ expressed by eqn (1.110) is shown in Fig. 1.19. For example, the values of p_{A} and p_{B} for the compound $\delta = 0.3$, which are 1.0 and 0.4 at absolute zero temperature, change along an arrow

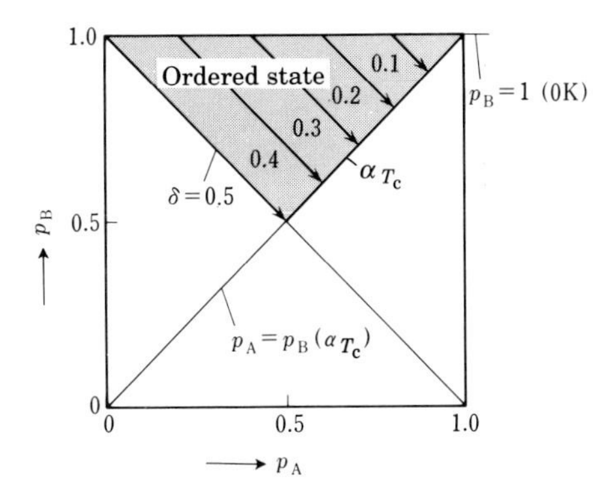

FIG. 1.19 Temperature dependence of p_A and p_B at fixed composition (eqn (1.110)). The direction of the arrows shows the increase of temperature.

on heating reaching 0.7 and 0.7 ($p_A = p_B$, random structure) at the order–disorder transition temperature α_{T_c}. The phase transition temperature α_{T_c} can be obtained as

$$\alpha_{T_c} = \lim_{p_A \to p_B} \alpha = \delta(1 - \delta) \tag{1.111}$$

This equation shows that α_{T_c} is a maximum at $\delta = 0.5$ and decreases with decreasing δ. After the treatment of order–disorder transition of metal alloy systems, the degree of order S is defined as,

$$S = p_B - p_A \tag{1.112}$$

Combined with eqn (1.110), we have

$$\left.\begin{aligned} p_A &= (1 - \delta) - \frac{S}{2} \\ p_B &= (1 - \delta) + \frac{S}{2} \end{aligned}\right\} \tag{1.113}$$

On substituting these for p_A and p_B in eqn (1.107), it follows that

$$\alpha = \frac{S}{\ln \dfrac{\left(\delta + \dfrac{S}{2}\right)\left(1 - \delta + \dfrac{S}{2}\right)}{\left(1 - \delta - \dfrac{S}{2}\right)\left(\delta - \dfrac{S}{2}\right)}} \tag{1.114}$$

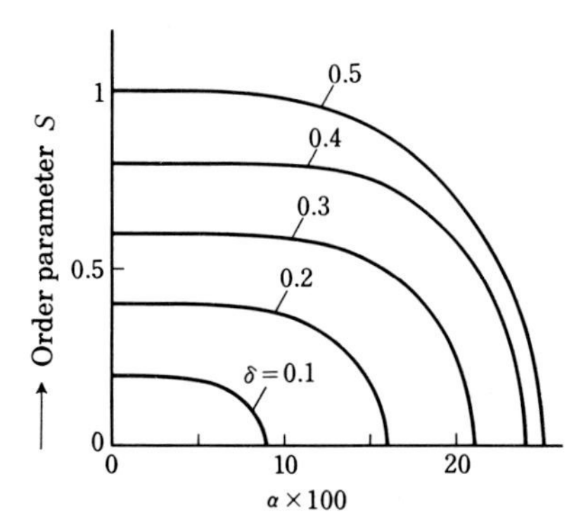

FIG. 1.20 Temperature dependence of order parameter S (eqn (1.114)).

In Fig. 1.20 the degree of order S is plotted as a function of temperature for a given δ.

The results depicted in Fig. 1.20 were originally obtained from eqn (1.105), which was derived from the condition that the activity of metal in the A-sites is equal to that in the B-sites at equilibrium. These results can also be deduced from the condition that the free energy is a minimum.

Equation (1.100) can be rewritten as a function of S, and by using eqns (1.109) and (1.113), we have

$$G(S, \delta, \alpha) \equiv \frac{G - N\{\mu'_{\mathrm{MX}} + \delta\varepsilon^{\mathrm{M}}_{\mathrm{v}} + 2\delta^2(\varepsilon^{\mathrm{M_1}}_{\mathrm{vv}} + \varepsilon^{\mathrm{M_2}}_{\mathrm{vv}} + \varepsilon^{\mathrm{M_3}}_{\mathrm{vv}})\}}{2\{\varepsilon^{\mathrm{M_1}}_{\mathrm{vv}} - (\varepsilon^{\mathrm{M_2}}_{\mathrm{vv}} + \varepsilon^{\mathrm{M_3}}_{\mathrm{vv}})\}N}$$

$$= -\frac{S^2}{4} + \alpha\left\{\left(\delta + \frac{S}{2}\right)\ln\left(\delta + \frac{S}{2}\right) + \left(1 - \delta - \frac{S}{2}\right)\ln\left(1 - \delta - \frac{S}{2}\right)\right.$$

$$\left. + \left(\delta - \frac{S}{2}\right)\ln\left(\delta - \frac{S}{2}\right) + \left(1 - \delta + \frac{S}{2}\right)\ln\left(1 - \delta + \frac{S}{2}\right)\right\} \qquad (1.115)$$

In Fig. 1.21 the function $G(S, 0.5, \alpha)$ is shown for a given temperature α. On each curve the minimum of the free energy (shown by an arrow) gives the S value at equilibrium, which must be coincident with the result obtained from eqn (1.114).

From eqns (1.101) and (1.102), we have an expression for μ_{X} as follows:

$$\mu_{\mathrm{X}} = \mu'_{\mathrm{MX}} + \varepsilon^{\mathrm{M}}_{\mathrm{v}} + 2(2\delta - \delta^2)(\varepsilon^{\mathrm{M_1}}_{\mathrm{vv}} + \varepsilon^{\mathrm{M_2}}_{\mathrm{vv}} + \varepsilon^{\mathrm{M_3}}_{\mathrm{vv}})$$

$$+ 2(\varepsilon^{\mathrm{M_1}}_{\mathrm{vv}} - \varepsilon^{\mathrm{M_2}}_{\mathrm{vv}} - \varepsilon^{\mathrm{M_3}}_{\mathrm{vv}})\frac{S^2}{4} + \tfrac{1}{2}kT\left[\ln\left(\delta^2 - \frac{S^2}{4}\right)\right] \qquad (1.116)$$

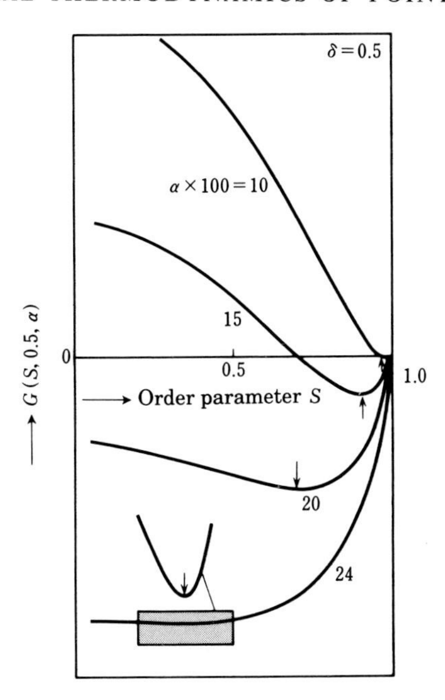

FIG. 1.21 Free energy $G(S, 0.5, \alpha)$ versus order parameter S as a function of temperature (eqn (1.115)). Arrows show the points of minimum free energy.

For simplicity we assume that $(\mu'_{MX} + \varepsilon_{v}^{M}) = -2\varepsilon_{vv}^{M_1} < 0$ and $\varepsilon_{vv}^{M_2} = \varepsilon_{vv}^{M_3} = 0$ (this assumption does not change the essence of the discussion). Then we get

$$\ln a_X = \frac{1}{2\alpha}\left\{(-\delta^2 + 2\delta - 1) + \frac{S^2}{4}\right\} + \tfrac{1}{2}\ln\left(\delta^2 - \frac{S^2}{4}\right) \tag{1.117}$$

considering the relation $\mu_X = kT \ln a_X$. Figure 1.22 shows the temperature variation of $\ln a_X$ for $\delta = 0.5$. In the temperature range $\alpha > \alpha_{T_c}$ (=0.25), eqn (1.117) becomes

$$\ln a_X = \frac{1}{2\alpha}(-\delta^2 + 2\delta - 1) + \ln \delta \tag{1.118}$$

because $S = 0$. This equation is parallel to eqn (1.93) in nature. In this model structure the degree of order S decreases continuously to zero with increasing temperature as shown in Fig. 1.20, this is called the second order phase change. This is the reason why we cannot observe an anomaly at α_{T_c} on the $\ln a_X$ versus α curves. It is to be noted, however, that by plotting $\ln a_X$ against inverse temperature α^{-1} we can estimate α_{T_c}, because at $\alpha > \alpha_{T_c}$ the $\ln a_X$ changes linearly with α^{-1} as shown in the insert of Fig. 1.22.

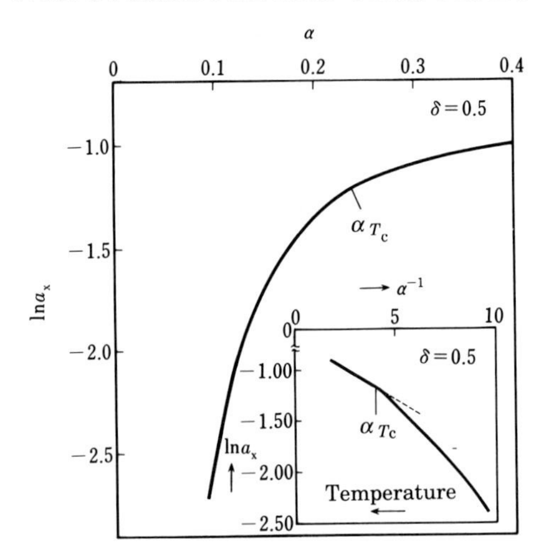

FIG. 1.22 Ln a_X versus α for $\delta = 0.5$ (eqn (1.117)). The insert shows the ln a_X versus α^{-1} curve, on which the order–disorder transition temperature, α_{T_c}, is clearly seen.

1.3.7 Point defects, electrons, and holes as chemical species

So far we have derived an expression for the free energy of the solid phase under consideration after Libowitz, and discussed the composition or temperature dependence of the activity derived from this expression of free energy. Let us now consider non-stoichiometry from another point of view.

As a typical example, we shall consider metal hydride systems MH_s, in which s depends on metal M: for example the maximum value of s is 3 in the U–H system. Metal hydrides are usually prepared by flowing hydrogen gas over fine metal powders at constant temperature. The quantity of absorbed hydrogen is controlled by changing the pressure of flowing hydrogen gas (the activity of H_2 in the atmosphere). Therefore if the pressure of atmospheric hydrogen gas, P_{H_2}, is less than that of equilibrium hydrogen gas for the solid, the hydrogen in the solid is desorbed as H_2 gas (strictly speaking, we should discuss this in terms of the activity of hydrogen in the solid and gas phases). By this process, vacant hydrogen sites in the solid phase are produced, but the number of metal atoms remains constant, and also point defects of metal sites are considered to be very rare. In this case we can express this state by the following chemical reaction, showing equilibrium between hydrogens in the solid and those of the atmosphere.

$$\text{H(lattice)} \rightleftharpoons \tfrac{1}{2}\text{H}_2\text{(gas)} + \text{V}_\text{H} \qquad (1.119)$$

where H(lattice) and V_H denote hydrogens sited in the lattice and vacant hydrogen sites, respectively. Hence we have

$$K = \frac{a_{V_H} a_{H_2}^{1/2}}{a_H} \tag{1.120}$$

where K is equilibrium constant, and a_{V_H}, a_{H_2}, and a_H are the activities for hydrogen vacancies, hydrogen gas, and hydrogen atoms in the solid, respectively. The activities in eqn (1.120) can be expressed as[7,8]

$$\left.\begin{aligned} a_{H_2}^{1/2} &= P_{H_2}^{1/2} \\ a_{V_H} &= X_{V_H} \exp(BX_H^2) \\ a_H &= X_H \exp(BX_{V_H}^2) \end{aligned}\right\} \tag{1.121}$$

where X_{V_H} and X_H are mole fractions of V_H and H. These relations were derived on the assumption that the metal hydride systems can be regarded as binary V_H–H systems. In the equation, B equals $2T_c/T$, and below T_c there appears a phase separation between V_H and H. On substituting the relations in eqn (1.121) for the activities in eqn (1.120) and after rearrangement, we get

$$\ln K = \ln(X_{V_H}/X_H) + \tfrac{1}{2}\ln P_{H_2} + B(X_H^2 - X_{V_H}^2) \tag{1.122}$$

Consider a metal hydride MH_s in which there are N lattice points for metals and sN for hydrogens, and hydrogen atoms occupy N_H in sN lattice points ($sN = N_H + N_V$, where N_V is the number of vacant hydrogen sites). Hence we have

$$\left.\begin{aligned} X_H &= \frac{N_H}{N_H + N_V} = \frac{N_H}{sN} \equiv \frac{n}{s} \\ X_{V_H} &= \frac{N_V}{N_H + N_V} = 1 - X_H = \frac{s-n}{s} \end{aligned}\right\} \tag{1.123}$$

Combining this with eqn (1.122), we get

$$\ln \frac{P_{H_2}}{P_{H_2}^0} = 2\ln\frac{n}{s-n} + \frac{4T_c}{sT}(s-2n) \tag{1.124}$$

where $P_{H_2}^0$ is the pressure of hydrogen gas, H_2, at the composition $n = N_H/N = s/2$. The equation shows the relation between the deviation from the stoichiometry δ in $MH_{s(1-\delta)}$ and the hydrogen pressure, P_{H_2}, where δ is defined as $(s-n)/s$ in this case. It is to be noted that eqn (1.124) has been derived by assuming the chemical reaction expressed by eqn (1.119) and treating the hydrogen vacancy as if it is a kind of chemical species. We can also get the same equation by the Libowitz method as follows. If it is assumed that there are no vacancies for metal sites, the total free energy of the solid may be expressed as

$$G = N\mu'_{\mathrm{MH}_s} + (sN - N_{\mathrm{H}})\varepsilon_{\mathrm{v}}^{\mathrm{H}} + \tfrac{1}{2}Z\varepsilon_{\mathrm{vv}}^{\mathrm{H}} \cdot \frac{(sN - N_{\mathrm{H}})^2}{sN} - kT \ln {}_{sN}\mathrm{C}_{sN - N_{\mathrm{H}}} \quad (1.125)$$

The chemical potential of hydrogen atoms in the solid, μ_{H}, is

$$\mu_{\mathrm{H}} = \left(\frac{\partial G}{\partial N_{\mathrm{H}}}\right) = -\varepsilon_{\mathrm{v}}^{\mathrm{H}} + \frac{Z\varepsilon_{\mathrm{vv}}^{\mathrm{H}}}{s}(n - s) + kT \ln \frac{n}{s - n} \quad (1.126)$$

By using the relation $\mu_{\mathrm{H}} = \frac{1}{2}\mu_{\mathrm{H}_2} = \frac{1}{2}kT \ln a_{\mathrm{H}_2} = \frac{1}{2}kT \ln P_{\mathrm{H}_2}$, we have

$$\ln \frac{P_{\mathrm{H}_2}}{P_{\mathrm{H}_2}^0} = 2 \ln \frac{n}{s - n} + \frac{Z\varepsilon_{\mathrm{vv}}^{\mathrm{H}}}{skT}(2n - s) \quad (1.127)$$

where $P_{\mathrm{H}_2}^0$ is the value of pressure for $n = s/2$. The equation is coincident with eqn (1.124), because $T_c = -Z\varepsilon_{\mathrm{vv}}^{\mathrm{H}}/4k$.

Next let us discuss the electronic defects associated with point defects in semiconductive or insulating compounds, which lead to non-stoichiometry. Consider a NiO crystal, which has a NaCl-type structure, as NiO can be regarded as an ionic crystal, the valence states of Ni and O are Ni^{2+} and O^{2-}, respectively. We assume that the non-stoichiometry originates only from metal vacancies. Generation of metal defects in NiO may be expressed by a chemical reaction similar to eqn (1.119), i.e.

$$\tfrac{1}{2}\mathrm{O}_2(\mathrm{gas}) \rightleftharpoons \mathrm{O}_0^{2-} + \mathrm{V}_{\mathrm{Ni}}^{2+} \quad (1.128)$$

Namely, oxygen gas is adsorbed by the solid on the surface, then it changes to O^{2-} ions by receiving electrons from the solid, and finally occupies the regular lattice sites for oxygen ions. By this reaction, vacancies for metal sites are produced, which are considered to have +2 charge (denoted by V_{Ni}^{2+}, see Fig. 1.23(b)). An alternative explanation for the charge of the metal vacancy is that Ni^{2+} changes to Ni^{3+} by giving electrons to adsorbed oxygen atoms, and these Ni^{3+} occupy sites close to metal vacancies at low temperatures (see Fig. 1.23(c)). In any case, plus electrons, e^+, i.e. holes, produced to maintain charge neutrality in the crystal, are trapped by metal vacancies at low temperatures, as expressed by eqn (1.128). With a rise in temperature the following reactions proceed,

$$\mathrm{V}_{\mathrm{Ni}}^{2+} \rightleftharpoons \mathrm{V}_{\mathrm{Ni}}^{+} + \mathrm{h} \quad (1.129)$$

$$\mathrm{V}_{\mathrm{Ni}}^{+} \rightleftharpoons \mathrm{V}_{\mathrm{Ni}}^{0} + \mathrm{h} \quad (1.130)$$

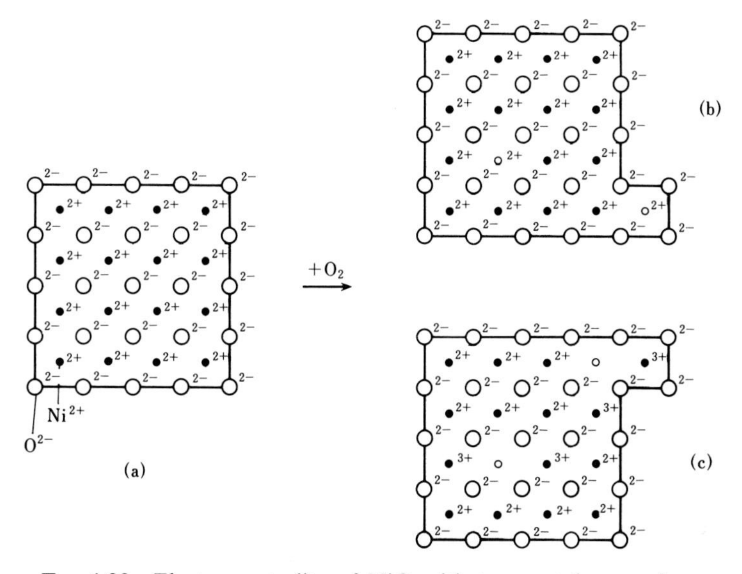

FIG. 1.23 Electro-neutrality of NiO with two metal vacancies. A perfect crystal (a) is oxidized to crystal (b) or (c). In crystal (b), there are two metal vacancies with +2 charge. In crystal (c), there are two metal vacancies with neutral charge and four metal ions with excess charge (+3). (b) and (c) are alternative representations of the oxidized crystal.

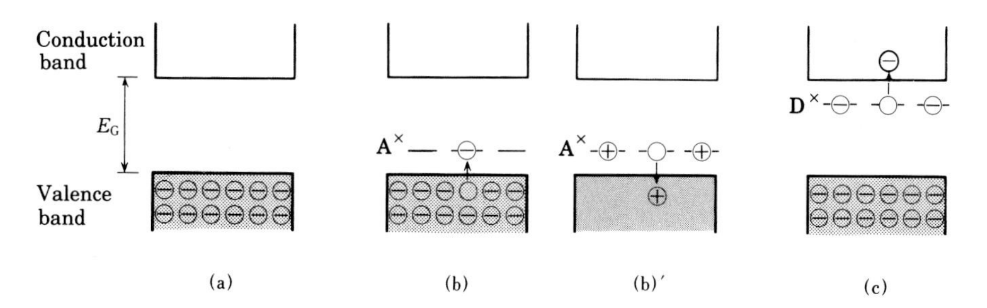

FIG. 1.24 Impurity levels in semiconductors. $A^{\times}$ and $D^{\times}$ denote the acceptor and donor levels.

This situation can be expressed in terms of the band model as shown in Fig. 1.24. Stoichiometric NiO is an intrinsic semiconductor, having an energy gap of E_G $(=E_C - E_V)$. Non-stoichiometric $Ni_{1-\delta}O$, which has metal vacancies or electronic defects, has an acceptor level $A^{\times}$ between the valence band and the conduction band.* By receiving electrons from the valence band, $A^\times$ creates holes successively in the valence band as follows

$$A^\times \rightarrow A^- + h \tag{1.131}$$

$$A^- \rightarrow A^{2-} + h \tag{1.132}$$

Concentration equilibrium among $A^\times$, A^-, A^{2-}, and h is discussed on the assumption that these equations can be treated as chemical equilibrium ones. (Similarly, $D^\times$, D^+, D^{2+} (donor levels), and e are regarded as chemical species, see Fig. 1.24(c).) We have a reasonable reason for regarding these species as chemical species. As is well known, the electrical properties of metals and alloys are independent of the concentration of point defects or imperfections existing in their crystals, because the number of electrons or holes in metals or alloys is roughly equal to that of the constituent atoms. For the case of semiconductors or insulators, however, the number of electrons or holes is much lower than that of the constituent atoms and is closely correlated to the concentration of defects. In the latter case, electrons and holes can be considered as kinds of chemical species, for a reason similar to that discussed above for the case of point defects. Let us consider the chemical potential, which is most characteristic of chemical species. Electrochemical potential of electrons μ_e is written as

$$\mu_e = \mu_e^0 + kT \ln \frac{n}{N} \tag{1.133}$$

where μ_e^0 is the value of μ_e at the standard state, N is the number of possible energy states, and n is the number of existing electrons. Therefore n/N denotes the concentration of electrons. This equation is valid for the concentration region $n/N < 0.1$.

On the other hand, the concentrations of electrons and holes, n and p, respectively, in the conduction band and valence band of the semiconductors, which contribute to the electrical conductivity, are expressed by the following equations[9]

$$\left.\begin{aligned} n &= N_c \exp\{(E_F - E_C)/kT\} \\ p &= N_v \exp\{(E_V - E_F)/kT\} \end{aligned}\right\} \tag{1.134}$$

* The acceptor $A^\times$ is regarded as neutral in charge for the following reason. To satisfy the charge neutrality locally, a metal vacancy is considered to have an effective charge of 2− (the site was for a metal bearing a 2+ charge). For this reason, the metal vacancy acts as a trapping centre for holes. In the field of semiconductor physics, $A^\times$ denotes a vacancy trapping two holes, which is neutral in charge. By following eqns (1.131) and (1.132), $A^\times$ changes to A^- and A^{2-} by releasing holes. Therefore $A^\times$, A^-, and A^{2-} correspond to V_{Ni}^{2+}, V_{Ni}^{+}, and V_{Ni}^{0}, respectively. Hereafter we use the effective charge notation. It is noted that in this notation O and Ni in regular lattice points are neutral in charge. For example, eqn (1.128) may be written as

$$\tfrac{1}{2}O_2(\text{gas}) = O_0^\times + A^\times$$

in the low concentration region, where E_F is the Fermi energy, and N_c and N_v are constant. N_c and N_v are given as $(m_n/m)^{3/2}N_0$ and $(m_p/m)^{3/2}N_0$, where m, m_n, and m_p are the mass of a free electron and the effective mass of an electron and a hole, respectively.

From eqn (1.134), we get

$$E_F = E_C + kT \ln \frac{n}{N_c} \tag{1.135}$$

As can be verified,[10] E_F equals μ_e. Therefore it has been shown that eqn (1.133) is identical to eqn (1.135). From this, the creation and annihilation of electrons and holes in semiconductors may be written as the following chemical equation

$$\mathrm{eh} \rightleftharpoons \mathrm{e} + \mathrm{h} \tag{1.136}$$

The equation can also be considered as the ionization of a valence bond, i.e. eh means the ground state of the valence bond. The rate of ionization is believed to be roughly equal to that of the dissociation of water ($H_2O \rightleftharpoons H^+ + OH^-$) and therefore we can put [eh] = 1. Hence, by using eqns (1.134) and (1.135), the equilibrium constant K of eqn (1.136) is given as

$$\begin{aligned} K = np &= N_c N_v \exp[(E_F - E_C)/kT] \exp[(E_V - E_F)/kT] \\ &= c \exp\{-(E_C - E_V)/kT\} \\ &= c \exp -\frac{E_G}{kT} \end{aligned} \tag{1.137}$$

Thus, lattice defects such as point defects and carriers (electrons and holes) in semiconductors and insulators can be treated as chemical species, and the mass action law can be applied to the concentration equilibrium among these species. Without detailed calculations based on statistical thermodynamics, the mass action law gives us an important result about the equilibrium concentration of lattice defects, electrons, and holes (see Section 1.4.5).

1.4 Case studies of non-stoichiometric compounds

1.4.1 Introduction

So far we have discussed the basic concept of non-stoichiometry without showing real examples. Here we shall consider case studies in order to understand the non-stoichiometry appearing in various kinds of substances. To construct phase diagrams which contain non-stoichiometric compounds, it is indispensable to know the relationship between the deviation from stoichiometry δ, the partial pressure of coexisted gas, P_{X_2} (for diatomic gases), and the temperature, T. This is the main theme of this section, how to determine these three parameters precisely. Not only X-ray diffraction but also neutron and electron diffraction techniques are usually adopted to identify phases.

In principle the deviation δ can be determined by the use of usual analytical chemistry or a highly sensitive thermo-balance. These methods, however, are not suitable for very small deviations. In these cases the following methods are often applied to detect the deviation: physico-chemical methods (ionic conductivity, diffusion constant, etc.), electro-chemical methods (coulometric titration, etc.), and physical methods (electric conductivity, nuclear magnetic resonance, electron spin resonance, Mossbauer effect, etc.), some of which will be described in detail.

There are many types of methods for controlling the atmospheric gas pressure as mentioned below.

1.4.2 Study of the V_2O_3–V_2O_4 system—control of oxygen partial pressure and thermogravimetry

It has been well established that there is a series of compounds (homologous series) expressed by the general formula V_nO_{2n-1} ($n = 3, 4, \ldots, 9$) between V_2O_3 and V_2O_4, called the Magnéli phase. These compounds are non-stoichiometric compounds derived from extended defects (see Section 2.2.1). We describe a study on the phase diagram of this system, carried out by controlling the oxygen partial pressure.

The method of controlling P_{O_2} depends on the pressure region: low pressure ($10^{-20} < P_{O_2} < 1$ atm) and high pressure (1 atm $< P_{O_2}$) regions. In the pressure region 1 atm $> P_{O_2} > 10^{-20}$ atm, the following two methods are usually adopted.

1. *Gas mixing* (for example oxygen gas + inert gas, such as Ar and N_2). This method is suitable for obtaining a constant P_{O_2} independent of temperature. The control region is relatively narrow, and the purity of the inert gas is a serious consideration.

2. *Dissociation equilibrium* (for example H_2O gas). The following chemical equilibrium is utilized,

$$H_2O(\text{gas}) \rightleftharpoons H_2 + \tfrac{1}{2}O_2 \qquad K = \frac{P_{O_2}^{1/2} P_{H_2}}{P_{H_2O}} \tag{1.138}$$

The partial pressure P_{O_2} depends on temperature, because the equilibrium constant K is a function of temperature. It is also possible to make use of a more complex gas mixing system, such as $H_2O + H_2$, $CO_2 + H_2$, $CO_2 + CO$.

For example, the gas mixing system ($CO_2 + H_2$) shows the following chemical equilibria:

$$CO_2 + H_2 \rightleftharpoons CO + H_2O(\text{gas}) \qquad K_1(T) = \frac{P_{CO}P_{H_2O}}{P_{CO_2}P_{H_2}} \tag{1.139}$$

$$CO_2 \rightleftharpoons CO + \tfrac{1}{2}O_2 \qquad K_2(T) = \frac{P_{CO}P_{O_2}^{1/2}}{P_{CO_2}} \tag{1.140}$$

P_{O_2} can be expressed as

$$\sqrt{P_{O_2}} = \frac{K_2}{2}\left[(\alpha - 1) + \sqrt{(\alpha - 1)^2 + \frac{4}{K_1}\alpha}\right] \tag{1.141}$$

where α is the mixing ratio of CO_2 to H_2 in moles at room temperature. Figure 1.25(a) shows P_{O_2} as a function of temperature for a given α calculated using eqn (1.141). The calculated value of P_{O_2} often differs from the observed one as shown in Fig. 1.26. The following factors are considered to be the reasons for the difference: inaccuracy of the mixing ratio, the flow rate of gases, impurity gas in the mixed gases, the local fluctuation of P_{O_2} due to the reaction between the solid and gas phases, and the thermal separation of the mixed gases. The method for measuring P_{O_2} will be described in Section

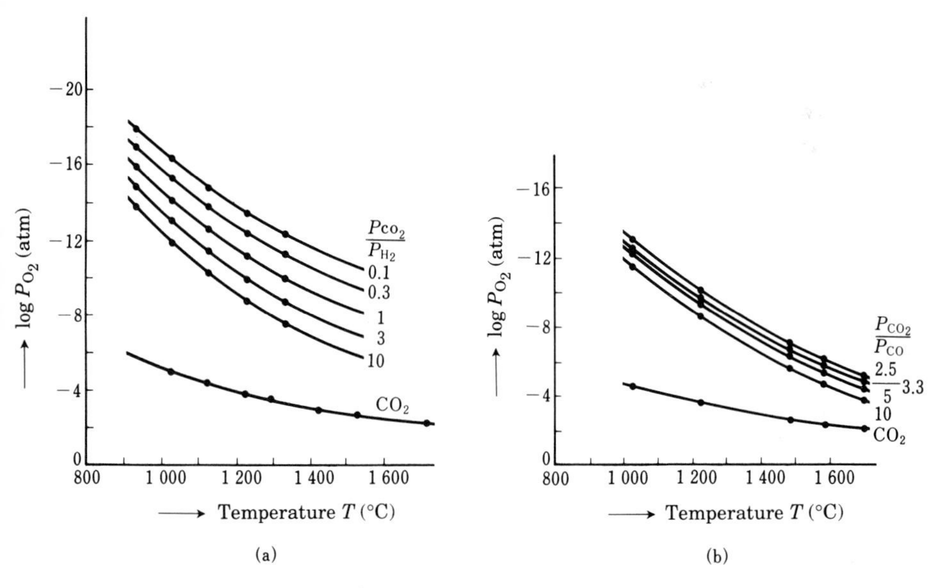

FIG. 1.25 Computed P_{O_2} versus temperature curves as a function of the mixing ratio of gases: (a) for the CO_2–H_2 system, calculated using eqn (1.141); (b) for the CO_2–CO system.

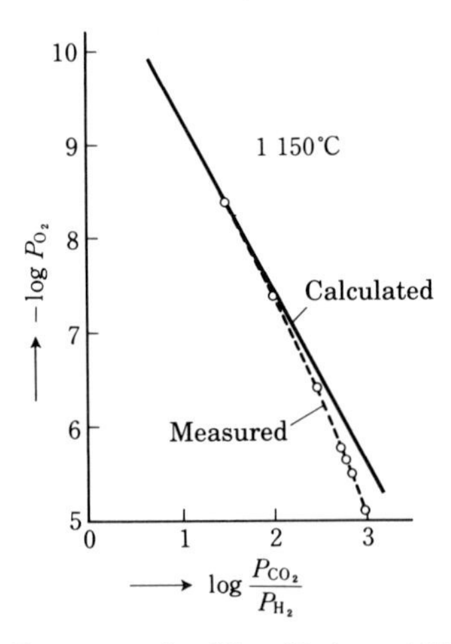

FIG. 1.26 Log P_{O_2} versus $\log(P_{CO_2}/P_{H_2})$ at 1150 °C. The solid curve was calculated using eqn (1.141) and the dotted curve is from the measured values.

1.4.7. Figure 1.25(b) shows the calculated P_{O_2} versus temperature for the CO_2–CO system.

For the high pressure region, $P_{O_2} > 1$ atm, the following three methods can be utilized.

1. *Compression of O_2 gas.* Oxygen gas compressed by a compressor is used up to about 4000 atm, this is a very popular method.

2. *Heating of liquid oxygen in a closed vessel.* By heating a closed vessel which contains samples and liquid oxygen, a high pressure oxygen atmosphere can be achieved.

3. *Thermal decomposition of oxides and peroxides.* High pressure oxygen can be produced by the thermal decomposition of oxides (for example, CrO_3, Mn_2O_7) and peroxides (for example, BaO_2, CaO_2) in a closed vessel, the pressure of which can reach up to 50 000 atm.

In order to determine the relation between the composition of the solid phase and the oxygen partial pressure P_{O_2} at higher temperatures, it is convenient to trace the weight change *in situ* by thermogravimetric analysis, changing P_{O_2} at a fixed temperature or changing temperature at a fixed P_{O_2}. Chemical, electric, and quartz balances etc. are used for this purpose. However, special care should be taken because of the flow of gases.

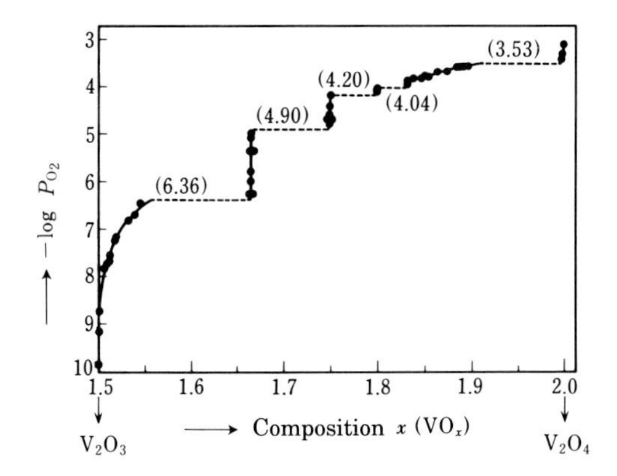

FIG. 1.27 Log P_{O_2} versus x in VO_x at 1600 K.[11]

Katsura and Hasegawa[11] studied the phase diagram of the V_2O_3–V_2O_4 system at 1600 K, to control P_{O_2}, the low pressure methods described above were utilized.* The starting compound was stoichiometric V_2O_3, which was obtained by reduction of V_2O_5 with flowing hydrogen gas at high temperature (~800 °C). The weight change was traced *in situ* as a function of P_{O_2} at 1600 K. Chemical analyses of quenched samples were also made, in order to take vaporization of samples into consideration. The results are shown in Fig. 1.27. This experiment was carried out under a total pressure of 1 atm. Because the method is, in principle, the same as that shown in Fig. 1.2, the compositional regions shown by closed circles (measured points) correspond to the single phase regions and those shown by dotted lines, parallel to the horizontal axis, correspond to the two-phase regions. Using this method, the two-phase regions are hardly observed, therefore the dotted regions were estimated by extrapolation from the single-phase regions at each side.

Table 1.3 shows the relation between the composition and the equilibrium oxygen partial pressure for each phase of V_nO_{2n-1}. For example, the phase V_2O_3 exists as a single phase in the composition range $VO_{1.500}$–$VO_{1.555}$.† The corresponding P_{O_2} ranges from 10^{-15} to $10^{-6.36}$ atm. Within experimental errors the phases V_3O_5, V_4O_7, and V_5O_9, which correspond to $n = 3$, 4, and 5 in V_nO_{2n-1}, have the compositions $VO_{1.667}$, $VO_{1.750}$, and $VO_{1.800}$, i.e. stoichiometric composition. Each of these phases, however, exists in some

* For the case that P_{O_2} is controlled by flowing mixed gases such as ($Ar + O_2$), ($H_2O + H_2$), and ($CO_2 + CO$) etc., i.e. an open system, the phase rule $F = c + 2 - p$ has to be altered to $F = c + 3 - p$.

† The V_2O_3 phase has a corundum (Al_2O_3) type structure, in which vanadium atoms occupy regularly $\frac{2}{3}$ of the octahedral sites formed by hexagonal close packing of oxygen atoms. Because non-stoichiometry of the V_2O_3 phase may be considered to originate from metal vacancies, the correct expression for $VO_{1.500}$–$VO_{1.555}$ has to be V_2O_3–$V_{1.929}O_3$. Generally, non-stoichiometric V_2O_3 must be written as $V_{2-\delta}O_3$.

Table 1.3

Phase equilibrium of V_2O_3–V_2O_4 system at 1600 K[11]

Phase	$-\log_{10} P_{O_2}$	Composition range
V_2O_3	~15.00–6.36	1.500–1.555
V_3O_5	6.36–4.90	1.677
V_4O_7	4.90–4.20	1.750
V_5O_9	4.20–4.04	1.800
V_6O_{11}	4.04–3.53	1.833–1.905[a]
V_2O_4	3.53–	2.000–

[a] This value is not correct.

P_{O_2} region, which means that these phases are strictly non-stoichiometric compounds.

Later the phase diagrams for this system at 1700 and 1307 K were determined by Taniguchi and co-workers.[12–13] An important result obtained in these studies was that the V_6O_{11}, V_7O_{13}, and V_8O_{15} phases were faintly revealed at 1307 K, although they were not confirmed at 1600 K, as shown in Fig. 1.27 and Table 1.3. Though all the phases from $n = 3$ to 9 in V_nO_{2n-1} were confirmed by electron diffraction techniques,[14] it remains to be seen whether or not phases corresponding to $n > 6$ exist at higher temperatures.

The vanadium–oxygen system will be further described in Sections 1.4.7 and 2.2.

1.4.3 Study of $CrO_{2+\delta}$—control of high pressure oxygen gas by thermal decomposition of oxides

CrO_2 is a useful magnetic material, because it shows metallic ferromagnetism. Here we show an investigation of the phase diagram of the Cr–O system, i.e. the relation between the non-stoichiometry, δ in $CrO_{2+\delta}$, and the equilibrium oxygen pressure, at rather low temperature by use of thermal decomposition of CrO_3, as an example of control of high pressure oxygen gas.[15]

As shown in Fig. 1.28, a reaction vessel (cylinder) was made of 18-8 Cr stainless steel. The inner room for sample preparation, into which a platinum crucible was fitted, was sealed by screwing up a stainless steel disc wrapped in aluminium foil. This vessel can be used under the conditions that the reaction temperature is below 600 °C and the oxygen pressure is lower than 1000 atm. The experimental procedure is as follows: place the weighed CrO_3 into the crucible, seal quickly, heat the vessel using an outer heater, and then keep it at a set temperature.

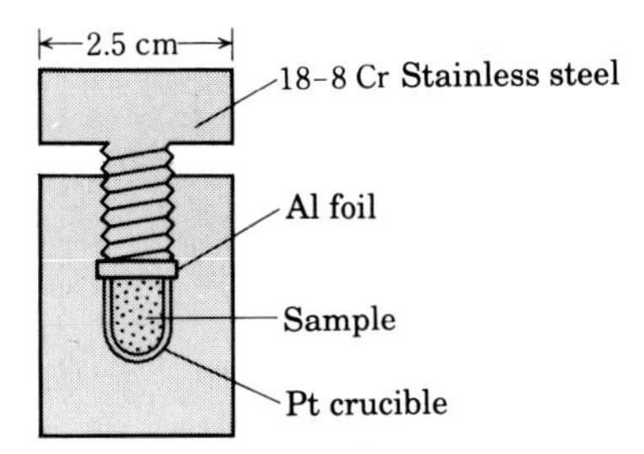

FIG. 1.28 A reaction vessel for high pressure oxygen.[15]

When heated, CrO_3 decomposes into CrO_{3-x} and O_2 by the following equation.

$$CrO_3 \rightarrow CrO_{3-x} + \frac{x}{2} O_2 \qquad (1.142)$$

The oxygen gas thus evolved is shut in the reaction vessel, and therefore a high pressure is obtained. At chemical equilibrium, the composition of the solid phase or phases and the pressure of O_2 are fixed at a fixed temperature following the phase rule. Phase identification and chemical analysis (composition) of the solid phase were performed for quenched samples. The oxygen content in the quenched samples CrO_{3-x} was determined by the reduction of CrO_{3-x} to the stoichiometric Cr_2O_3 in air. By combining the data of the weight of starting CrO_3 and of the composition of CrO_{3-x} at equilibrium, the oxygen pressure in the reaction vessel can be calculated, because the volume of the inner room is known. At a fixed temperature, the oxygen pressure can be controlled by controlling the weight of starting CrO_3. In Fig. 1.29, the phase diagram thus determined is shown, focusing on the CrO_2 phase. The CrO_2 phases, which lies between the Cr_2O_5 and the Cr_2O_3

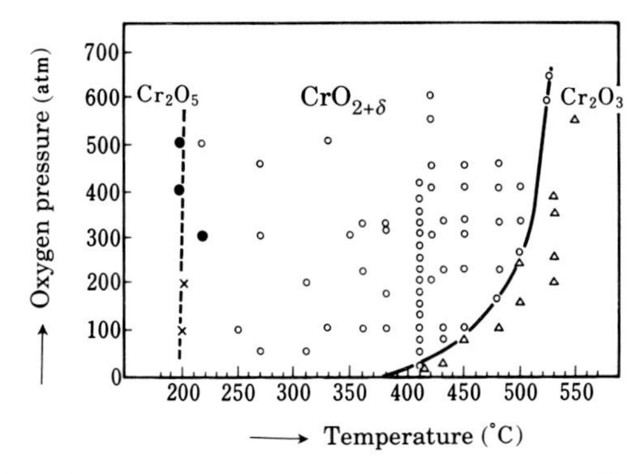

FIG. 1.29 An oxygen pressure versus temperature diagram for the CrO_2 phase region in the Cr–O system.[15]

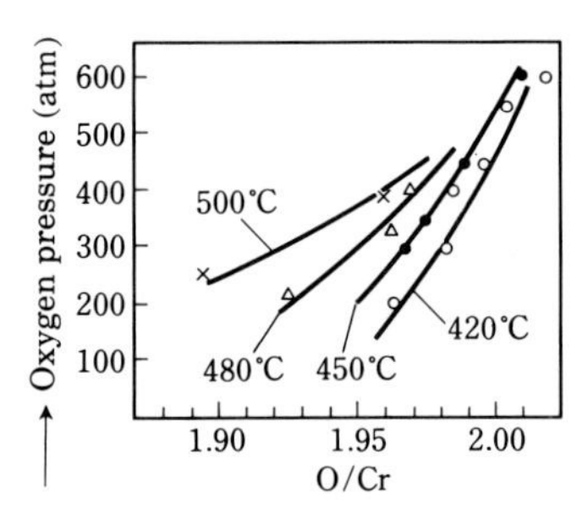

FIG. 1.30 Relation between P_{O_2}, T, and composition (O/Cr) in the CrO_2 phase region.[15]

phases, is stable over a wide temperature and oxygen pressure range. Figure 1.30 shows the isothermal curves for the relationship between the composition (the ratio of O to Cr) and the pressure. It has been found that to produce stoichiometric CrO_2, the preparation has to be performed at the lowest temperature and highest oxygen pressure possible. The non-stoichiometric $CrO_{2+\delta}$ phase is stable in the composition region $-0.11 < \delta < +0.02$. A similar study, which gave similar results, was independently performed by Kubota.[16]

Following these studies, Fukunaga and Saito[17] investigated the phase diagram up to 35 000 atm and to 1400 °C using high pressure piston-cylinder-type apparatus.

It is possible to use the O_2 gas evolved by the thermal decomposition of CrO_3 for determination of the phase diagrams of other oxides at high oxygen pressure. For this purpose it is necessary to divide the reaction room into two parts. For reference, Fig. 1.31 shows the relation between a temperature,

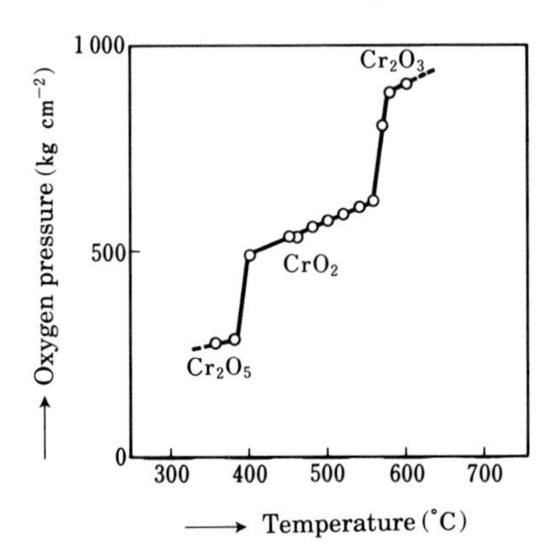

FIG. 1.31 An oxygen pressure versus temperature curve for CrO_3 (50 g) sealed in a 36 cm^3 high pressure cylinder.[16]

pressure, and phases in the Cr–O system when the weight of the starting CrO_3 is 50.0 g and the volume of the reaction room is 36 cm^3.[18]

1.4.4 *Study of $Ni_{1-\delta}S$—control of sulfur gas pressure*

There are many non-stoichiometric compounds in the 3*d* transition metal chalcogenides. Among them, the NiS phase with the NiAs-type structure has been investigated in detail from both chemical and physical viewpoints. Let us adopt this compound as a typical example of a non-stoichiometric compound.

First, we describe the methods of partial pressure control of sulfur gas.

(a) *Mixed gas of H_2S and H_2.* In the sulfur pressure range 10^{-3} to 10^{-15} atm, the following chemical reaction is utilized,

$$H_2S \rightleftharpoons H_2 + \tfrac{1}{2}S_2 \qquad K_{H_2S} = \frac{P_{H_2} P_{S_2}^{1/2}}{P_{H_2S}} \tag{1.143}$$

It is noted that the sulfur gas contains a variety of molecular species S_n ($n = 2, 3, \ldots, 8$). Figure 1.32 shows the ratio of the total pressure of the sulfur

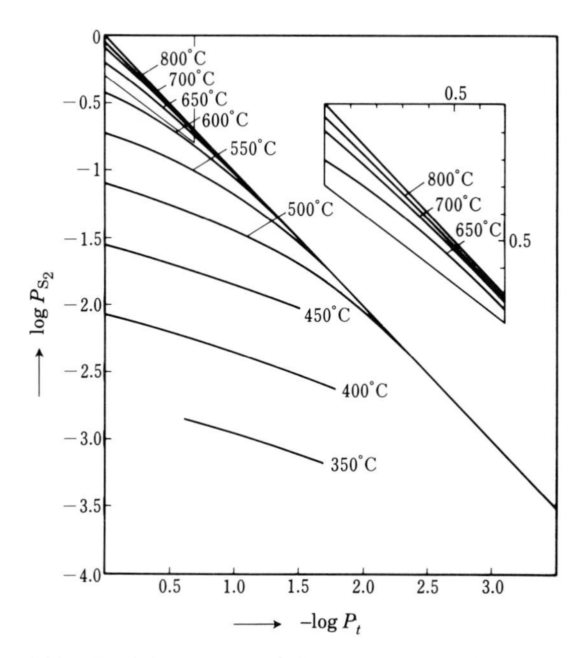

FIG. 1.32 Partial pressure of S_2, P_{S_2}, versus total pressure of sulfur gas, P_t, at different temperatures.[19]

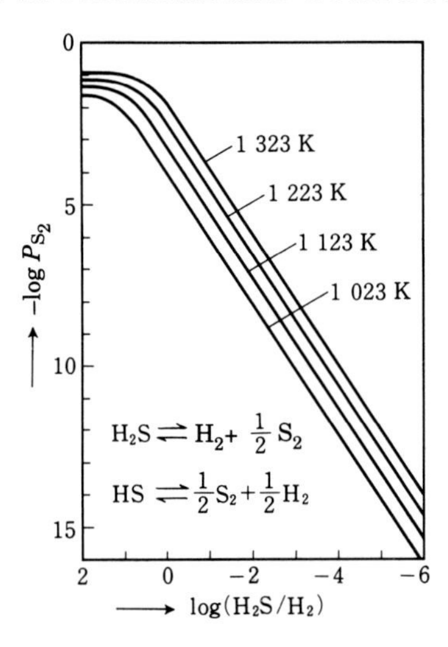

FIG. 1.33 Log P_{S_2} versus log(H_2S/H_2) curves at different temperatures (eqns (1.143) and (1.144)).[20]

gas P_t to the partial pressure of the dimolecular sulfur P_{S_2}.[19] This indicates that above 650 °C the S_2 molecule prevails overwhelmingly over the other species. Since the mixture of H_2S and H_2 gases also produces HS gas, the following chemical equilibrium has to be taken into consideration (in addition to eqn (1.143))

$$HS \rightleftharpoons \tfrac{1}{2}H_2 + \tfrac{1}{2}S_2 \qquad K_{HS} = \frac{P_{H_2}^{1/2} P_{S_2}^{1/2}}{P_{HS}} \tag{1.144}$$

In Fig. 1.33 the P_{S_2} against the mixing ratio H_2S/H_2 curves, calculated from eqns (1.143) and (1.144), are shown as a function of temperature. At high sulfur pressures ($P_{S_2} > 10^{-4}$ atm), the HS gas formation expressed by eqn (1.144) plays an important role in the determination of the S_2 pressure.[20] From Fig. 1.33, it seems to be possible to control P_{S_2} down to 10^{-15}, which necessitates a high mixing ratio of H_2/H_2S of 10^6. To do this it is necessary to use H_2 and H_2S starting gases of highest purity and to adopt a multi-step mixing method, as described by Young *et al.*[21]

(b) *Equilibrium vapour pressure of liquid sulfur.* By this method the P_{S_2} can be controlled in the pressure range $10^{-4} < P_{S_2} < 1$ atm.[22]

These methods, (a) and (b), have been used for both the closed and open (gas flowing) systems.

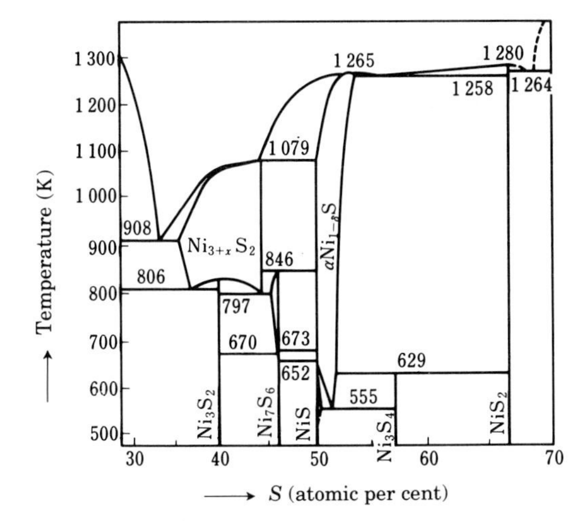

FIG. 1.34 Phase diagram of the Ni–S system.[23]

Many papers on the $Ni_{1-\delta}S$ system have been published. Figure 1.34 shows the phase diagram of the Ni–S system determined by Kullerud and Yund;[23] the region denoted $\alpha Ni_{1-\delta}S$ is discussed here. The $Ni_{1-\delta}S$ phase has the NiAs-type structure, in which cations occupy all of the octahedral voids formed by hexagonal close packed anions.

Rau studied the phase diagram of the $Ni_{1-\delta}S$ phase in detail, i.e. the relation between the composition (the deviation from stoichiometry δ), the equilibrium sulfur pressure P_{S_2}, and the temperature.[24,25] He measured P_{S_2} by following two methods. The procedure of the first method is as follows:

1. $Ni_{1-\delta}S$, with a known value of δ_s, and a fixed quantity of H_2 gas are shut in a vessel of known volume.

2. The vessel is heated to a set temperature, T_1.

3. After the system has attained chemical equilibrium, it is quenched to room temperature.

4. By analysing the mole ratio of H_2S to H_2 in the vessel, the P_{S_2} is calculated using eqn (1.143).

5. The deviation, δ_e, which is not equal to the starting δ_s, can be calculated from procedure 4.

Thus the relationship between δ and P_{S_2} at T_1 is obtained. By repeating this procedure but changing variables such as temperature, starting composition δ_s, quantity of starting H_2 gas, volume of reaction vessel etc., the phase diagram can be established.

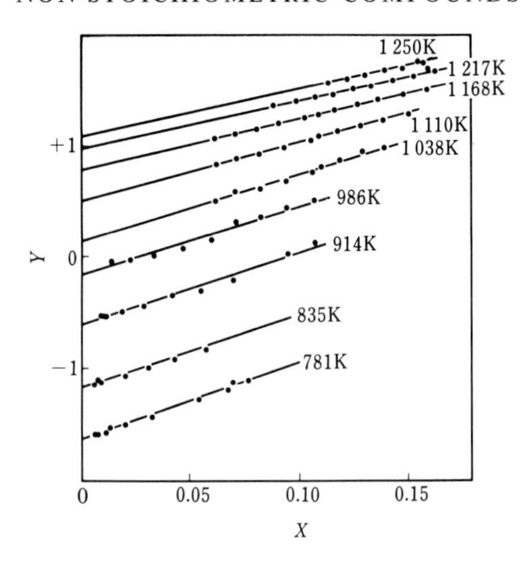

FIG. 1.35 $Y = \ln a_S(\gamma/\gamma - 1)$ versus $X = (\gamma^2 - 1)/\gamma^2$ plotted as a function of temperature (eqn (1.145)).[25]

In the second method P_{S_2} was measured directly against temperature by use of a Bourdon gauge made of quartz. The composition change during measuring could be neglected by shutting a large quantity of the sample with known composition in the gauge system.

Rau analysed these data on the assumption that $Ni_{1-\delta}S$ has defects only in the metal sites, i.e. metal vacancies, and these vacancies are distributed randomly with vacancy–vacancy interaction ε_{vv}^{M}, see Section 1.3.5. Because this assumption is the same as that adopted in Section 1.3.5, we can apply eqn (1.90) to this problem. On replacing Z_M and N/N_M by 8 (in the NiAs type structure, a metal has 8 near neighbour metals separated by the same metal–metal distance) and γ (by doing this, δ in eqn (1.90) equals $(\gamma - 1)/\gamma$) in the equation, we have

$$\ln a_S \frac{\gamma}{\gamma - 1} = \frac{\mu'_{NiS} + \varepsilon_v^M}{RT} + \frac{4\varepsilon_{vv}^M}{RT} \cdot \frac{\gamma^2 - 1}{\gamma^2} \qquad (1.145)$$

In this equation the parameters μ'_{NiS}, ε_v^M, and ε_{vv}^M are expressed in units per mole.

If the above assumption is correct, plotting $Y = \ln a_S(\gamma/\gamma - 1)$ against $X = (\gamma^2 - 1)/\gamma^2$ at a fixed temperature must give a straight line. As shown in Fig. 1.35, there is an excellent linear relation between X and Y, which suggests that the 'ε_{vv}^M model' is plausible in the $Ni_{1-\delta}S$ phase. From this figure, the temperature dependence of $(\mu'_{NiS} + \varepsilon_v^M)$ and ε_{vv}^M have been obtained

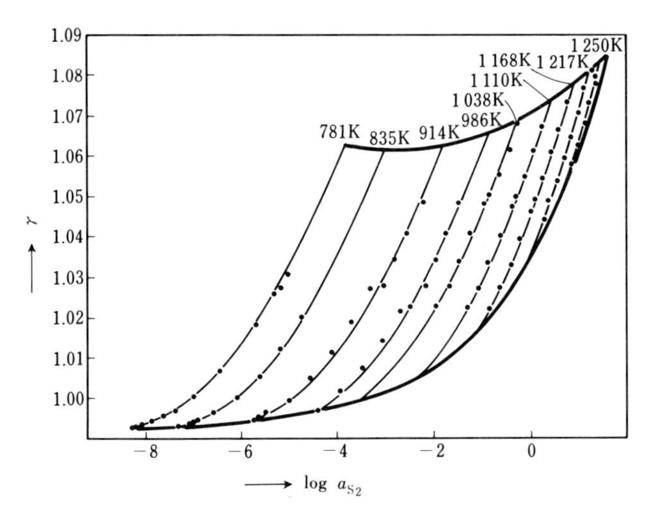

FIG. 1.36 Log a_{S_2} versus γ curves at different temperatures.[25] Closed circles are measured points and solid curves are calculated from eqns (1.145), (1.146), and (1.147). The thicker solid curves for the mixed phases are referred to in Refs (26) and (27).

as follows

$$(\mu'_{NiS} + \varepsilon_v^M) = -26\,270 + 26.041T \quad (\text{cal mol}^{-1}) \tag{1.146}$$

$$\varepsilon_{vv}^M = -12\,905 + 39.43T - 0.019\,56T^2 \quad (\text{cal mol}^{-1}) \tag{1.147}$$

The value of ε_{vv}^M is almost constant (6–7 kcal mol^{-1}) in the measured temperature range and the positive value means that the vacancy–vacancy interaction is repulsive. On the other hand, the value of $(\mu'_{NiS} + \varepsilon_v^M)$ changes sign from minus to plus with increasing temperature. Upon substituting eqns (1.145) for $(\mu'_{NiS} + \varepsilon_v^M)$ and ε_{vv}^M, from eqns (1.146) and (1.147), eqn (1.145) can be rewritten as the relation between γ, a_S, and T, as shown in Fig. 1.36. The curves for phase boundaries (thicker curves), i.e. the upper curve for coexistent condensed phases ($Ni_{1-\delta}S$ phase + adjacent sulfur rich phase) and the lower curve for coexistent condensed phases ($Ni_{1-\delta}S$ phase + adjacent sulfur poor phase), were taken from Refs 26 and 27, in which the temperature dependence of P_{S_2} for coexistent samples was investigated in detail. (As mentioned in Section 1.2, the relationship between the equilibrium sulfur pressure for coexistent condensed phases and temperature must show one to one correspondence. Rau calculated δ in $Ni_{1-\delta}S$ for the coexistent phases by substitution of the data from refs 26 and 27 for a_S and T into eqn (1.145).)

Thus, we can prepare $Ni_{1-\delta}S$ with a predetermined composition by controlling the sulfur pressure, using the phase diagram shown in Fig. 1.36.

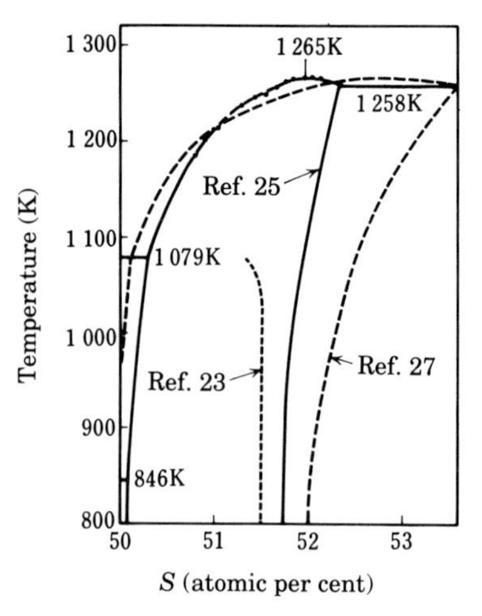

FIG. 1.37 Phase diagram of $Ni_{1-\delta}S$.[23,25,27]

In Fig. 1.37, the relation between the composition and temperature obtained from Fig. 1.36 is shown, together with the data of other investigations.[23,27]

1.4.5 Study of $Ni_{1-\delta}O$—estimation of type of point defects by measuring electrical conductivity

The MO compounds with NaCl-type structure, where M is a divalent metal of 3*d*, are non-stoichiometric compounds except for the Cr compound. The width of non-stoichiometry becomes narrower with increasing number of *d*-electrons. For example, the TiO phase shows a wide homogeneity range from $Ti_{0.80}O$ to $Ti_{1.30}O$ at higher temperatures, whereas the CoO phase shows a narrow homogeneity range from $Co_{0.988}O$ to $Co_{1.000}O$. For the case of non-stoichiometric compounds showing an extremely narrow homogeneity range, such as NiO, it is very difficult to determine precisely the width of non-stoichiometry by usual chemical analysis (see Table 1.7 below).

The value of non-stoichiometry δ in $Ni_{1-\delta}O$ is 1×10^{-3} at most, that is, there is only one vacancy in 1000 lattice points of Ni. Osburn and Vest[28] measured the electrical conductivity, σ, of high purity NiO (single crystal) as a function of temperature (1000–1400 °C) and oxygen partial pressure (1–10^{-4} atm), to elucidate the conduction mechanism. Figure 1.38 shows σ versus temperature curves at fixed P_{O_2} values. The following relation between σ and T is roughly satisfied for the measured temperature, T, and oxygen

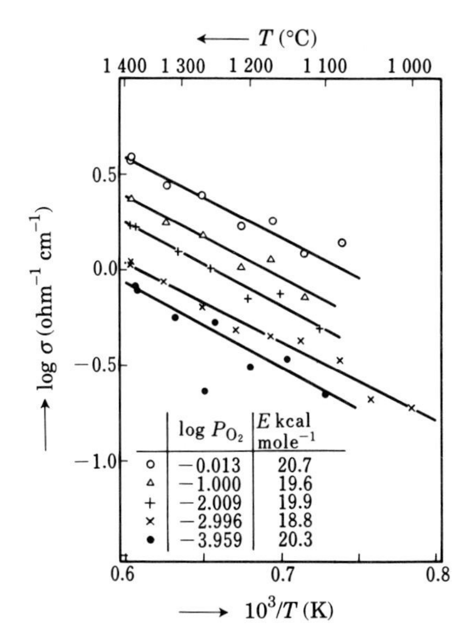

FIG. 1.38 Electrical conductivity of NiO single crystal as a function of temperature at different P_{O_2} values.[28]

pressure, P_{O_2}, range

$$\sigma = A \exp -\frac{E}{RT} \tag{1.148}$$

This equation does not explicitly include P_{O_2} as a variable, and the activation energy E is about 20 kcal mol^{-1} ($\sim$0.87 eV), independent of P_{O_2}. Figure 1.39, on the other hand, shows the σ versus P_{O_2} relation at fixed temperatures.

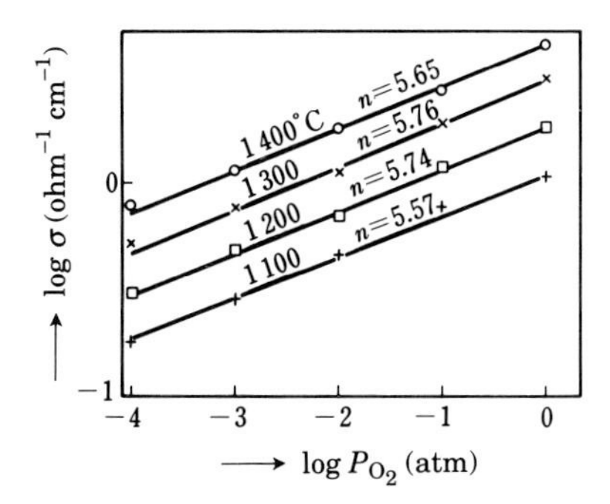

FIG. 1.39 Electrical conductivity of NiO single crystal as a function of P_{O_2} at different temperatures.[28]

This figure suggests the following relationship

$$\sigma = BP_{O_2}^{1/n} \tag{1.149}$$

where the value of n is about 5.5–5.8, independent of temperature. From eqns (1.148) and (1.149), we have

$$\sigma = CP_{O_2}^{1/n} \exp -\frac{E}{RT} \tag{1.150}$$

where n and E are constants, and C is also a constant which determines the absolute value of σ.

As described in Section 1.1, both the procedures of changing P_{O_2} at fixed temperature and changing temperature at fixed P_{O_2} control the non-stoichiometry in $Ni_{1-\delta}O$. Let us consider the relation between σ and δ in $Ni_{1-\delta}O$, in which the non-stoichiometry δ is believed to originate from metal vacancies. By use of the notation based on the effective charge, described in Section 1.3.7, the chemical equilibrium between the oxygen gas in the atmosphere and the oxygen in the solid may be expressed as

$$\tfrac{1}{2}O_2(\text{gas}) \rightleftharpoons O_0^{\times} + V_{Ni}^{\times} \qquad K_0 = \frac{[V_{Ni}^{\times}]}{P_{O_2}^{1/2}} \tag{1.151}$$

where $O_0^{\times}$ denotes the neutral oxygen on the regular oxygen sites in the solid phase and $V_{Ni}^{\times}$ denotes the neutral impurity level (acceptor) trapped by two holes. With increasing temperature, the acceptors are successively ionized by receiving electrons from the valence band, this process is expressed by the following equations

$$V_{Ni}^{\times} \rightleftharpoons V_{Ni}^{-} + h \qquad K_1 = \frac{[V_{Ni}^{-}]p}{[V_{Ni}^{\times}]} \tag{1.152}$$

$$V_{Ni}^{-} \rightleftharpoons V_{Ni}^{2-} + h \qquad K_2 = \frac{[V_{Ni}^{2-}]p}{[V_{Ni}^{-}]} \tag{1.153}$$

In the case of eqn (1.152) being dominant, we get the following chemical equation as a total equation,

$$\tfrac{1}{2}O_2(\text{gas}) \rightleftharpoons O_0^{\times} + V_{Ni}^{-} + h \qquad K_0K_1 = \frac{[V_{Ni}^{-}]p}{P_{O_2}^{1/2}} \tag{1.154}$$

In the case of eqn (1.153) being dominant, we have

$$\tfrac{1}{2}O_2(\text{gas}) \rightleftharpoons O_0^{\times} + V_{Ni}^{2-} + 2h \qquad K_0K_1K_2 = \frac{[V_{Ni}^{2-}]p^2}{P_{O_2}^{1/2}} \tag{1.155}$$

The following three cases are considered, depending on which equation dominates for electrical conduction of $Ni_{1-\delta}O$.

1. *Equation (1.151) dominates.* Electrical conduction due to non-stoichiometry is impossible. This situation may be realized at lower temperatures.

2. *Equation (1.152) dominates.* Holes contribute to the electrical conduction. As is obvious from eqn (1.154), the concentration of ionized acceptor $[V_{Ni}^-]$ equals p. Then we have

$$p = \sqrt{K_{01}}\, P_{O_2}^{1/4} \tag{1.156}$$

   where $K_{01} = K_0 K_1$. This situation is realized at intermediate temperatures.

3. *Equation (1.153) dominates.* From eqn (1.155), we get

$$p = \sqrt[3]{K_{012}}\, P_{O_2}^{1/6} \tag{1.157}$$

   where $K_{012} = K_0 K_1 K_2$. This situation is realized at higher temperatures.

As described in Section 1.2.1, the equilibrium constant K is related to the free energy change at standard state ΔG^0. Applying eqn (1.39) to eqns (1.154) and (1.155), the electrical conductivities σ_{01} and σ_{012} are expressed as

$$\sigma_{01} \propto P_{O_2}^{1/4} \exp -\frac{\Delta G_{01}^0}{2RT} \tag{1.158}$$

$$\sigma_{012} \propto P_{O_2}^{1/6} \exp -\frac{\Delta G_{012}^0}{3RT} \tag{1.159}$$

noting that σ is proportional to p. Comparing these theoretically derived equations with the experimentally derived one, eqn (1.150), it is apparent that eqn (1.159) is more appropriate to describe the electrical conductivity of $Ni_{1-\delta}O$. The equation shows that the carriers for electrical conductivity are holes originating from metal vacancies, the acceptors (impurity levels) are perfectly ionized and hence σ is roughly proportional to $P_{O_2}^{1/6}$.

Generally, by measuring σ as a function of P_{O_2} and T for metal oxides, we can gain knowledge on the character of defects and also on the mechanism of the electrical conductivity. For the case of $Ni_{1-\delta}O$, in fact, it has been confirmed by these measurements that the lattice defects are mainly metal vacancies charged with $-2e$ in the effective charge description and that the ionization energy of acceptors is about 0.87 eV. Similar investigations have been done on many compounds, by combination with other measurements such as the Hall effect, Zeebeck coefficient etc., a more conclusive discussion is possible.

Osburn and Vest[28] also measured the weight change as a function of P_{O_2} at fixed temperatures, the results of which are shown in Fig. 1.40. The

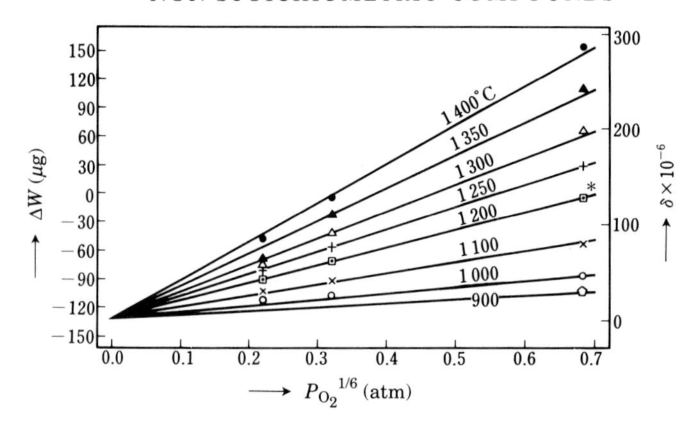

FIG. 1.40 Weight change (ΔW) of NiO powder as a function of P_{O_2} at different temperatures.[28] The standard point of weight change is marked with ▣* ($T = 1200$ °C, $P_{O_2} = 0.21$ atm).[28]

results can be expressed by the following equation,

$$\delta = 0.107 P_{O_2}^{1/6} \exp\{(-18\,600 \pm 1300)/RT\} \tag{1.160}$$

By taking it into consideration that $\delta \propto [V_{Ni}^{2-}] \propto p$, the equation is in good agreement with eqn (1.150) regarding the dependence of P_{O_2} and temperature (in the equation, $n = 5.7$ and $E = 20$ kcal mol^{-1}), although these equations were obtained experimentally in different ways.

1.4.6 Study of $(ZrO_2)_{0.85}(CaO)_{0.15}$*—electronic and ionic conduction*

ZrO_2 shows the following successive phase transitions on heating:

$$\text{Monoclinic} \xrightarrow{\sim 1000\,^\circ\text{C}} \text{Tetragonal} \xrightarrow{\sim 2370\,^\circ\text{C}} \text{Cubic}$$

ZrO_2 has been useful to the manufacturing industry mainly as firebricks, although breaking by heat cycling was a problem, due to the phase transitions mentioned above. It was found that addition of CaO or MgO to ZrO_2 is effective in stabilizing the cubic phase at lower temperatures, and the material has been put to practical uses. Figure 1.41 shows a part of the phase diagram of the pseudo-binary ZrO_2–CaO system,[29] in which the shaded area is the cubic phase denoted C_{ss}. The cubic phase containing more than 10 mol per cent CaO, with a Fluorite (CaF_2) type structure (Fig. 1.42), can be quenched to room temperature.

Figure 1.43 shows the electrical conductivity σ versus oxygen pressure P_{O_2} curves at fixed temperatures for $Zr_{0.85}Ca_{0.15}O_{1.85}$ or $(ZrO_2)_{0.85}(CaO)_{0.15}$ (called stabilized zirconia), as an example.[30] In the oxygen pressure range

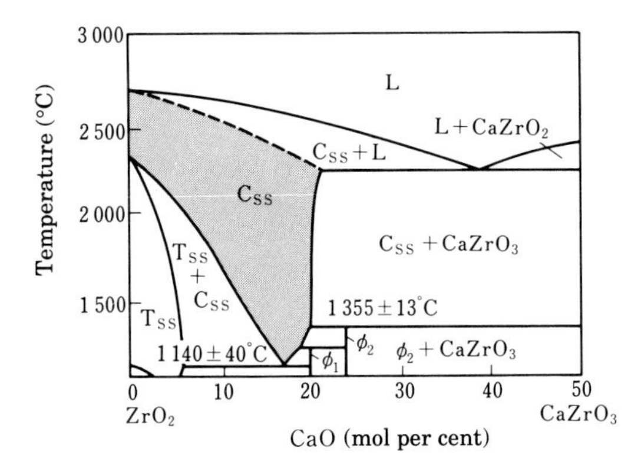

FIG. 1.41 Phase diagram of the pseudo-binary ZrO_2–CaO system.[29]

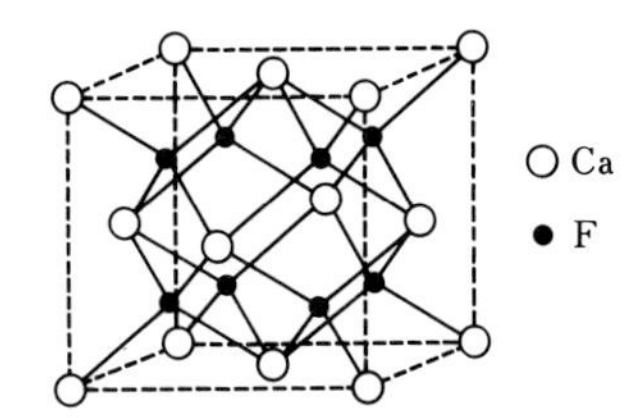

FIG. 1.42 Crystal structure of CaF_2 (Fluorite).

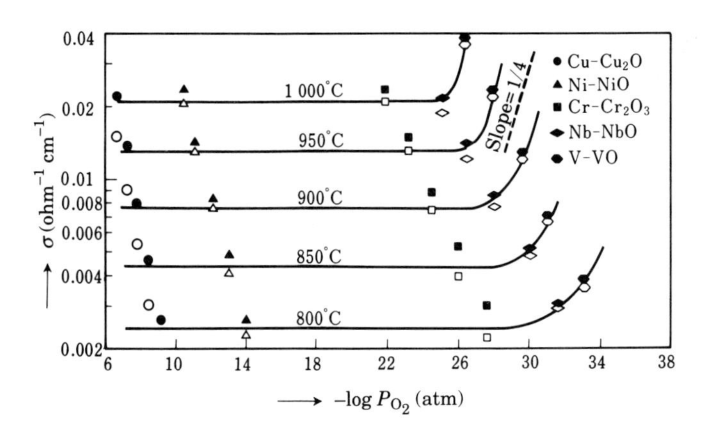

FIG. 1.43 Electrical conductivity of the stabilized zirconia $(ZrO_2)_{0.85}(CaO)_{0.15}$ as a function of P_{O_2} at different temperatures. P_{O_2} was controlled by the use of a two phase mixture such as Cu–Cu_2O, Ni–NiO etc. Open and closed marks are data obtained from different samples.[30]

$P_{O_2} > 10^{-26}$ atm, σ depends only on temperature, i.e. σ is constant at fixed temperature in spite of changing P_{O_2}. If the conduction mechanism was similar to that of $Ni_{1-\delta}O$, described in Section 1.4.5, σ would have to depend on P_{O_2} at fixed temperature.

Since the ratio of number of anions to cations in a unit cell for the Fluorite structure is 1 to 2, the compound $Zr_{0.85}Ca_{0.15}O_{1.85}$ can be said to be non-stoichiometric. The possible defect types are anion vacancies or interstitial cations. X-ray diffraction studies have definitely confirmed that the former type of defect structure is dominant; therefore, there exist oxygen vacancies up to 7.5 per cent. The concentration of oxygen vacancies must depend on P_{O_2}, as is usual for the metal oxides.

Consider the compound $Zr_{0.85}Ca_{0.15}O_{1.85-\delta}$, where the value of δ depends on P_{O_2} and T. In the case $\delta = 0$, there exist Ca_{Zr}^{2-} and V_O^{2+} as lattice defects. Namely, the substitution of Zr with Ca produces the atomic defects Ca_{Zr}^{2-} and also the same number of oxygen vacancies, due to the principle of charge neutrality in a crystal. It is impossible for δ to be less than 0, because both the cations have their highest valences, Zr^{4+} and Ca^{2+}, in $Zr_{0.85}Ca_{0.15}O_{1.85}$. In the case $\delta > 0$, the concentration of oxygen vacancies increases with increasing P_{O_2}, and accordingly Zr and Ca take lower valences.

Let us consider the defect equilibrium, with the condition $P_{O_2} < P_{O_2}^{\delta=0}$ at fixed temperature, where $P_{O_2}^{\delta=0}$ denotes the equilibrium oxygen pressure at $\delta = 0$. In this condition, oxygen escapes from the solid phase by the following reactions

$$O_O^{\times} \rightleftharpoons \tfrac{1}{2}O_2 + V_O^{\times} \qquad K_0 = P_{O_2}^{1/2}[V_O^{\times}] \tag{1.161}$$

$$V_O^{\times} \rightleftharpoons V_O^{+} + e \qquad K_1 = \frac{[V_O^{+}]n}{[V_O^{\times}]} \tag{1.162}$$

$$V_O^{+} \rightleftharpoons V_O^{2+} + e \qquad K_2 = \frac{[V_O^{2+}]n}{[V_O^{+}]} \tag{1.163}$$

where $O_O^{\times}$ denotes oxygen atoms on lattice points. In eqn (1.161), $V_O^{\times}$ represents the situation of (oxygen vacancy + two trapped electrons). Equations (1.162) and (1.163) show the ionization processes of donors, which donate electrons to the conduction band. It is noted that $[V_O^{2+}]$ in eqn (1.163) denotes the sum of $[V_O^{2+}]_{\delta=0}$ (7.5 per cent for the composition $Zr_{0.85}Ca_{0.15}O_{1.85}$) and $[V_O^{2+}]_{\delta>0}$. Because the concentration of defects derived from changing P_{O_2}, $[V_O^{2+}]_{\delta>0}$, is much lower than that derived from the intrinsic defects, i.e. the substitution of Zr with Ca, we can replace $[V_O^{2+}]$ with $[V_O^{2+}]_{\delta=0}$.

Let us consider the chemical equilibria expressed by eqns (1.161), (1.162), and (1.163). If eqn (1.162) is dominant, we have

$$n = \sqrt{K_{01}}\, P_{O_2}^{-1/4} \qquad K_{01} = K_0K_1 \tag{1.164}$$

since $[V_0^{+}] = n$. On the other hand, if eqn (1.163) is dominant, we get

$$n = \sqrt{K_{012}}\,[V_0^{2+}]^{-1/2} P_{O_2}^{-1/4} \qquad K_{012} = K_0 K_1 K_2 \tag{1.165}$$

noting that $[V_0^{2+}]$ nearly equals the concentration of Zr ions substituted by Ca ions, as mentioned above. Thus σ must be proportional to $P_{O_2}^{-1/4}$ at fixed temperatures, in these cases, if the conduction mechanism is very similar to that of $Ni_{1-\delta}O$. As seen from Fig. 1.43, this occurs in the lower oxygen pressure region ($P_{O_2} < 10^{-26}$ atm), if $P_{O_2} > 10^{-26}$ atm, σ is independent of P_{O_2} at fixed temperatures. This suggests that the conduction mechanism of the stabilized zirconia differs in origin from that of $Ni_{1-\delta}O$. From various kinds of measurements it was concluded that the carriers for electricity are oxygen ions, i.e. ionic conduction by diffusion of ions, as is often observed in solution systems. Solid substances which show ionic conductivity are called solid electrolytes. In these, the total electrical conductivity, σ_{total}, is expressed as

$$\sigma_{total} = \sigma_{ion} + \sigma_{|e|} \tag{1.166}$$

where σ_{ion} and $\sigma_{|e|}$ denote the ionic and electronic conductivities, respectively (|e| denotes both electrons and holes). The transference number t_{ion} is defined as $t_{ion} = \sigma_{ion}/\sigma_{total}$. In stabilized zirconia, oxygen ions can carry charges via oxygen ion vacancies V_0^{2+} (in terms of the effective charge expression, oxygen in the crystal has a neutral charge $O_0^{\times}$), as oxygen diffuses, charged V_0^{2+} moves in the opposite direction. The ionic conductivity σ_{ion} may be expressed by

$$\sigma_{ion} = 2|e|\mu_{ion}[V_0^{2+}] \tag{1.167}$$

where μ_{ion} is the mobility of oxygen ions. On the other hand, the electronic conductivity $\sigma_{|e|}$, which is caused by electrons derived from lattice defects, is expressed as

$$\sigma_{|e|} = |e|\mu_{|e|} P_{O_2}^{-1/4} \tag{1.168a}$$

or

$$\sigma_{|e|} = |e|\mu_{|e|} P_{O_2}^{-1/4}[V_0^{2+}]^{-1/2} \tag{1.168b}$$

In both cases, $\sigma_{|e|}$ depends on P_{O_2}. Figure 1.43 shows that eqn (1.167) is dominant in the pressure range $P_{O_2} > 10^{-26}$ atm and eqn (1.168) is dominant in the pressure range $P_{O_2} < 10^{-26}$ atm. In the former case, σ_{ion} depends on temperature because ion mobility is temperature dependent. The relation between the ionic and electronic conductivity for solid electrolytes is shown schematically in Fig. 1.44. Since ionic conductivity originates from diffusion of ions in the solid phase, σ_{ion} is closely related to the coefficient of self diffusion of ions (D_{ion}) shown by the following equation

$$D_{ion} = \frac{kT\sigma_{ion}}{Z^2 e^2 n} \tag{1.169}$$

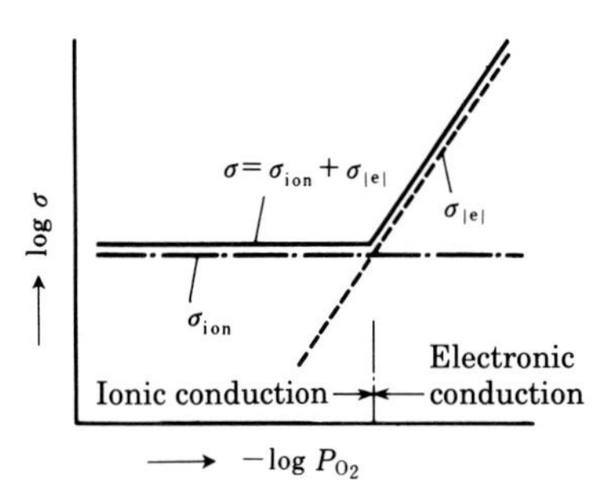

FIG. 1.44 Schematic drawing of log σ versus log P_{O_2} curves for ionic and electronic conductivity.

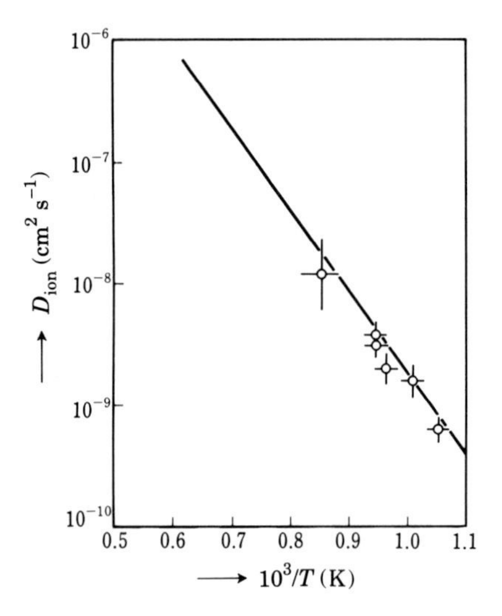

FIG. 1.45 Diffusion coefficient of ^{18}O in stabilized zirconia, D_{ion}, as a function of temperature.[31] The solid line was calculated by use of eqn (1.169).

where Z is the number of ionic charges and n is the number of ions per unit volume. If a crystal shows perfect ionic conductivity ($t_{ion} = 1$), we can calculate the value of D_{ion} from eqn (1.169) by use of the measured value of σ_{ion}. On the other hand, D_{ion} can be directly measured by a tracer experiment. By making a comparison between the directly measured and the calculated D_{ion} values, the transference number t_{ion} ($=\sigma_{ion}/\sigma_{total}$) can be estimated. Figure 1.45 shows the temperature dependence of D_{ion} for ^{18}O in stabilized zirconia, measured by Kingerly *et al.*[31] The straight line in the figure was calculated using eqn (1.169). This figures indicates that stabilized zirconia can be regarded as a perfect ionic conductor.

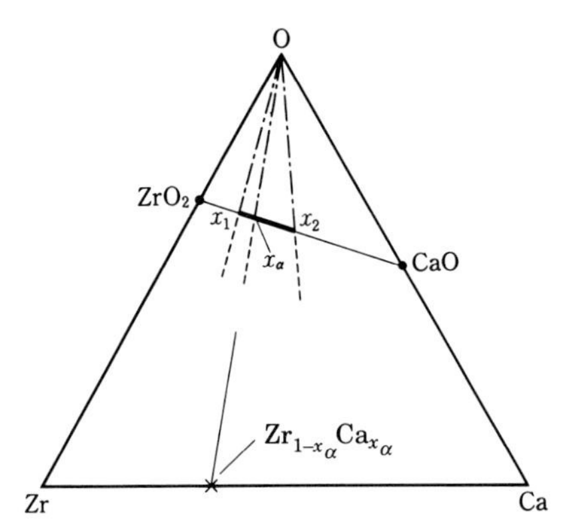

FIG. 1.46 Stabilized zirconia, $(ZrO_2)_{1-x}(CaO)_x$, in the ternary Zr–Ca–O system.

Stabilized zirconia has special or complex characteristics as a non-stoichiometric compound, compared with those mentioned above. The cubic phase of zirconia (ZrO_2), which is stable only at high temperatures (more than 2300 °C), is stabilized at lower temperature by adding CaO. In other words, the solid solution system $xZrO_2$–$(1-x)$CaO, i.e. $(Zr_xCa_{1-x}O_{1+x})$, has a cubic phase with the Fluorite structure in a wide composition range at lower temperatures. The homogeneous region depends on temperature: $0.80 < x < 0.90$ at 1750 °C and $0.78 < x < 1.0$ at 2300 °C. The solid solution $(Zr_xCa_{1-x}O_{1+x})$ is a phase appearing in the ternary Zr–Ca–O system, depicted in Fig. 1.46. If $(Zr_xCa_{1-x}O_{1+x})$ is stable in the composition region $x_1 < x < x_2$ at a fixed temperature T_1, the composition of the compound is expressed by the line x_1–x_2 in this figure, which is a part of the tie line ZrO_2–CaO. From the viewpoint that the composition changes continuously along the line x_1–x_2 but that the structure type stays the same, this compound can be said to be non-stoichiometric. The equilibrium oxygen pressure changes with x at constant temperature and pressure, because the ZrO_2–CaO system can be regarded as a pseudo-binary system in terms of the phase rule.

Next we consider the compound $Zr_{x_\alpha}Ca_{1-x_\alpha}O_{1+x_\alpha}$, in which the ratio of Zr to Ca is fixed and therefore both the concentrations of defects Ca_{Zr}^{2-} and V_0^{2+} are constant. This compound is also regarded as a compound on the tie line $Zr_{x_\alpha}Ca_{1-x_\alpha}$ (an alloy of the Zr–Ca system) –O. By changing P_{O_2}, the composition of oxygen in the compound changes as expressed by $Zr_{x_\alpha}Ca_{1-x_\alpha}O_{1+x_\alpha-\delta}$. At fixed temperatures there is a one-to-one correspondence between P_{O_2} and δ. The chemical reactions in this process are expressed by eqns (1.161), (1.162), and (1.163). It is noted, as mentioned above, that the chemical reaction corresponding to $\delta < 0$ does not occur. Thus the compound $Zr_{x_\alpha}Ca_{1-x_\alpha}O_{1+x_\alpha-\delta}$ can also be said to be non-stoichiometric.

1.4.7 Application of solid electrolytes. I. Measurement of oxygen partial pressure

Solid electrolytes, such as described in Section 1.4.6, have been utilized in a wide variety of fields, and here we describe their application to the chemistry of non-stoichiometric oxides (see Section 3.2 for industrial uses).

Consider a chamber which is partitioned into two rooms (I) and (II) by a wall made of solid electrolyte, as shown in Fig. 1.47(a). $P_{O_2}^{(I)}$ and $P_{O_2}^{(II)}$ represent the oxygen partial pressure of each room, where $P_{O_2}^{(I)}$ is higher than $P_{O_2}^{(II)}$. This can be regarded as a kind of reversible cell. When the cell operates the following chemical reactions take place at each surface of the electrolyte (denoted by electrode (I) and (II)):

$$\text{Electrode (I):} \quad O_2(\text{gas}) + 4e = 2O^{2-}(\text{electrolyte}) \tag{1.170}$$

$$\text{Electrode (II):} \quad 2O^{2-}(\text{electrolyte}) = O_2(\text{gas}) + 4e \tag{1.171}$$

The electromotive force (EMF) of the cell is expressed at temperature T as[32]

$$E = \frac{RT}{4F} \ln \frac{P_{O_2}^{(I)}}{P_{O_2}^{(II)}} \tag{1.172}$$

where R is the gas constant and F is the Faraday constant. If one of the P_{O_2} values is known (for example, P_{O_2} of air $= 0.21$ atm), by using this equation, the unknown P_{O_2} can be calculated after measuring E. Let us describe some experimental examples based on this principle.

As mentioned in Section 1.4.2, the oxygen partial pressure is controlled by changing the ratio of mixed gases, and the value of the pressure can be calculated, in principle, by the known equilibrium constants. The value

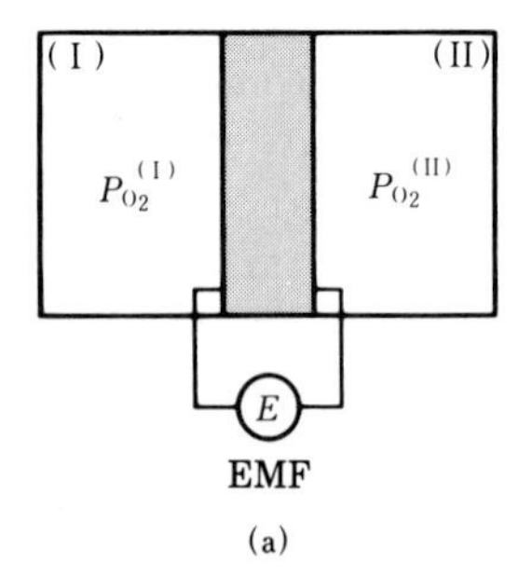

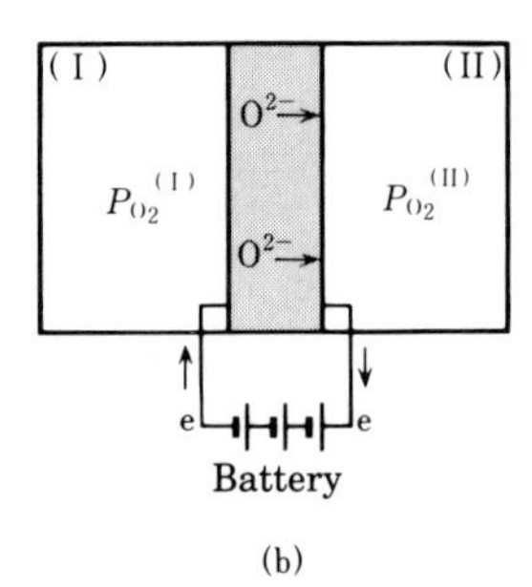

FIG. 1.47 Use of solid electrolyte (a) for P_{O_2} measurement, (b) for coulometric titration.

determined by experiments often deviates from the calculated one, the reasons for which may be attributed to the realization of a non-chemical equilibrium state due to the inadequacies of the ratio and the flowing rate of mixed gases, and the thermal separation of mixed gases. The open circles on the dotted line in Fig. 1.26 (which shows the measured values of P_{O_2} as a function of mixing ratio of CO_2 to H_2 at a constant flow rate ($=1\ \mathrm{cm\ s^{-1}}$) and at 1150 °C) are the values measured using stabilized zirconia, taking the air as a standard. As seen from the figure, the measured values gradually deviate from the calculated ones with increasing mixing ratio. Accordingly, in the case of controlling the oxygen pressure by the use of mixed gases, it is necessary to monitor the oxygen pressure *in situ* using the electrolyte cell.

The Magnéli compound V_nO_{2n-1} shows many discrete phases ($n = 3 - 9$) in the composition range V_2O_3–V_2O_4, each of which is non-stoichiometric, although the phases with higher values of n have a rather narrow range of homogeneity. In Section 1.4.2, a phase diagram study on the system was introduced, in which the thermogravimetric method was used, in an atmosphere of controlled oxygen pressure. Using this method it is difficult to obtain the P_{O_2} value for a two-phase region for solids directly, and therefore the value is usually obtained by extrapolation of the composition versus P_{O_2} curves of the neighbouring phases to the intercept. The temperature dependence of P_{O_2} for the coexisting phases of solids is a very important consideration in the preparation of Magnéli phases. It is, however, laborious to get it by measuring P_{O_2} versus composition at various temperatures, to give curves similar to those shown in Fig. 1.27. By use of the electrolyte cell, as shown in Fig. 1.47(a), we can easily get the data for the P_{O_2} values of the coexisting phases of solids, the procedure of which is, in principle, as follows: room (I) is filled with air as a standard, a mixture of two neighbouring phases $\{V_nO_{2n-1} + V_{(n+1)}O_{2(n+1)-1}\}$ is placed in room (II) and then the EMF is measured at various temperatures.

Figure 1.48 shows the P_{O_2} versus $1/T$ curves for each pair of neighbouring phases, measured using a stabilized zirconia test-tube in which the mixture was placed.[33] Filled circles (●) indicate the data from the EMF, open circles (○) indicate the data from Fig. 1.27, and crosses (×) indicate the points where the samples were prepared directly by controlling P_{O_2} and T, in order to confirm the EMF results. This figure suggests that each of the two-phase regions is on a straight line in the expression of $\log P_{O_2}$ versus $1/T$, from which the standard free energy change ΔG^0 for each pair can be calculated (eqn (1.39)). In this figure, one phase regions are marked 'V_nO_{2n-1}'. For example, 'V_2O_3' exists in the wide P_{O_2} and T region, which suggests the compound has to be written as the non-stoichiometric compound $V_{2-\delta}O_3$.

As noted above, we have to use solid electrolytes as the cell when $t_{ion} = 1$, i.e. the restricted regions of P_{O_2} and T.

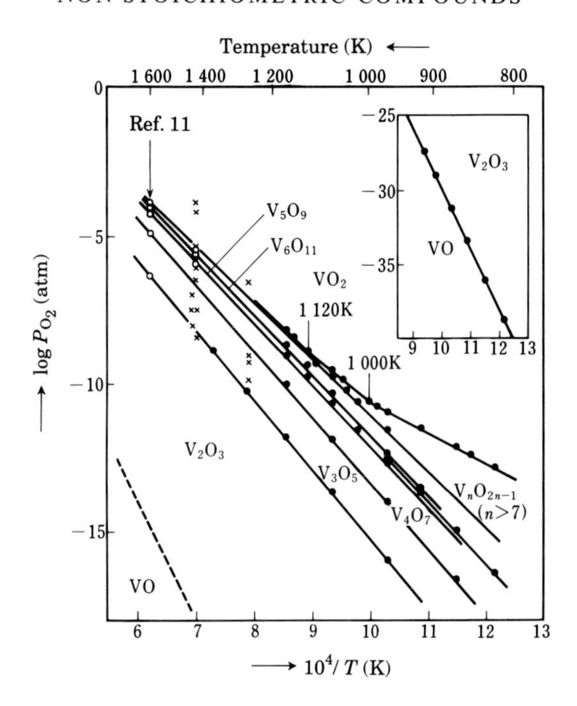

FIG. 1.48 Log P_{O_2} versus inverse temperature curves for the adjacent two-phase mixtures in the V_nO_{2n-1} system.[33]

1.4.8 Application of solid electrolytes. II. Coulometric titration

As mentioned in Section 1.4.7, the EMF for the electrolyte cell depicted in Fig. 1.47(a) is given by eqn (1.172). On the application of a voltage from an external battery, as shown in Fig. 1.47(b), on the other hand, we can forcibly transfer oxygen molecules from room (I) to room (II) by the following process: as in eqn (1.170), oxygen gas, O_2, in room (I) in the vicinity of the electrode is ionized to O^{2-}, which migrates towards room (II). At the surface of the electrode in room (II), O^{2-} is oxidized to O_2 as in eqn (1.171). The quantity of oxygen gas transferred from room (I) to (II) by this process, Δn_{O_2}, is described as

$$\Delta n_{O_2} = \frac{It}{4F} \tag{1.173}$$

where I is the electric current, t is the time duration, and F is the Faraday constant. Thus we can transfer oxygen gas from room (I) to room (II) quantitatively. By reversing the polarity of the battery, oxygen gas moves from room (II) to (I). This technique is usually referred to as coulometric titration, by analogy with ordinary titrations.

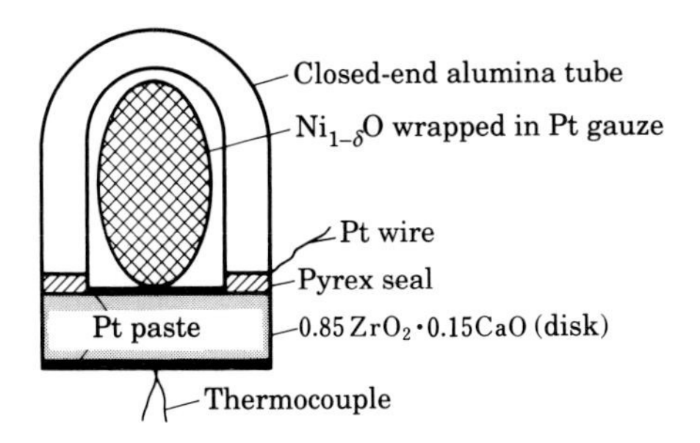

FIG. 1.49 Apparatus for coulometric titration of $Ni_{1-\delta}O$.[34]

Here let us introduce another report on the non-stoichiometry of NiO,[34] as a typical example of coulometric titration using stabilized zirconia. Figure 1.49 shows the schema of the experimental arrangement, i.e. a solid electrolyte cell. The cap of the cell is a closed-end alumina tube. A disc made of calcia-stabilized zirconia serves as the solid electrolyte. A Pyrex ring is placed between the cap and the electrolyte. Chunks of poorly sintered NiO are wrapped in Pt gauze and placed in the cell. Both the inner and outer faces of the electrolyte tablet are painted with platinum paste. The cell is heated in air up to 1050 °C; during heating the Pyrex ring melts and provides an excellent viscous seal. A platinum wire, which is connected to the inner electrode, is placed over the Pyrex ring.

In this experiment the non-stoichiometric nickel oxide is expressed as $NiO_{1+\gamma}$. Since $\gamma < 10^{-3}$, the concentration of nickel vacancies $[V_{Ni}]$ (this denotes the total concentration nickel vacancies having different electronic states, such as $V_{Ni}^{\times}$, V_{Ni}^{-}, and V_{Ni}^{2-}) can be described as

$$[V_{Ni}] = 1 - \frac{1}{1+\gamma} \simeq \gamma = \delta \tag{1.174}$$

where δ is in the usual expression $Ni_{1-\delta}O$. The experimental procedure and the analysis of the results are as follows. At the start of an experiment, the cell is shorted to equilibrate the $NiO_{1+\gamma}$ at $P_{O_2} = 0.21$ atm (air). Then oxygen is coulometrically titrated at a constant current, I, for a fixed period of time, t. After equilibration the cell potential becomes constant, and then the procedure is repeated. The number of moles of oxygen Δn_{O_2} removed from or added to the closed chamber by the titration is given by

$$\Delta n_{O_2} = \Delta n_{O_2}(\text{sp}) + \Delta n_{O_2}(\text{gas}) \tag{1.175}$$

where $\Delta n_{O_2}(\text{sp})$ is the change in the number of moles of oxygen due to oxidation or reduction of the solid sample and $\Delta n_{O_2}(\text{gas})$ is the change in the number of moles of oxygen gas in the chamber. The pressure used is sufficiently low for the oxygen gas to be regarded as an ideal gas. Hence we have

$$\Delta n_{O_2}(\text{gas}) = (V/RT)\,\Delta P_{O_2}$$

where V is the volume of the chamber (known value) and ΔP_{O_2} is the difference between the oxygen pressure before and after titration. Because ΔP_{O_2} can be measured by the change of the EMF, $\Delta n_{O_2}(\text{gas})$ can be calculated. Subtraction of $\Delta n_{O_2}(\text{gas})$ from Δn_{O_2} gives $\Delta n_{O_2}(\text{sp})$.

If $NiO_{1+\gamma_1}$ and $NiO_{1+\gamma_2}$ are the compositions before and after titration, respectively, then we get

$$\begin{aligned}\Delta\gamma = \gamma_2 - \gamma_1 &= \frac{2[NiO_{1+\gamma_1}]\,\Delta n_{O_2}(\text{sp})}{m_{NiO_{1+\gamma_1}}}\\ &\simeq \frac{2[NiO]}{m_{NiO}}\,\Delta n_{O_2}(\text{sp})\end{aligned} \tag{1.176}$$

where [NiO] and m_{NiO} denote the molecular weight of NiO and the mass of the sample used in the experiment, respectively. The change of P_{O_2} is measured at fixed temperatures, by repeating this process ($\gamma_1 \to \gamma_2 \to \gamma_3 \cdots$). Because $\gamma = \delta = [V_{Ni}]$, as noted in Section 1.4.5, the following assumption is plausible

$$\gamma = kP_{O_2}^{1/n} \tag{1.177}$$

Then we have

$$\begin{aligned}\Delta\gamma &= \gamma_2 - \gamma_1\\ &= k[P_{O_2}^{1/n}(\gamma_2) - P_{O_2}^{1/n}(\gamma_1)]\end{aligned} \tag{1.178}$$

where $\Delta P_{O_2} = P_{O_2}(\gamma_2) - P_{O_2}(\gamma_1)$. On combining eqn (1.173) with eqns (1.175) and (1.176), we get

$$\Delta\gamma = \frac{2[NiO]}{m_{NiO}}\left[\frac{It}{4F} - \Delta n_{O_2}(\text{gas})\right] \tag{1.179}$$

Comparing this equation with (1.178), it is clear that we can get the value of n by plotting $[It - 4F\,\Delta n_{O_2}(\text{gas})]$ against $[P_{O_2}^{1/n}(\gamma_2) - P_{O_2}^{1/n}(\gamma_1 = \text{ref})]$ to fit a straight line. To illustrate the dependence of the curvature of such plots on the value of n, the same data at 836 °C are plotted for $n = 6, 5, 4$ in Fig. 1.50(a). It is apparent that the best linear fit is obtained for $n = 6$. In Fig. 1.50(b) the data for other temperatures are plotted using $n = 6$, which gives the straightest lines. These results definitely suggest that the chemical equilibrium of lattice defects in $Ni_{1-\delta}O$ can be expressed by eqn (1.155).

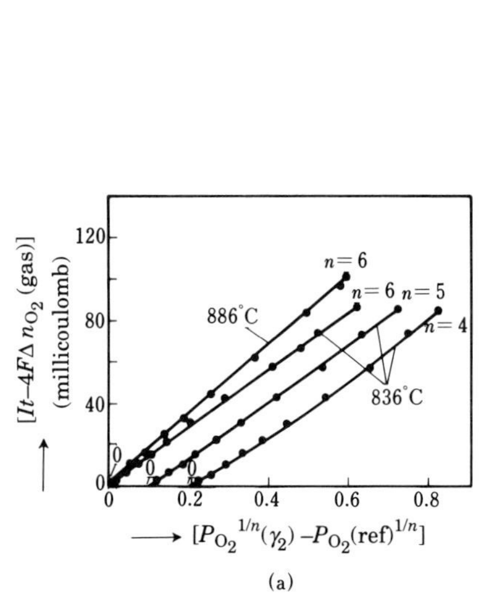

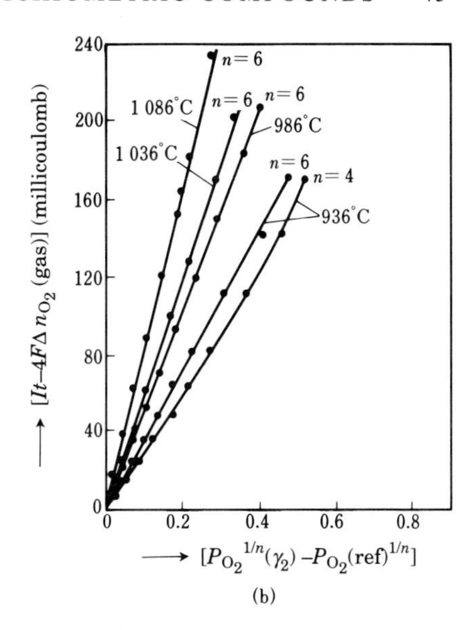

FIG. 1.50 $[It - 4F\,\Delta n_{O_2}(\text{gas})]$ versus $[P_{O_2}^{1/n}(\gamma_2) - P_{O_2}^{1/n}(\text{ref})]$ for $NiO_{1+\gamma}$ at different temperatures (eqns (1.178) and (1.179)).[34]

Also, the results are in good agreement with those mentioned in Section 1.4.5. The value of k in eqn (1.177), which depends on temperature, corresponds to the value of γ $(=\delta)$ at $P_{O_2} = 1$ atm. The temperature dependence of k, which is calculated from the slopes of the straight lines in Fig. 1.50, can be expressed as

$$\gamma(P_{O_2} = 1\,\text{atm}) = k = \left(0.51 \pm \begin{matrix}0.21\\0.15\end{matrix}\right) \exp\left(\frac{-19\,000 \pm 8700}{RT}\right) \tag{1.180}$$

Combining this equation with eqn (1.177), we get

$$\gamma = \left(0.51 \pm \begin{matrix}0.21\\0.15\end{matrix}\right) P_{O_2}^{1/6} \exp\left(\frac{-19\,000 \pm 8700}{RT}\right) \tag{1.181}$$

taking into consideration that $n = 6$ in eqn (1.177).

Thus, the temperature and P_{O_2} dependence of the non-stoichiometry γ in $NiO_{1+\gamma}$ could be measured precisely by use of the solid electrolyte, without chemical analysis. This result is shown in Fig. 1.51, as $\log P_{O_2}$ versus T^{-1} curves, in which the dotted lines indicate the iso-γ values. This figure shows that in order to get stoichiometric NiO, the sample has to be prepared at low temperature and low oxygen partial pressure in the NiO phase region.

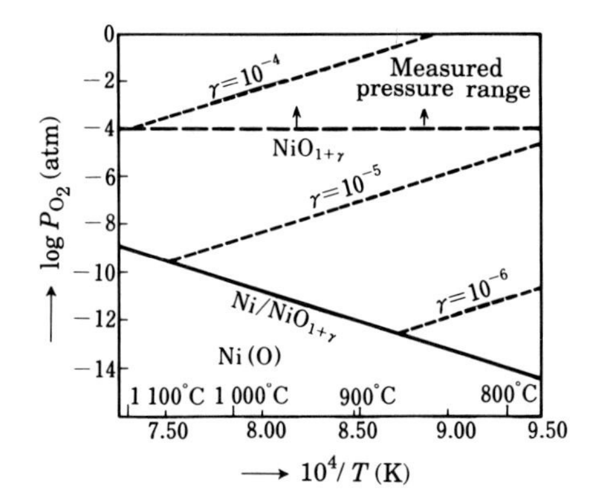

FIG. 1.51 Log P_{O_2} versus inverse temperature T^{-1} for $NiO_{1+\gamma}$.[34]

It is hard to say whether the absolute value of γ $(=\delta)$ calculated from eqn (1.181) agrees with that from eqn (1.160), although the dependency of oxygen partial pressure is consistent.

1.4.9 *Study of $Cu_{2-\delta}O$—measurement of diffusion coefficient*

The non-stoichiometry δ of $Cu_{2-\delta}O$ has been extensively studied by various methods such as chemical analysis of a quenched sample, thermogravimetry, electrical resistivity measurement, and coulometric titration, but the results obtained are not consistent.

Recently Peterson and Wiley[35] discussed the non-stoichiometry of $Cu_{2-\delta}O$, based on their measurement of the diffusion coefficient of ^{64}Cu (radioactive isotope of Cu) in $Cu_{2-\delta}O$ as a function of temperature and oxygen partial pressure, P_{O_2}. Here let us review the results in some detail.

The crystal structure of Cu_2O (cuprite) is shown in Fig. 1.52, oxygen

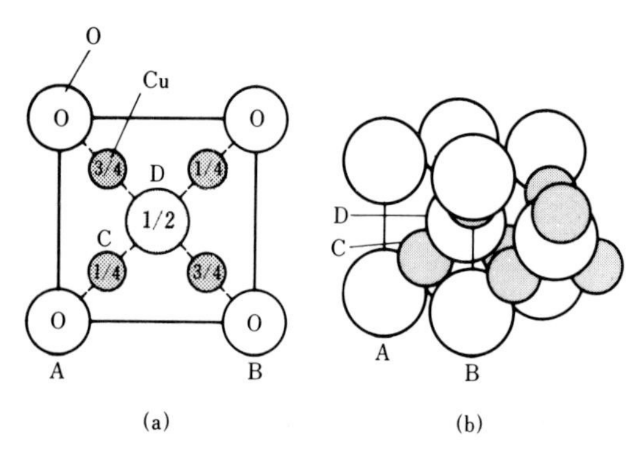

FIG. 1.52 Crystal structure of Cu_2O (cuprite).

atoms form a body-centered cubic lattice and each oxygen atom is surrounded by a tetrahedron of copper atoms.

The studies on $Cu_{2-\delta}O$ mentioned above concluded that Cu_2O is a metal-deficient p-type semiconductor with cation vacancies. It was not established, however, which kinds of defects ($V_{Cu}^{\times}$, V_{Cu}^{-}) were dominant and what the effect of O_i (interstitial oxygen) was on non-stoichiometry. To clarify these points, Peterson and Wiley measured the diffusion coefficient, D^*, of Cu in $Cu_{2-\delta}O$, by use of ^{64}Cu as a tracer over the temperature range 700–1153 °C and for oxygen partial pressures, P_{O_2}, greater than 10^{-6} atm. It has been widely accepted that lattice defects play an important role in the diffusion of atoms or ions. Accordingly it can be expected that the measurement of D^* gives important information on the lattice defects.

The 'tracer-sectioning technique' was employed for the diffusion measurement. The distribution of specific activity C of a radioactive tracer at fixed temperature and oxygen partial pressure is expressed as

$$C = \frac{M}{(\pi D^* t)^{1/2}} \exp\left(\frac{-x^2}{4D^* t}\right) \tag{1.182}$$

where x is the penetration distance, D^* is the tracer diffusion coefficient, M is the activity per unit area deposited at $t = 0$ on the surface $x = 0$ (see Fig. 1.53(a)), and t is the time elapsed for diffusion. The concentration profile of the radioactive tracer was determined by a serial sectioning technique. The radioactivity in each section was measured using a scintillation counter. The value of D^* (hereafter, D_{Cu}^*) can be determined by the measurement of the dependency of C on x, as is clear from this equation.

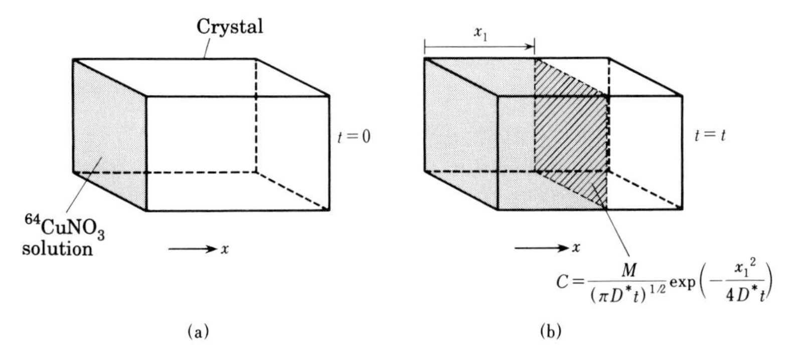

FIG. 1.53 Measurement of the diffusion coefficient by use of a radioactive isotope. (a) At $t = 0$, radioactive isotope M ($^{64}CuNO_3$ solution in this experiment) is deposited on the surface of the crystal ($x = 0$). (b) At $t = t$ and constant P_{O_2}, C(radioactivity at x_1) = $[M/(\pi D^* t)^{1/2}] \exp[-x_1^2/(4D^* t)]$.

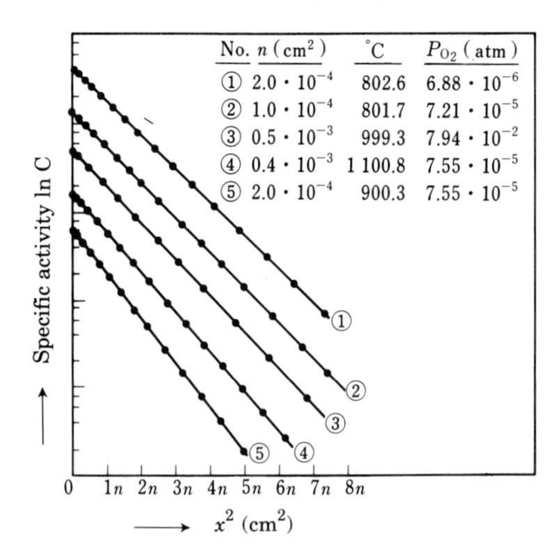

FIG. 1.54 Specific activity, C, of $Cu_{2-\delta}O$ versus x^2 at different P_{O_2} values and temperatures (eqn (1.182)).[34]

Figure 1.54 shows the typical penetration curves for ^{64}Cu diffusion in Cu_2O, plotted as log C versus x^2. Experimental conditions (temperature, °C, partial oxygen pressure, P_{O_2}, and thickness of each section, n) are also tabulated in this figure. Because all the plots are linear under the conditions, the distribution of radioactive tracer follows eqn (1.182). The errors in D^*_{Cu} from the least-square fits of data to eqn (1.182) are less than 0.5 per cent.

Figure 1.55 shows the log D^*_{Cu} versus the inverse temperature relation at $P_{O_2} = 7.55 \times 10^{-5}$ atm, from which we get the relation (in the Arrhenius type form)

$$D^*_{Cu} = D^*_0 \exp\left(-\frac{Q}{RT}\right) \tag{1.183}$$

where $D^*_0 = (9.1 \pm 1.9) \times 10^{-4}\ \mathrm{cm^2\ s^{-1}}$ and $Q = (28.47 \pm 0.47)\ \mathrm{kcal\ mol^{-1}}$. Figure 1.55(b) shows the deviation of the measured D^*_{Cu} value from the values calculated from eqn (1.183), indicating that the deviation is less than ± 10 per cent, and that the deviation varies systematically with temperature. The deviation gives meaningful information, as discussed below.

Figure 1.56 shows the P_{O_2} dependence of D^*_{Cu} at fixed temperatures. The slopes of the curves, log D^*_{Cu}/log P_{O_2}, are quite close to $\frac{1}{6}$ and decrease slightly with decreasing temperature.

Let us consider the simple defect model for $Cu_{2-\delta}O$, proposed by Peterson and Wiley, based on the assumption that the concentration of defects is low and therefore the interaction among defects is negligible. Because it has been discovered, as mentioned above, that the dominant lattice defects are metal

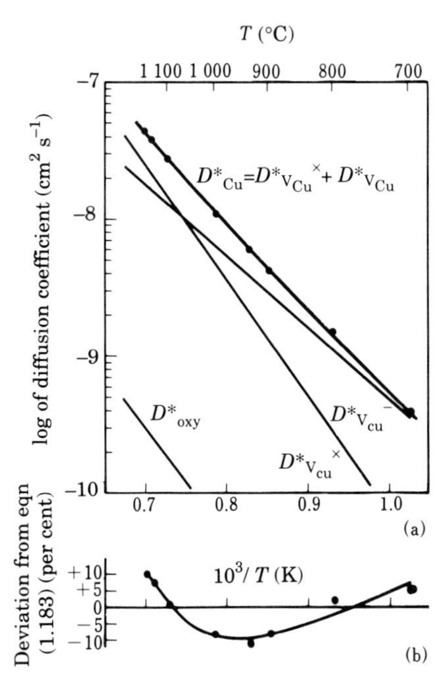

FIG. 1.55 (a) Diffusion coefficients of $Cu_{2-\delta}O$ as a function of temperature at $P_{O_2} = 7.55 \times 10^{-5}$ atm.[34] (b) Deviation from eqn (1.183) in per cent.[34] Solid circles are measured points.

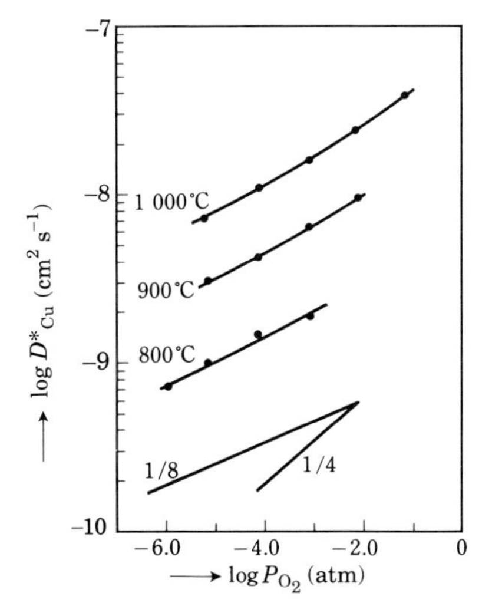

FIG. 1.56 Diffusion coefficient of $Cu_{2-\delta}O$ as a function of P_{O_2} at various temperatures.[34]

vacancies, the formation of these vacancies can be described by the following chemical reactions

$$\tfrac{1}{4}O_2(\text{gas}) \rightleftharpoons \tfrac{1}{2}O_O^{\times} + V_{Cu}^{\times} \qquad K_{V_{Cu}^{\times}} = \frac{[V_{Cu}^{\times}]}{P_{O_2}^{1/4}} \tag{1.184}$$

$$V_{Cu}^{\times} \rightleftharpoons V_{Cu}^{-} + h \qquad K_{V_{Cu}^{-}} = \frac{[V_{Cu}^{-}]p}{[V_{Cu}^{\times}]} \tag{1.185}$$

where $O_O^{\times}$, $V_{Cu}^{\times}$, V_{Cu}^{-}, and h denote a neutral anion on an anion lattice site (regular site for anion), a neutral cation vacancy, a singly ionized cation vacancy, and an electron hole, respectively. From these equations we have

$$[V_{Cu}^{-}] = \frac{K_{V_{Cu}} P_{O_2}^{1/4}}{p} \qquad K_{V_{Cu}} = K_{V_{Cu}^{\times}} K_{V_{Cu}^{-}} \tag{1.186}$$

On the other hand, oxygen diffusion measurements,[36,37] by use of ^{18}O as a tracer, indicate that singly charged interstitials, O_i^{-}, are the dominant defects of anions. (Anion defects O_i^{-} are considered as minority defects compared to the cation defects.) The formation of O_i^{-} is described by the chemical reaction

$$\tfrac{1}{2}O_2(\text{gas}) \rightleftharpoons O_i^{-} + h \qquad K_{O_i^{-}} = \frac{[O_i^{-}]p}{P_{O_2}^{1/2}} \tag{1.187}$$

If the chemical reactions described by eqns (1.184), (1.185), and (1.187) are in chemical equilibrium, the following charge neutrality condition in a crystal must hold

$$p = [V_{Cu}^{-}] + [O_i^{-}] \tag{1.188}$$

We can get the values of the unknown quantities $[V_{Cu}^{\times}]$, $[V_{Cu}^{-}]$, $[O_i^{-}]$, and p as functions of P_{O_2} by solving eqns (1.184), (1.185), (1.187), and (1.188) if all the equilibrium constants are known. First let us try to get answers in the extreme conditions: $p \doteqdot [V_{Cu}^{-}] \gg [O_i^{-}]$ (Region I) and $p \doteqdot [O_i^{-}] \gg [V_{Cu}^{-}]$ (Region II). The results are tabulated in Table 1.4 and plotted as log of defect

Table 1.4

Approximated solution of eqns (1.184)–(1.188) (see Fig. 1.57)

Concentration of defects	Region (I) $p \doteqdot [V_{Cu}^{-}] \gg [O_i^{-}]$	Region (II) $p \doteqdot [O_i^{-}] \gg [V_{Cu}^{-}]$
p	$(K_{V_{Cu}})^{1/2} P_{O_2}^{1/8}$	$(K_{O_i^{-}})^{1/2} P_{O_2}^{1/4}$
$[V_{Cu}^{\times}]$	$K_{V_{Cu}^{\times}} P_{O_2}^{1/4}$	$K_{V_{Cu}^{\times}} P_{O_2}^{1/4}$
$[V_{Cu}^{-}]$	$(K_{V_{Cu}})^{1/2} P_{O_2}^{1/8}$	$(K_{O_i^{-}})^{-1/2} K_{V_{Cu}}$
$[O_i^{-}]$	$(K_{V_{Cu}})^{-1/2} K_{O_i^{-}} P_{O_2}^{3/8}$	$(K_{O_i^{-}})^{1/2} P_{O_2}^{1/4}$

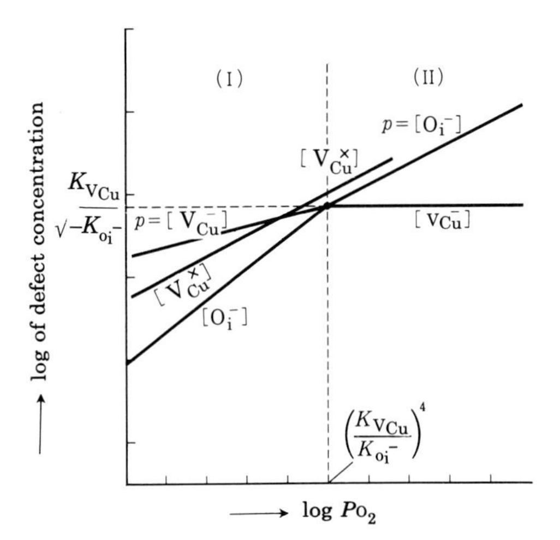

FIG. 1.57 Graphic solution of defect concentration change with P_{O_2} (eqns (1.184)–(1.188)).

concentration versus log P_{O_2} in Fig. 1.57. In Region I, the concentration of holes p, which contributes to electrical conductivity σ, is given as

$$p \doteqdot (K_{V_{Cu}})^{1/2} P_{O_2}^{1/8} \tag{1.189}$$

namely $\sigma \propto P_{O_2}^{1/8}$. In Region II, p is expressed as

$$p \doteqdot (K_{O_i^-})^{1/2} P_{O_2}^{1/4} \tag{1.190}$$

namely $\sigma \propto P_{O_2}^{1/4}$.

Here let us try to solve eqns (1.184), (1.185), (1.187), and (1.188) correctly. Using these equations, we obtain

$$p = \{K_{V_{Cu}} P_{O_2}^{1/4} + K_{O_i^-} P_{O_2}^{1/2}\}^{1/2} \tag{1.191}$$

If we assume that $[V_{Cu}^-] \gg [O_i^-]$, for example, it can be shown that a power-series expansion of eqn (1.191) yields a P_{O_2} dependence of hole concentration, which contains a $P_{O_2}^{1/8}$ term, a $P_{O_2}^{3/8}$ term, and higher order terms (see eqn (1.189)).

The electrical conductivity σ is expressed by the relation

$$\sigma = |e| \mu_h p \tag{1.192}$$

where μ_h and $|e|$ denote the hole mobility and the magnitude of the charge on the electron. Combining this equation with eqn (1.191), we get

$$\begin{aligned} \sigma &= |e| \mu_h \{K_{V_{Cu}} P_{O_2}^{1/4} + K_{O_i^-} P_{O_2}^{1/2}\}^{1/2} \\ &\equiv \{\kappa_{V_{Cu}} P_{O_2}^{1/4} + \kappa_{O_i^-} P_{O_2}^{1/2}\}^{1/2} \end{aligned} \tag{1.193}$$

where

$$\kappa_{V_{Cu}} = (\mu_h e)^2 K_{V_{Cu}} \equiv (\kappa_{V_{Cu}})_0 \exp(-Q\kappa_{V_{Cu}}/RT) \qquad (1.194)$$

$$\kappa_{O_i^-} = (\mu_h e)^2 K_{O_i^-} \equiv (\kappa_{O_i^-})_0 \exp(-Q\kappa_{O_i^-}/RT) \qquad (1.195)$$

Equations (1.194) and (1.195) can be accepted, within reason, because both the chemical equilibrium constants and the hole mobility μ_h for semi-conductors have an Arrhenius-type temperature dependence. It has been shown, by a least-square fitting of the electrical conductivity data of Maruenda *et al.*[38] to eqn (1.193), that 85 per cent of the data points are within 1.5 per cent of the calculated values, as shown in Fig. 1.58. This indicates that the model proposed here gives an accurate description of the data. The fitting parameters are listed in Table 1.5.

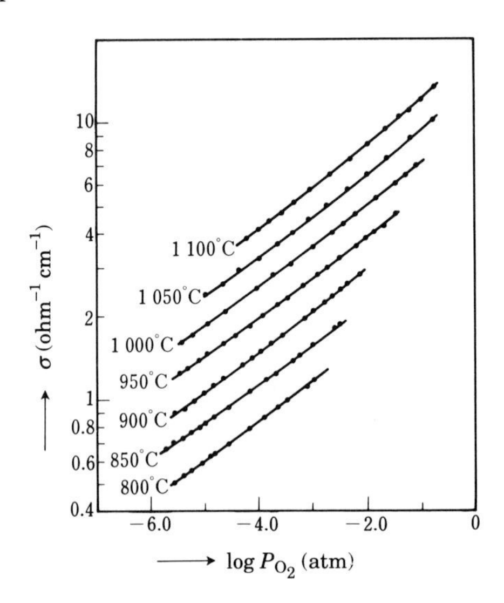

FIG. 1.58 Electrical conductivity of $Cu_{2-\delta}O$ as functions of log P_{O_2} at various temperatures.[38] Solid lines are calculated ones (eqn (1.193)).

Table 1.5

Value of constants for equations in Section 1.4.9

eqn (1.194)	$(\kappa_{V_{Cu}})_0$	$= (6.78 \pm 0.78) \times 10^6$	$Q_{\kappa V_{Cu}}$	$= (29.53 \pm 0.25)$ kcal mol^{-1}
eqn (1.195)	$(\kappa_{O_i^-})_0$	$= (1.6 \pm 1.0) \times 10^7$	$Q_{\kappa O_i^-}$	$= (31.6 \pm 1.7)$ kcal mol^{-1}
eqn (1.197)	D^*_{18}	$= (3.7 \pm 1.1) \times 10^{-4}$ cm^2 s^{-1}	Q_{18}	$= (24.46 \pm 0.64)$ kcal mol^{-1}
eqn (1.197)	D^*_{14}	$= (4.1 \pm 2.2) \times 10^{-1}$ cm^2 s^{-1}	Q_{14}	$= (40.6 \pm 1.4)$ kcal mol^{-1}
eqn (1.200)	$(d_{V_{Cu}})_0$	$= (1.13 \pm 0.36) \times 10^{-7}$	$Q_{dV_{Cu}}$	$= (9.28 \pm 0.69)$ kcal mol^{-1}
eqn (1.201)	$(\kappa_{V_{Cu}^{\times}})_0$	$= (2.1 \pm 1.4) \times 10^6$	$Q_{\kappa V_{Cu}}$	$= (29.6 \pm 1.6)$ kcal mol^{-1}

The diffusion constant for Cu can also be discussed using this model. Because diffusion by singly charged cation vacancies, $D^*_{V^-_{Cu}}$, and by neutral cation vacancies, $D^*_{V^\times_{Cu}}$, are both important, we may write

$$\begin{aligned} D^*_{Cu} &= D^*_{V^-_{Cu}} + D^*_{V^\times_{Cu}} \\ &= D^*_1[V^-_{Cu}] + D^*_2[V^\times_{Cu}] \\ &= D^*_1\sqrt{K_{V_{Cu}}}\, P^{1/8}_{O_2} + D^*_2 K_{V^\times_{Cu}} P^{1/4}_{O_2} \end{aligned} \qquad (1.196)$$

where it is assumed, $[O^-_i] \ll [V^\times_{Cu}], [V^-_{Cu}]$. D^*_{Cu} varies with P_{O_2} as the sum of a $P^{1/8}_{O_2}$ term and a $P^{1/4}_{O_2}$ term. Considering that the equilibrium constants show an Arrhenius-type behaviour, we can rewrite the equation as

$$D^*_{Cu} = D^*_{18} P^{1/8}_{O_2} \exp\left(-\frac{Q_{18}}{RT}\right) + D^*_{14} P^{1/4}_{O_2} \exp\left(-\frac{Q_{14}}{RT}\right) \qquad (1.197)$$

The tracer diffusion data shown in Fig. 1.55 correlate well with this equation. The thick curves in this figure, (a) and (b), are calculated ones using fitting parameters listed in Table 1.5. The difference between eqn (1.183) and eqn (1.197) is in the number of exponential terms. The deviation from eqn (1.183) indicates that the correct form of the diffusion equation must contain more than one exponential term.

We shall now develop an exact expression for D^*_{Cu} which includes the coupling of V^-_{Cu} and O^-_i through the charge neutrality condition given by eqn (1.188). Starting with the first equation in eqn (1.196) and following a similar procedure to that used in developing eqn (1.193), we can get*

$$D^*_{Cu} = d_{V_{Cu}} \left\{ \frac{\kappa_{V_{Cu}}}{[\kappa_{V_{Cu}} P^{1/4}_{O_2} + \kappa_{O^-_i} P^{1/2}_{O_2}]^{1/2}} + \kappa_{V^\times_{Cu}} \right\} P^{1/4}_{O_2} \qquad (1.198)$$

$$d_{V_{Cu}} \equiv \frac{D_{V_{Cu}} f_V}{\mu_h |e|} \qquad (1.199)$$

where both $d_{V_{Cu}}$ and $\kappa_{V^\times_{Cu}}$ also have an Arrhenius-type temperature dependence given by

$$d_{V_{Cu}} \equiv (d_{V_{Cu}})_0 \exp(-Q_{dV_{Cu}}/RT) \qquad (1.200)$$

$$\kappa_{V^\times_{Cu}} = (\mu_h |e|) K_{V^\times_{Cu}} \equiv (\kappa_{V^\times_{Cu}})_0 \exp(-Q_{\kappa V^\times_{Cu}}/RT) \qquad (1.201)$$

* The first line in eqn (1.196) is rewritten as

$$\begin{aligned} D^*_{Cu} &= D^*_{V^-_{Cu}} + D^*_{V^\times_{Cu}} \\ &= D_V f_v([V^-_{Cu}] + [V^\times_{Cu}]) \end{aligned}$$

where D_V and f_v denote the uncorrelated diffusion coefficient and the correlation factor, respectively (see ref. 36). On combining eqn (1.186) with eqn (1.194), $[V^-_{Cu}]$ is expressed as

$$[V^-_{Cu}] = \kappa_{V_{Cu}} P^{1/4}_{O_2} / (p \cdot (\mu_n e)^2)$$

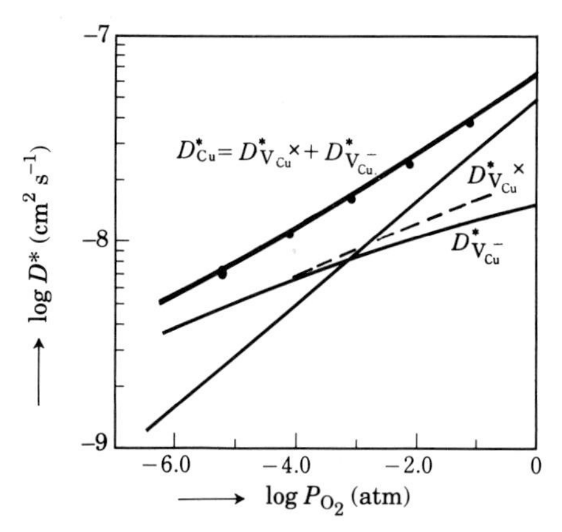

FIG. 1.59 Diffusion coefficient of $Cu_{2-\delta}O$ as functions of P_{O_2} at 1000 °C (see Fig. 1.56).[34] The dotted line shows the deviation from a linear relation between log of defect concentration $D^*_{V_{Cu}^-}$ and $\log P_{O_2}$.

The parameters $\kappa_{V_{Cu}}$ and $\kappa_{O_i^-}$ are already known from the least-squares fitting of the electrical conductivity data to eqn (1.193) (see Fig. 1.58). By a least-squares fitting of the tracer diffusion data to eqn (1.198) for all temperatures and P_{O_2} values, the values of $d_{V_{Cu}}$ and $\kappa_{V_{Cu}^{\times}}$, i.e. $(d_{V_{Cu}})_0$, $Q_{dV_{Cu}}$, $(\kappa_{V_{Cu}^{\times}})_0$, and $Q_{\kappa V_{Cu}^{\times}}$, have been determined, as listed in Table 1.5. The temperature dependence of D^*_{Cu} (thick line), $D^*_{V_{Cu}^{\times}}$, and $D^*_{V_{Cu}^-}$ for $P_0 = 7.55 \times 10^{-5}$ atm are shown in Fig. 1.55. In Fig. 1.59 is shown the P_{O_2} dependence of these values at 1000 °C. The slight deviation from the linear relation (dotted line) for $D^*_{V_{Cu}^-}$ at higher values of P_{O_2} results from the coupling of V_{Cu}^- with O_i^-, see eqn (1.188). (The diffusion constant of neutral copper vacancies $D^*_{V_{Cu}^{\times}}$, shows a correct linear relation in the graph of $\log D^*_{V_{Cu}^{\times}}$ versus $\log P_{O_2}$, because $[V_{Cu}^{\times}]$ does not depend on $[O_i^-]$.)

The relationship between eqns (1.197) and (1.198) is as follows: It can be easily seen that the expression for D^*_{Cu} in eqn (1.198) contains the sum of more than two exponential terms, each with a different activation energy. Considering the P_{O_2} dependence, it is straightforward to show that D^*_{Cu} in eqn (1.198) has a $P_{O_2}^{1/4}$ term, a $P_{O_2}^{1/8}$ term, and higher-order terms by a power-series expansion, assuming $[O_i^-] \ll [V_{Cu}^-]$. Thus eqn (1.197) has a similar content, in essence, to eqn (1.198), although the latter equation has the higher-order terms of P_{O_2}.

The absolute values of the various defect concentrations, p, $[V_{Cu}^{\times}]$, $[V_{Cu}^-]$, and $[O_i^-]$, can now be determined from measurements of δ in $Cu_{2-\delta}O$. Similar to the calculation of σ and D^*_{Cu} (eqns (1.193) and (1.198)), the defect concentration δ in a mole fraction of $Cu_{2-\delta}O$ is written as

$$\delta = [V_{Cu}^{-}] + [V_{Cu}^{\times}] + 2[O_i^{-}] \qquad (1.202)^*$$

$$= \frac{1}{\mu_h|e|}\left\{\frac{\kappa_{V_{Cu}^{-}}P_{O_2}^{1/4} + 2\kappa_{O_i^{-}}P_{O_2}^{1/2}}{[\kappa_{V_{Cu}^{-}}P_{O_2}^{1/4} + \kappa_{O_i^{-}}P_{O_2}^{1/2}]^{1/2}} + \kappa_{V_{Cu}^{\times}}P_{O_2}^{1/4}\right\} \qquad (1.203)$$

The parameters $\kappa_{V_{Cu}^{-}}$, $\kappa_{O_i^{-}}$, and $\kappa_{V_{Cu}^{\times}}$ in eqn (1.203) are already known, as shown in Table 1.5. Only the value of $\mu_h|e|$ is unknown, but can be calculated using the experimental data for $Cu_{2-\delta}O$. Adopting the data ($\delta = 1.37 \times 10^{-3}$, $T = 1000\,°C$, $P_{O_2} = 1.32 \times 10^{-2}$ atm) of O'Keeffe and Moore[39] and the values of κs determined by the present study, the value of $\mu_h|e|$ obtained is 9.10×10^{-3} (ohm cm)$^{-1}$. Thus we could calculate the value of δ in $Cu_{2-\delta}O$ at any temperature and oxygen pressure. And moreover the values of $[V_{Cu}^{\times}]$, $[V_{Cu}^{-}]$, and $[O_i^{-}]$ could be calculated as functions of temperature and oxygen pressure. Figures 1.60 and 1.61 show the temperature dependence, at $P_{O_2} = 10^{-3}$ atm, and oxygen pressure dependence, at $T = 1000\,°C$, of these defects, together with δ, p ($=[V_{Cu}^{-}] + [O_i^{-}]$), and $[V_{Cu}]$ ($=[V_{Cu}^{\times}] + [V_{Cu}^{-}]$). It is interesting to compare Fig. 1.61 with Fig. 1.57. It was predicted in Fig. 1.57 that the values of $\log[V_{Cu}^{-}]/\log P_{O_2}$ and $\log[O_i^{-}]/\log P_{O_2}$ have to vary with P_{O_2}, whereas that of $\log[V_{Cu}^{\times}]/\log P_{O_2}$ has to be constant ($=\frac{1}{4}$) over all the P_{O_2} range. Such a tendency is clearly observed in Fig. 1.61.

It has been shown that the accurate measurement of the diffusion constant D^*, combined with such measurements as electrical conductivity and quantity of non-stoichiometry as functions of temperature and oxygen pressure, afford us significant knowledge on lattice defects.

---

* Supposing that the number of lattice sites for copper is $2N$ and that for oxygen is N, the number of copper atoms in a crystal is $2N - ([V_{Cu}^{\times}] + [V_{Cu}^{-}])$ and that of oxygen atoms is $(N + [O_i^{-}])$, i.e. $Cu_{(2N-\{[V_{Cu}^{\times}]+[V_{Cu}^{-}]\})}O_{(N+[O_i^{-}])}$. Thus $(2-\delta)$ in $Cu_{2-\delta}O$ can be calculated as

$$2 - \delta = (2N - \{[V_{Cu}^{\times}] + [V_{Cu}^{-}]\})/(N + [O_i^{-}])$$

$$= (2 - \{[V_{Cu}^{\times}] + [V_{Cu}^{-}]\}/N)/(1 + [O_i^{-}]/N)$$

$$\sim \langle(2 - \{[V_{Cu}^{\times}] + [V_{Cu}^{-}]\}/N)\rangle\langle(1 - [O_i^{-}]/N)\rangle$$

$$\sim 2 - \{[V_{Cu}^{\times}] + [V_{Cu}^{-}] + 2[O_i^{-}]\}/N$$

$$= 2 - \{[V_{Cu}^{\times}] + [V_{Cu}^{-}] + 2[O_i^{-}]\} \quad \text{(for } N = 1 \text{ mole)}$$

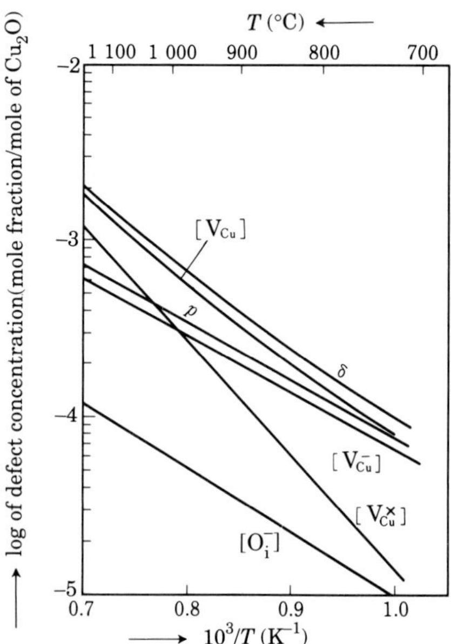

FIG. 1.60 Defect concentrations of $Cu_{2-\delta}O$ as functions of temperature at $P_{O_2} = 10^{-3}$ atm.[34]

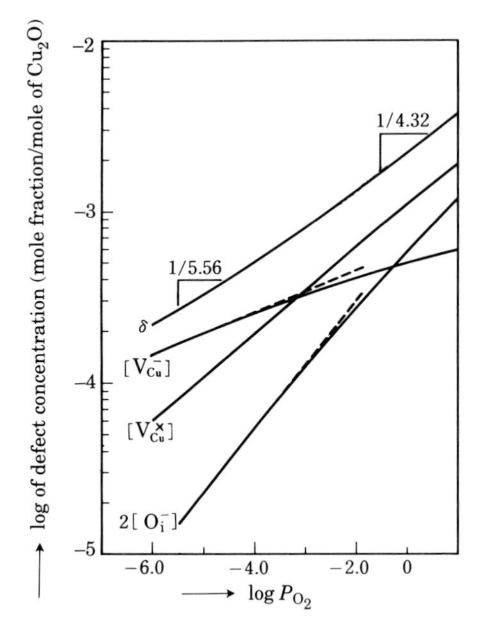

FIG. 1.61 Defect concentrations of $Cu_{2-\delta}O$ as functions of P_{O_2} at 1000 °C.[34] Dotted lines show the deviation from a linear relation between log of defect concentration and log P_{O_2}.

1.4.10 *Study of $Pb_{1-\delta}S$—Kröger–Vink diagram*

It has been shown in Section 1.3.7 that in semiconductors or insulators the lattice defects and electronic defects (electrons and holes), derived from non-stoichiometry, can be regarded as chemical species, and that the creation of non-stoichiometry can be treated as a chemical reaction to which the law of mass action can be applied. This method was demonstrated for $Ni_{1-\delta}O$, $Zr_xCa_{1-x}O_{1+x-\delta}$, and $Cu_{2-\delta}O$ in Sections 1.4.5, 1.4.6, and 1.4.9, as typical examples. We shall now introduce a general method based on the above-mentioned principle after Kröger,[40] and then discuss the impurity effect on the electrical properties of PbS as an example. This method is very useful in investigating the relation between non-stoichiometry and electrical properties of semiconductive compounds.

Consider a crystal $M^{2+}X^{2-}$ with semiconducting properties. A small fraction of both sites are unoccupied, i.e. V_M and V_X. Therefore V_M acts as an acceptor and V_X as a donor as shown in Fig. 1.62, which is the band structure of a typical semiconductor. The lattice defects existing in the crystal are assumed to be $V_M^\times$, V_M^-, $V_X^\times$ and V_X^+.

In the crystal the following chemical equilibria are established:

$$\mathrm{M(g)} \rightleftharpoons \mathrm{M_M^\times} + \mathrm{V_X^\times} \qquad E_{MV} \qquad K_{MV} = [\mathrm{V_X^\times}]/P_M \tag{1.204}$$

$$\tfrac{1}{2}\mathrm{X_2(g)} \rightleftharpoons \mathrm{X_X^\times} + \mathrm{V_M^\times} \qquad E_{X_2V} \qquad K_{X_2V} = [\mathrm{V_M^\times}]/P_{X_2}^{1/2} \tag{1.205}$$

$$\mathrm{V_X^\times} \rightleftharpoons \mathrm{V_X^+} + \mathrm{e} \qquad E_X \qquad K_X = [\mathrm{V_X^+}]n/[\mathrm{V_X^\times}] \tag{1.206}$$

$$\mathrm{V_M^\times} \rightleftharpoons \mathrm{V_M^-} + \mathrm{h} \qquad E_M \qquad K_M = [\mathrm{V_M^-}]p/[\mathrm{V_M^\times}] \tag{1.207}$$

$$0 \rightleftharpoons \mathrm{V_M^\times} + \mathrm{V_X^\times} \qquad E_S \qquad K_S = [\mathrm{V_M^\times}][\mathrm{V_X^\times}] \tag{1.208}$$

$$0 \rightleftharpoons \mathrm{V_M^-} + \mathrm{V_X^+} \qquad E_S' \qquad K_S' = [\mathrm{V_M^-}][\mathrm{V_X^+}] \tag{1.209}$$

$$0 \rightleftharpoons \mathrm{e} + \mathrm{h} \qquad E_I \qquad K_I = np \tag{1.210}$$

$$\mathrm{MX} \rightleftharpoons \mathrm{M(g)} + \tfrac{1}{2}\mathrm{X_2(g)} \qquad E_{MX} \qquad K_{MX}' = P_M P_{X_2}^{1/2} \tag{1.211}$$

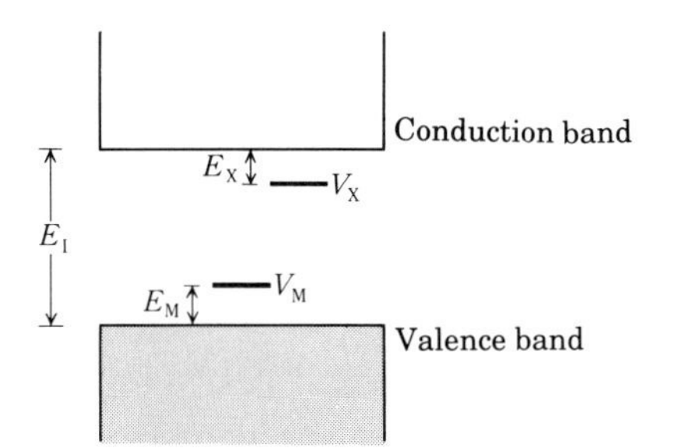

FIG. 1.62 Band structure of semiconductive compounds.

Equation (1.204) shows an equilibrium between the solid and gas phases for metal M, assuming that the molecular species of metal gas is M(g). Towards the right-hand side, the reaction gives an excess metal atom (neutral), $M_M^\times$, with a vacancy at an anion site, $V_X^\times$. Equation (1.205) shows an equilibrium between the solid and gas phases for anion X, assuming that the molecular species of anion gas is $X_2(g)$. Towards the right-hand side, the reaction gives an excess anion (neutral), $X_X^\times$, with a vacancy at a cation site, $V_M^\times$. In place of these two reactions, we can choose the following reactions:

$$M_M^\times \rightleftharpoons M(g) + V_M^\times \qquad K_1 = P_M[V_M^\times]/[M_M^\times] \qquad (1.212)$$

$$X_X^\times \rightleftharpoons \tfrac{1}{2}X_2(g) + V_X^\times \qquad K_2 = P_{X_2}^{1/2}[V_X^\times]/[X_X^\times] \qquad (1.213)$$

Equation (1.212) shows the vaporization of a neutral metal atom sited in the crystal, leaving a vacancy, $V_M^\times$, in the metal sites. Equation (1.213) describes a similar reaction for anion X. These equations are, however, not independent of eqns (1.204)–(1.211), because the equilibrium constants K_1 and K_2 can be expressed in terms of the other Ks as K_S/K_{MV} and K_S/K_{X_2V}, respectively.

Equations (1.206) and (1.207) describe the ionization of neutral vacancies ($V_X^\times, V_M^\times$). We assume here that the ionization of V_X^+ and V_M^- to V_X^{2+} and V_M^{2-} does not take place. In a crystal in thermal equilibrium, electrons and holes will be formed by thermal excitation of electrons from the valence band to the conduction band, and the reverse process is also possible. This process can be expressed by eqn (1.210) as a chemical reaction. (see eqn (1.136)). Such reactions are called creation–annihilation reactions. Equations (1.208) and (1.209) describe the creation–annihilation reactions of neutral vacancies and charged vacancies in a crystal. Equation (1.211) shows the formation reaction of MX from constituent gases. It is to be noted that of these eight equations two are not independent. For example, the equilibrium constants K'_S and K'_{MX} in eqns (1.209) and (1.211) are expressed in terms of the other Ks as

$$K'_S = \frac{K_M K_X K_S}{K_I} \qquad (1.214)$$

$$K'_{MX} = \frac{K_S}{K_{XV} K_{MV}} \qquad (1.215)$$

In addition to these six independent equations, we must take the condition of electroneutrality into consideration:

$$n + [V_M^-] = p + [V_X^+] \qquad (1.216)$$

We, therefore, have seven equations with eight unknown quantities, viz. n, p, $[V_X^\times]$, $[V_X^+]$, $[V_M^\times]$, $[V_M^-]$, P_M, and P_{X_2}. Hence, we can express all of

Table 1.6

Approximated solution of eqns (1.204)–(1.216) (Kröger–Vink diagram, see Figs 1.63 and 1.64)

	Region (I)	Region (II)		Region (III)
	$n \doteqdot [V_X^+]$	$n \doteqdot p$	$[V_M^-] \doteqdot [V_X^+]$	$p \doteqdot [V_M^-]$
n	$\left(\dfrac{K_I K_S'}{K_M R}\right)^{1/2}$	$K_I^{1/2}$	$\dfrac{K_I K_S'^{1/2}}{K_M R}$	$\dfrac{K_I}{(K_M R)^{1/2}}$
p	$\left(\dfrac{K_I K_M R}{K_S'}\right)^{1/2}$	$K_I^{1/2}$	$\dfrac{K_M R}{K_S'^{1/2}}$	$(K_M R)^{1/2}$
$[V_M^-]$	$\left(\dfrac{K_S' K_M R}{K_I}\right)^{1/2}$	$\dfrac{K_M R}{K_I^{1/2}}$	$K_S'^{1/2}$	$(K_M R)^{1/2}$
$[V_X^+]$	$\left(\dfrac{K_I K_S'}{K_M R}\right)^{1/2}$	$\dfrac{K_S' K_I^{1/2}}{K_M R}$	$K_S'^{1/2}$	$\dfrac{K_S'}{(K_M R)^{1/2}}$
$[V_M^\times]$	R [a]	R	R	R
$[V_X^\times]$	$\dfrac{K_I K_S'}{K_M K_X R}$	$\dfrac{K_I K_S'}{K_M K_X R}$	$\dfrac{K_I K_S'}{K_M K_X R}$	$\dfrac{K_I K_S'}{K_M K_X R}$

[a] $R = K_{X_2V} P_{X_2}^{1/2}$.

them in terms of one unknown quantity (for example, P_{X_2}) and the equilibrium constants. Whereas the exact calculation of all the concentrations is a tedious affair, approximate solutions are easily obtained if eqn (1.216) is approximated by its dominant members. This method, after Kröger,[40] gives a general view of this problem.

Under a strongly reducing atmosphere, viz. excess cation, eqn (1.216) can be approximated as

$$n \doteqdot [V_X^+] \gg [V_M^-], p \tag{1.217}$$

In this region (I), the unknown variables can be calculated, as tabulated in Table 1.6, in terms of Ks and $R = K_{X_2V} P_{X_2}^{1/2}$. Thus n $(=[V_X^+])$ decreases proportional to $R^{-1/2}$ and p and $[V_M^-]$ increase proportional to $R^{1/2}$. The above approximation, $n \doteqdot [V_X^+]$, is no longer valid when either p or $[V_M^-]$ becomes larger than n.

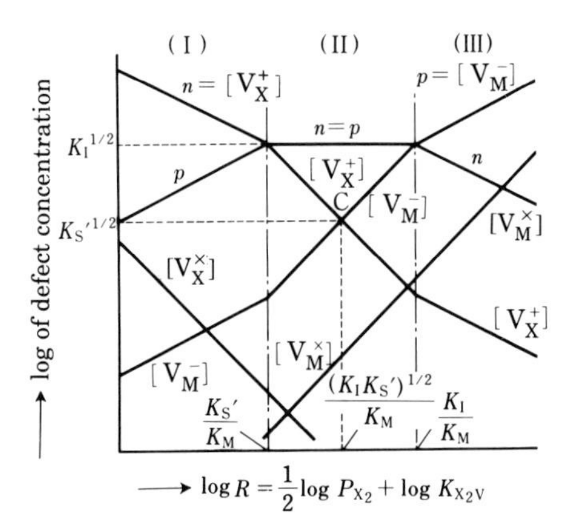

FIG. 1.63 Kröger–Vink diagram of semiconductive compound MX for the case $K'_S < K_I$ (see Table 1.6).

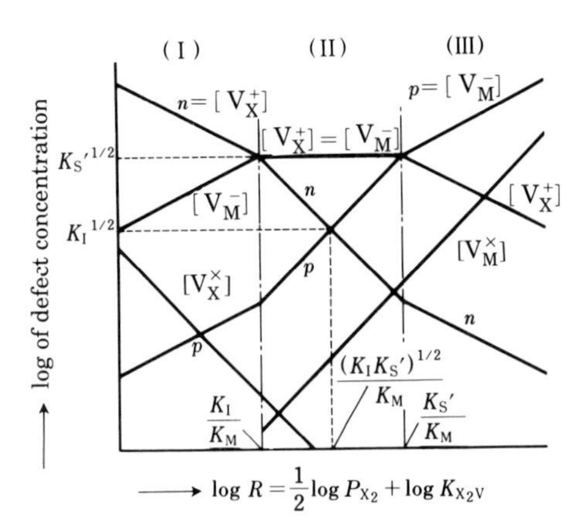

FIG. 1.64 Kröger–Vink diagram of semiconductive compound MX for the case $K'_S > K_I$ (see Table 1.6).

For the case of $p > [V_M^-]$, viz. $K_I > K'_S$, we get a new region (II) when p reaches n at $R = K'_S/K_M$. In this region, the neutral condition can be approximated to $n \doteqdot p$ ($\gg [V_M^-], [V_X^+]$). The electron concentration n does not depend on R (P_{X_2}), and $[V_M^-]$ increases proportional to R and $[V_X^+]$ decreases proportional to R^{-1} (see Table 1.6). When $[V_M^-]$ becomes larger than n, we get a new region (III), in which the neutrality condition is approximated to $p \doteqdot [V_M^-]$ ($\gg n, [V_X^+]$). In this region, the compound has an anion excess composition, because the conditions correspond to a strongly oxidizing atmosphere.

For the case of $p < [V_M^-]$, viz. $K_S' > K_I$, the neutrality condition in region (II) is approximated to $[V_M^-] \doteqdot [V_X^+]$ ($\gg n, p$). The calculated expressions for each concentration are listed in Table 1.6. In this region under this approximation the concentrations of charged vacancies $[V_M^-]$ and $[V_X^+]$ are dominant, which may be expected to induce ionic conduction. When p becomes larger than $[V_X^+]$, we get a new region (III), in which the neutrality condition is approximated to $p \doteqdot [V_M^-]$.

The calculated values of the various imperfections in the various regions are listed in Table 1.6. It is to be noted that the concentrations of the neutral imperfections, $[V_M^\times]$ and $[V_X^\times]$, are calculated from the same expressions in all regions. Figures 1.63 and 1.64 show the plots of the logarithms of the concentrations of the defects against the logarithms of R. The former and the latter correspond to $p > [V_M^-]$ and $p < [V_M^-]$, viz. $K_I > K_S'$ and $K_I < K_S'$, respectively. We call figures such as Figs 1.63 and 1.64 Kröger–Vink diagrams.

Let us examine Fig. 1.63 in detail. As is well understood, this figure has been obtained using extreme approximations. It is to be emphasized, therefore, that this figure reflects the correct solutions of the simultaneous equations only in the left side of region (I) (lower P_{X_2}), around the centre of region (II), and the right side of region (III) (higher P_{X_2}). Between these, the figure shows qualitatively the P_{X_2} dependence of the concentrations. For example, the concentration of electronic carriers $(n - p)$, which can be determined by measurement of the Hall coefficient, is expressed as

$$n - p = [V_X^+] - [V_M^-] \tag{1.218}$$

Therefore, the dominant carriers in regions (I) and (III) are electrons and holes, respectively. In region (II), electrons are dominant in the left-half and holes are dominant in the right-half. Roughly speaking, the carrier concentration $(n - p)$ is expressed by $[V_X^+]$ in the region $R < \dfrac{(K_I K_S')^{1/2}}{K_M}$ and by $[V_M^-]$ in the region $R > \dfrac{(K_I K_S')^{1/2}}{K_M}$, as shown in Fig. 1.65(a).

On the other hand, the deviation from stoichiometry is expressed as

$$\begin{aligned} \delta &= [M_M^\times] - [X_X^\times] \\ &= \{[V_X^\times] + [V_X^+]\} - \{[V_M^\times + V_M^-]\} \\ &\simeq [V_X^+] - [V_M^-] \end{aligned} \tag{1.219}$$

considering the condition of equality of the number of M and X sites. (This equation is derived on the assumption that most of the vacancies are ionized.) In Fig. 1.65 the P_{X_2} dependence of carrier concentration and δ are schematically shown.

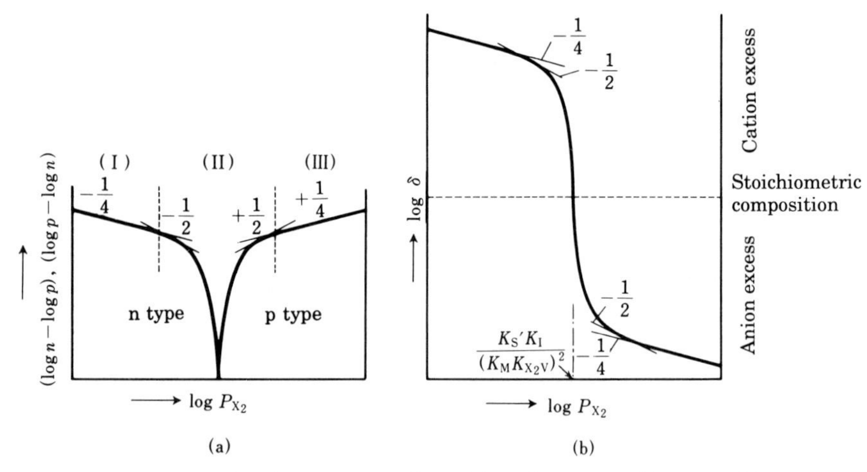

FIG. 1.65 Carrier concentrations (a) and non-stoichiometry (b) of compound MX as a function of P_{X_2}.

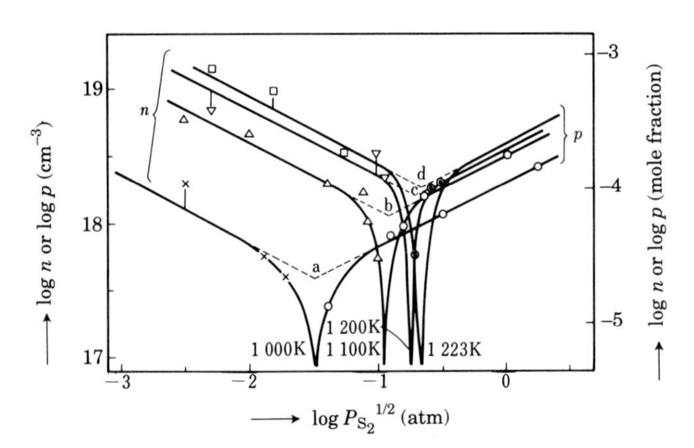

FIG. 1.66 Carrier concentration of PbS as a function of P_{S_2} at different temperatures.[41]

Thus, it has been shown that the electrical properties of semiconductive compounds depend on the chemical composition, viz. non-stoichiometry, and therefore the control of the composition is indispensable in the control of the semiconductive properties of the compounds.

Figure 1.66 shows the carrier concentration (n or p) versus P_{S_2} isotherms for semiconductive PbS, measured by Bloem.[41] This figure is very similar to Fig. 1.65(a); the slope at low P_{S_2} values (electron conduction region) is $-\frac{1}{4}$ and that at high P_{S_2} values (hole conduction region) is $+\frac{1}{4}$ in the log[carrier concentration] versus log P_{S_2} diagram. The point **a** corresponds to the stoichiometric composition at 1000 K. The value of $P_{S_2}^{\delta=0}$, corresponding to the stoichiometric composition, increases with increasing temperature. By using the data shown in Fig. 1.66, Bloem and Kröger calculated various thermodynamic quantities and discussed the nature of the electrical properties of PbS.[42]

Next, let us consider the dependence of carrier concentration on P_{S_2} for Bi- or Ag-doped PbS. This problem is related to the method of controlled valency, developed by Verwey[43] and Kröger and Vink.[44] It is assumed, as observed, that Bi substitutes for Pb in the crystal and acts as a donor. The equilibrium equations for this case are the six independent ones, eqn (1.204) to eqn (1.211), which describe the chemical equilibrium in pure PbS, and the following two equations:

$$\mathrm{Bi(g)} + \mathrm{V}_{\mathrm{Pb}}^{\times} \rightleftharpoons \mathrm{Bi}_{\mathrm{Pb}}^{\times} \qquad K_{\mathrm{Bi}} = [\mathrm{Bi}_{\mathrm{Pb}}^{\times}]/P_{\mathrm{Bi}}[\mathrm{V}_{\mathrm{Pb}}^{\times}] \tag{1.220}$$

$$\mathrm{Bi}_{\mathrm{Pb}}^{\times} \rightleftharpoons \mathrm{Bi}_{\mathrm{Pb}}^{+} + \mathrm{e} \qquad K_{\mathrm{FD}} = [\mathrm{Bi}_{\mathrm{Pb}}^{+}]n/[\mathrm{Bi}_{\mathrm{Pb}}^{\times}] \tag{1.221}$$

The electroneutrality condition is expressed as

$$n + [\mathrm{V}_{\mathrm{Pb}}^{-}] = p + [\mathrm{V}_{\mathrm{S}}^{+}] + [\mathrm{Bi}_{\mathrm{Pb}}^{+}] \tag{1.222}$$

For the case of Ag-doped PbS, we can write equations similar to eqns (1.220), (1.221), and (1.222). It is noted that Ag acts as an acceptor. Figure 1.67(a) and (c) show the experimental results (thick curves) of the carrier concentration dependence on P_{S_2} for the case of Bi- and Ag-doped PbS, respectively, together with the calculated ones (thin lines) based on the above equations. This figure is obtained under the condition that the concentration of foreign donors (Bi) or acceptors (Ag) is nearly equal to p or n. For comparison, Fig. 1.67(b) is shown for pure PbS.

Characteristic features seen from the figure are summarized as follows:

1. In the Bi-doped PbS (donor doping), the transition of electron to hole conduction is shifted to higher P_{S_2}, and the region of n-type conduction extends to a higher P_{S_2}. In region (II), the concentration of carrier n ($\doteqdot [\mathrm{Bi}_{\mathrm{Pb}}^{+}]$) is weakly dependent on P_{S_2}.

2. In the Ag-doped PbS (acceptor doping), the transition of electron to hole conduction is shifted to lower P_{S_2}, and the region of p-type conduction extends to a lower P_{S_2}. In region (III), the concentration of carrier p ($\doteqdot [\mathrm{Ag}_{\mathrm{Pb}}^{-}]$) is weakly dependent on P_{S_2}.

The stoichiometric composition, i.e. when the ratio Pb/S equals 1/1, is realized at the value of P_{S_2} where the transition from electron to hole-type conduction takes place, similar to the case of pure PbS. For the Ag-doped

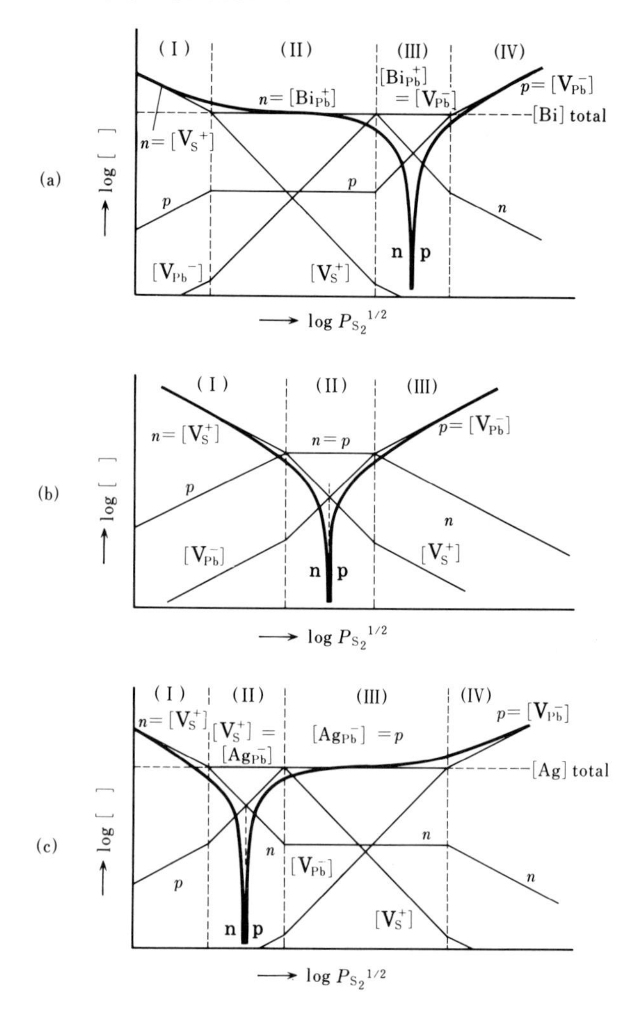

FIG. 1.67 Kröger–Vink diagrams of: (a) Bi-doped PbS, (b) pure PbS, and (c) Ag-doped PbS.[42] Log[ ] denotes log of defect concentration. Thick curves show the P_{S_2} dependence of carrier concentrations (measured). Thin lines are calculated ones.

PbS, the quantity δ can be expressed as

$$\begin{aligned}\delta &= [Pb_{Pb}^{\times}] - [S_S^{\times}] \\ &= \{[V_S^{\times}] + [V_S^{+}]\} - \{[Ag_{Pb}^{\times}] + [Ag_{Pb}^{-}] + [V_{Pb}^{\times}] + [V_{Pb}^{-}]\}\end{aligned}$$

and the neutrality condition is

$$n + [Ag_{Pb}^{-}] + [V_{Pb}^{-}] = p + [V_S^{+}]$$

In region (II), the neutral condition is approximated to $[Ag_{Pb}^-] = [V_S^+] \gg n, p, [V_{Pb}^-]$. If we assume that the concentrations of the neutral defects, such as $[V_S^\times]$, $[Ag_{Pb}^\times]$, and $[V_{Pb}^\times]$, are negligibly small (as observed), we get $\delta = 0$ at the transition from the electron to hole conduction.

The method of control of lattice defects, viz. the control of the concentration of electronic carriers, is called the principle of controlled vacancy[43] or controlled electronic imperfections,[44] and is very useful in the fabrication of practical materials with semiconductive properties.

1.4.11 Study of chalcogenides with the NiAs type structure (V–S system)—order–disorder phase transition of vacancies

Most 3*d*-transition metal chalcogenides MX_n take the NiAs-type structure at the composition MX ($n = 1$) and the CdI_2-type structure at the composition MX_2 ($n = 2$). In these structures X ions form a hexagonal closed packed structure and M ions occupy the octahedral holes of the X lattices. In the NiAs-type structure, M ions are octahedrally coordinated by six X ions, and X ions are sited in a trigonal prism formed by six M ions, as shown in Fig. 1.68(a). Figure 1.68(b) shows the structure of CdI_2, in which the coordination of M by X is the same as that in the NiAs structure, but X is sited at a corner of the tetrahedron formed by (X + 3M). In other words, the structure can be regarded as that formed if metal layers were removed alternately from along the *c*-axis of the NiAs structure.

Generally it is well known that many phases originate from the ordering of metal vacancies. If $A_1A_2A_1A_2A_1\cdots$ denotes the packing of metal layers along the *c*-axis of the NiAs structure (in this model, it was assumed that the anion lattice forms the hexagonal packing without vacancies and that only the metal layer A_1 includes vacancies), the occupation probabilities, p_1 and p_2, of A_1 and A_2 ideally change linearly with x (of MX_{1+x}), as shown in Fig. 1.69. In the M_3X_4 phase, for example, 50 per cent of metal sites on the A_1 plane are vacant. At the phase transition the metal vacancies can be

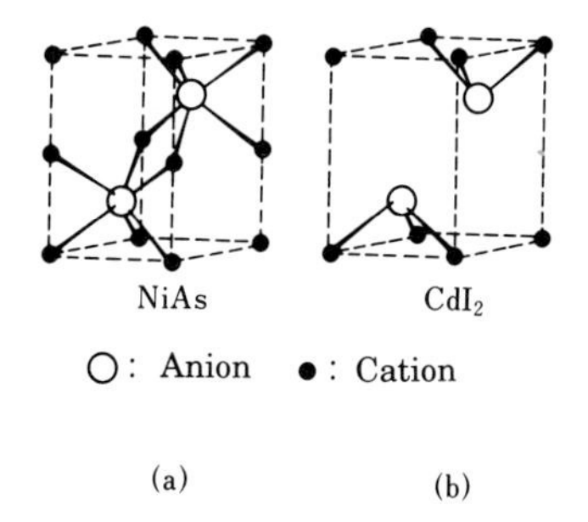

FIG. 1.68 Crystal structures of NiAs (a) and CdI_2 (b).

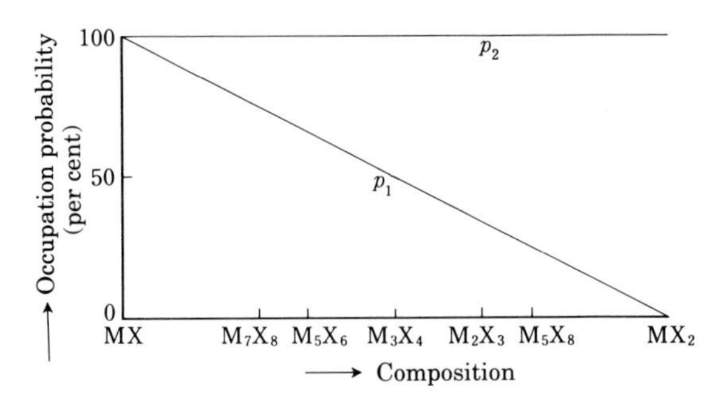

FIG. 1.69 Occupation probabilities (p_1, p_2) of the MX–MX_2 system as a function of composition at absolute zero temperature.

expected to change from a high-temperature disordered to a low-temperature ordered structure, the origin of which is due to the interaction between vacancies. In Fig. 1.69 are shown the vacancy-ordered compositions which have been discovered, and in Fig. 1.70 the ordered arrangements of vacancies on the A_1 plane for typical compositions are shown. These ordered structures are expected to show the following successive phase transitions at higher

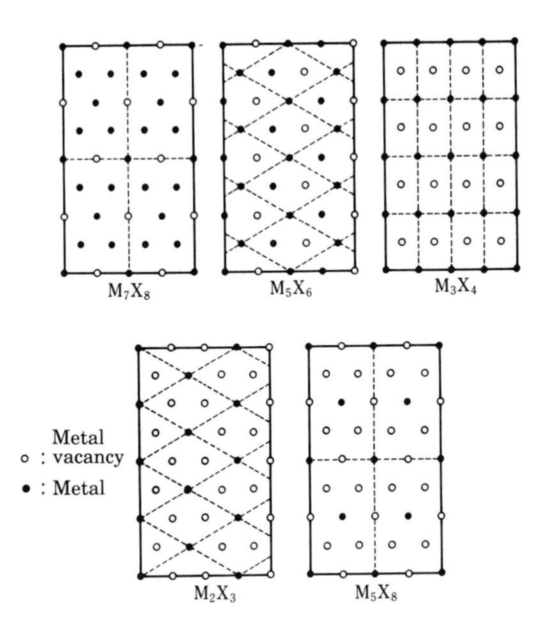

FIG. 1.70 Typical arrangement of metal vacancies on the A_1 (A_1' and A_1'': see Fig. 1.73) planes for the MX–MX_2 system. Dotted lines show the unit lattice for each compound.

temperatures:

vacancy-ordered structures → CdI_2-type → NiAs-type

where the structure type is defined by use of p_1 and p_2 as follows:

$$\begin{aligned} &\text{NiAs-type:} \quad p_1 = p_2 \\ &\text{CdI}_2\text{-type:} \quad p_1 < p_2 \end{aligned} \tag{1.223}$$

Accordingly over all the composition range from MX to MX_2, either the NiAs- or CdI_2-type structure is stable at higher temperatures. A number of papers have been published on 3*d*-transition metal chalcogenides from the viewpoint of order–disorder of metal vacancies. A study of the V–S system from this point of view is described here.

We shall first discuss the phase transition from the CdI_2- to the NiAs-type structure on heating based on a similar treatment to that in Section 1.3. Starting from the composition MX, metal vacancies are introduced in the metal layers A_1 and A_2 (vacancies on the metal layers are randomly distributed. In this case the metal layer A_2 also includes vacancies). The free energy, G, of the system having the composition $M_{1-\delta}X$ can be expressed as

$$\begin{aligned} G = N\mu'_{\mathrm{MX}} + \{N - (N_1 + N_2)\}\varepsilon_{\mathrm{v}}^{\mathrm{M}} + (N - N_{\mathrm{X}})\varepsilon_{\mathrm{v}}^{\mathrm{X}} + H_{\mathrm{vv}}^{\mathrm{M}} \\ - kT \ln {}_{N/2}C_{N_1} \cdot {}_{N/2}C_{N_2} \cdot {}_{N}C_{N_{\mathrm{X}}} \end{aligned} \tag{1.224}$$

$$\begin{aligned} H_{\mathrm{vv}}^{\mathrm{M}} = \frac{1}{2}\left(\frac{N}{2} - N_1\right)\left\{6 \times \left(1 - \frac{2N_1}{N}\right)\varepsilon_{\mathrm{vv}}^{\mathrm{C}\parallel} + 2 \times \left(1 - \frac{2N_2}{N}\right)\varepsilon_{\mathrm{vv}}^{\mathrm{C}\perp}\right\} \\ + \frac{1}{2}\left(\frac{N}{2} - N_2\right)\left\{6 \times \left(1 - \frac{2N_2}{N}\right)\varepsilon_{\mathrm{vv}}^{\mathrm{C}\parallel} + 2 \times \left(1 - \frac{2N_1}{N}\right)\varepsilon_{\mathrm{vv}}^{\mathrm{C}\perp}\right\} \end{aligned} \tag{1.225}$$

where the term $H_{\mathrm{vv}}^{\mathrm{M}}$ denotes the interaction energy between vacancies in the planes (interaction energy $\varepsilon_{\mathrm{vv}}^{\mathrm{C}\parallel}$) and perpendicular to the planes ($\varepsilon_{\mathrm{vv}}^{\mathrm{C}\perp}$). For simplicity, only the first nearest interactions (six sites in the planes and two sites perpendicular to the planes) are taken into consideration (see Figs 1.70 and 1.71).

The chemical potentials μ_1 and μ_2 of metals on the A_1 and A_2 layers are calculated as follows:

$$\mu_1 = \frac{\partial G}{\partial N_1} = -\varepsilon_{\mathrm{v}}^{\mathrm{M}} + 6\varepsilon_{\mathrm{vv}}^{\mathrm{C}\parallel}(p_1 - 1) + 2\varepsilon_{\mathrm{vv}}^{\mathrm{C}\perp}(p_2 - 1) - kT \ln \frac{1 - p_1}{p_1} \tag{1.226}$$

$$\mu_2 = \frac{\partial G}{\partial N_2} = -\varepsilon_{\mathrm{v}}^{\mathrm{M}} + 6\varepsilon_{\mathrm{vv}}^{\mathrm{C}\parallel}(p_2 - 1) + 2\varepsilon_{\mathrm{vv}}^{\mathrm{C}\perp}(p_1 - 1) - kT \ln \frac{1 - p_2}{p_2} \tag{1.227}$$

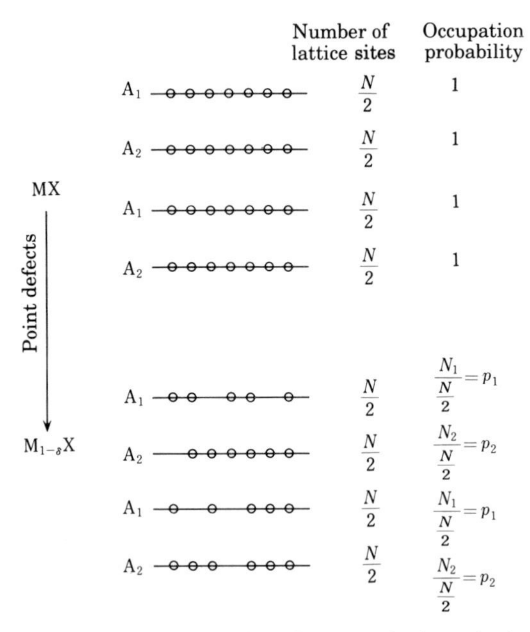

FIG. 1.71 Phases $M_{1-\delta}X$ derived from MX by introducing metal vacancies into the A_1 and A_2 planes.

From the condition of chemical equilibrium, viz. $\mu_1 = \mu_2$, we have

$$2(p_1 - p_2)(\varepsilon_{vv}^{C\perp} - 3\varepsilon_{vv}^{C\parallel}) = kT \ln \frac{p_1(1-p_2)}{p_2(1-p_1)} \tag{1.228}$$

The relation between δ in $M_{1-\delta}X$ and p_1 and p_2 can be obtained as

$$p_1 + p_2 = 2(1-\delta) \qquad (0 \leq \delta \leq \tfrac{1}{2}) \tag{1.229}$$

Thus, eqn (1.228) gives the temperature variation of $p_1(p_2)$ at fixed composition (by fixing the composition, only one of the occupation probabilities, p_1 or p_2, becomes an independent variable as is easily seen from eqn (1.229). If we define

$$\beta = \frac{kT}{2(\varepsilon_{vv}^{C\perp} - 3\varepsilon_{vv}^{C\parallel})} \tag{1.230}$$

β corresponds to (reduced) temperature (it is noted that the condition $(\varepsilon_{vv}^{C\perp} - 3\varepsilon_{vv}^{C\parallel}) > 0$ is necessary for $\beta > 0$). Then, from eqn (1.228), we have

$$\beta = \frac{p_1 - p_2}{\ln \dfrac{p_1(1-p_2)}{p_2(1-p_1)}} \tag{1.231}$$

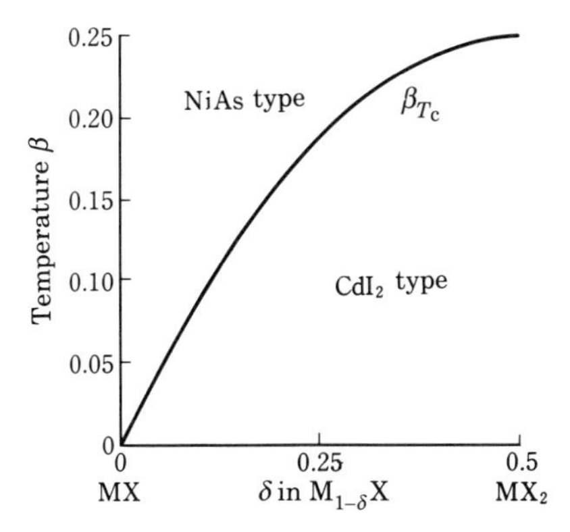

FIG. 1.72 Phase transition temperature, β_{T_c}, for the transition from the CdI_2- to NiAs-type structure as a function of δ in $M_{1-\delta}X$ (eqn (1.232)).

Then the phase transition temperature, β_{T_c}, from the CdI_2- to the NiAs-type structure is calculated as

$$\beta_{T_c} = \lim_{p_1 \to p_2} \beta = \delta(1 - \delta) \tag{1.232}$$

i.e. $T_c = \{2(\varepsilon_{vv}^{C\perp} - 3\varepsilon_{vv}^{C\parallel})/k\}\,\delta(1 - \delta)$. As shown in Fig. 1.72, T_c increases with δ.

In the composition range from VS to VS_2 of the V–S system, the vacancy-ordered phases V_3S_4 and V_5S_8 have been found to exist as typical examples of non-stoichiometric compounds. In the V_3S_4 phase, the occupation probabilities p_1 and p_2 on the metal layers should be 50 (at the stoichiometric composition of the phase) and 100 per cent, respectively. The ordered arrangement of vacancies in the A_1 layer (referred to as the vacant layer, hereafter) is just the same as that of the M_3X_4-type structure shown in Fig. 1.70. The length of the c-axis of this type of structure is twice that of the NiAs-type one, because of the phase shift of the stacking of the A_1 layers along the c-axis ($\cdots A_2A_1'A_2A_1''A_2\cdots$; see also Fig. 1.73). It is noted that the ordered structures shown in Fig. 1.70 are realized only at zero Kelvin for each of the stoichiometric compositions.

As shown later, both the phases V_3S_4 and V_5S_8 show non-stoichiometry in a wide composition range. A cation-poor region (compared to the stoichiometric composition) in the non-stoichiometric phase is realized when metal atoms are missed statistically from the regular metal sites on the A_1 layers. A cation-rich region, on the other hand, is realized when some sites of the regular metal vacant sites are occupied preferentially by metal atoms. Here we define the M_3X_4- and M_5X_8-type structures, in a similar way to the NiAs- and CdI_2-type structures (eqn (1.223)), the basic structure is shown

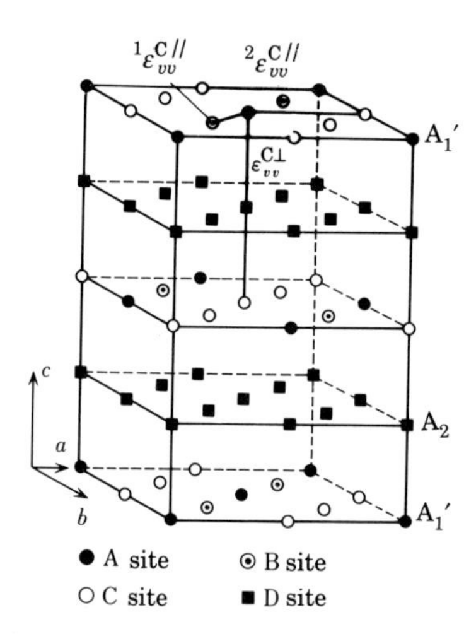

FIG. 1.73 Basic model structure for the definition of vacancy-ordered structures in the MX–MX_2 system.[45,46]

in Fig. 1.73. The cation sites on the A_1 planes are classified as A, B, and C sites and those on the A_2 planes as D sites. By using the occupation probabilities, *a*, *b*, *c*, and *d*, of each site, the above-mentioned four structure types can be defined as

$$\begin{aligned} &\text{NiAs-type:} && a = b = c = d < 1 \\ &M_3X_4\text{-type:} && 1 > d > a = b > c \\ &M_5X_8\text{-type:} && 1 > d > a > b > c \\ &CdI_2\text{-type:} && a = b = c < d < 1 \end{aligned} \tag{1.233}$$

This definition gives us four structure types for the whole composition range of MX–MX_2. For example, the phase transition from V_5S_8 (M_5X_8-type) to V_3S_4 (M_3X_4-type) corresponds to the phase transition from $a > b > c$ to $a = b > c$.

We shall now discuss the phase transition from the viewpoint of statistical thermodynamics.[45,46] The total free energy G can be expressed as a function of N (total number of cation sites = total number of anion sites), N_X (total number of anions), N_A (number of cations on the A sites), N_B, N_C, and N_D as

$$G = G(N, N_X, N_A, N_B, N_C, N_D) \tag{1.234}$$

The interaction between vacancies is taken into consideration as follows:

1. Intra-plane interaction: the first ($^1\varepsilon_{vv}^{C\parallel}$) and the second ($^2\varepsilon_{vv}^{C\parallel}$) nearest neighbour in the A_1' or A_1'' planes.

2. Inter-plane interaction: the first ($\varepsilon_{vv}^{C_\perp}$) nearest neighbour sites in the A_1' or A_1'' planes.

Here, we show only the outline of the process and the result of the calculation, the details can be found in the original papers.[45,46] At chemical equilibrium, we get

$$\left(\frac{\partial G}{\partial N}\right)_{N_X, N_A, N_B, \ldots} = 0 \tag{1.235}$$

The chemical potentials of anion and cation are calculated as

$$\mu_X = \left(\frac{\partial G}{\partial N_X}\right) \tag{1.236}$$

$$\mu_M^A = \left(\frac{\partial G}{\partial N_A}\right), \quad \mu_M^B = \left(\frac{\partial G}{\partial N_B}\right), \quad \mu_M^C = \left(\frac{\partial G}{\partial N_C}\right), \quad \mu_M^D = \left(\frac{\partial G}{\partial N_D}\right) \tag{1.237}$$

It is reasonably acceptable that the chemical potentials of cations on four different sites are equal to each other. Therefore we have

$$\left.\begin{aligned}(x - 3a - b) + \alpha(x - 2a - 2b) + t \ln \frac{a(2 - x + a + b)}{(1 - a)(x - a - b)} &= 0 \\ (x - a - 3b) + \alpha(x - 2a - 2b) + t \ln \frac{b(2 - x + a + b)}{(1 - b)(x - a - b)} &= 0\end{aligned}\right\} \tag{1.238}$$

where

$$\left.\begin{aligned} x &= a + b + 2c \\ \alpha &= \frac{\varepsilon_{vv}^{C_\perp}}{{}^1\varepsilon_{vv}^{C_\parallel} + {}^2\varepsilon_{vv}^{C_\parallel}} \equiv \frac{\xi_{vv}^{M'}}{\xi_{vv}^{M}} \\ t &= \frac{kT}{{}^1\varepsilon_{vv}^{C_\parallel} + {}^2\varepsilon_{vv}^{C_\parallel}} \equiv \frac{kT}{\xi_{vv}^{M}} \end{aligned}\right\} \tag{1.239}$$

The parameter x corresponds to the chemical composition, since Y in the notation MX_Y is equal to $8/(4 + x)$. α is the ratio of the inter-layer to intra-layer vacancy–vacancy interaction and t is the measure of temperature.

Assuming that the chemical species of the gas equilibrated with the solid phase of VS_X is S_2 only, the partial pressure of the gas phase (P_{X_2}) is calculated as

$$\left.\begin{aligned} P_{X_2} &= Q(t)\exp\left\{\frac{1}{t}\left[\frac{3+\alpha}{2} - \tfrac{1}{4}[ab + (2+\alpha)(a+b)c + c^2]\right]\right. \\ &\qquad \left. + \tfrac{1}{8}\ln(1-a)(1-b)(1-c)^2\right\} \\ Q(t) &= \tfrac{1}{2}\exp\frac{\mu'_{MX} + \frac{1}{2}\varepsilon_v^M}{t} \end{aligned}\right\} \tag{1.240}$$

We can therefore find the temperature and composition dependence of the occupation probabilities, and also $P_{X_2}(t, \alpha, a, b, c)$. The occupation probabilities which have to satisfy the condition of minimization of G, have been calculated numerically from eqn (1.238). Equation (1.238) was changed to simultaneous equations of a and b (which include parameter α) by fixing the value of x (the composition), and then the values of a, b, and c were obtained as functions of temperature, t. Figure 1.74 shows the occupation probability versus temperature curves for the composition $MX_{1.60}$ ($x = 1$, the stoichiometry of the M_5X_8 phase) at fixed values of α. In the case of $\alpha = 0$, i.e. non-existence of the inter-plane interaction, the M_5X_8 type structure is transformed into the CdI_2-type structure, this transformation is first order. With increasing α, the order of the phase transition gradually changes from

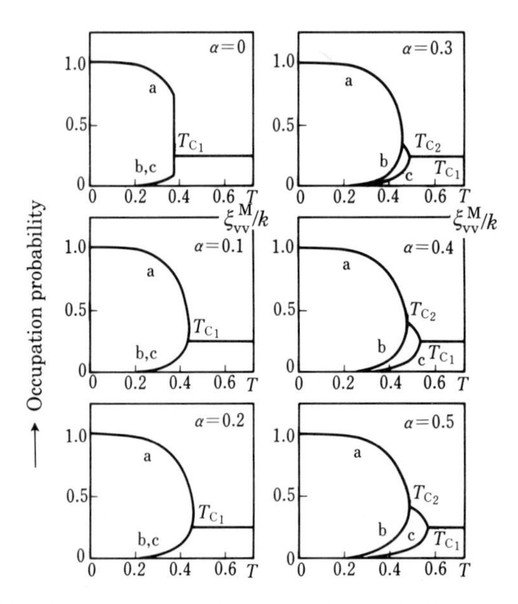

FIG. 1.74 Occupation probabilities (a, b, c) of $MX_{1.60}$ as a function of temperature for different α (eqn (1.238)).[45]

first to second order. If $0 < \alpha < 0.24$, the M_3X_4-type structure never appears. As shown later, the V–S and V–Se systems show successive phase changes on heating: $M_5X_8 \rightarrow M_3X_4 \rightarrow CdI_2$. To explain the transitions using this model, it is necessary that the inter-plane interaction ($\xi_{vv}^{M'}$) is more than 0.24 times the intra-plane interaction (ξ_{vv}^{M}).

Figure 1.75 shows the α dependency of T_{c_1} and T_{c_2} calculated by this

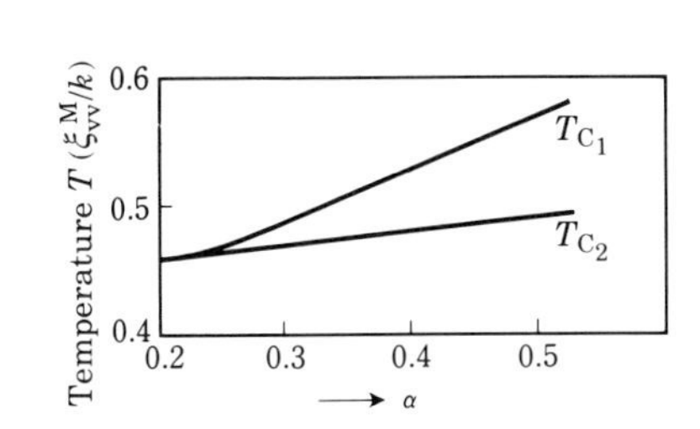

FIG. 1.75 Phase transition temperature (T_{c_1}, T_{c_2}) of $MX_{1.60}$ as a function of α.[45]

model for the compound M_5X_8, where T_{c_1} and T_{c_2} denote the phase transition temperatures from the M_5X_8- to M_3X_4-type structure and the M_3X_4- to CdI_2-type structure, respectively. By fitting the experimental values of T_{c_1} and T_{c_2} to the calculated values for the composition $VS_{1.58}$ (the V_5S_8 phase), we get the following parameter values: $\alpha = 0.30$, $\xi_{vv}^{M} = 4.65$ kcal mol^{-1}. Supposing that these values are constant in the composition range from VS to VS_2, the phase diagram of the VS–VS_2 system is calculated as shown in Fig. 1.76, it exhibits stability regions of M_5X_8-, M_3X_4-, and CdI_2-type structures qualitatively. The calculated and experimentally determined phase diagrams of the V–S system are compared in Fig. 1.77. The reason why the

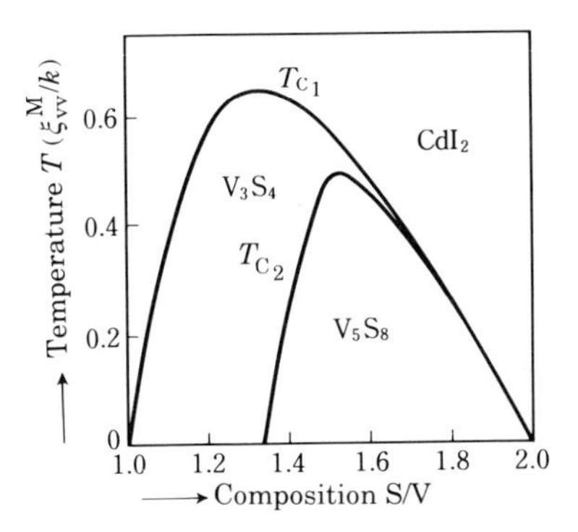

FIG. 1.76 Calculated phase diagram of the V–S system, on the assumption $\alpha = 0.30$ and $\xi_{vv}^{M} = 4.65$ kcal mol^{-1}.[46]

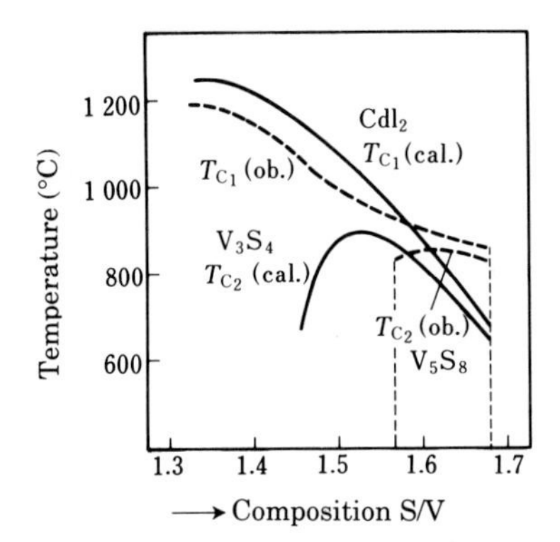

FIG. 1.77 Phase diagram of the V–S system: calculated, solid curve; observed, dotted curve.[45,46]

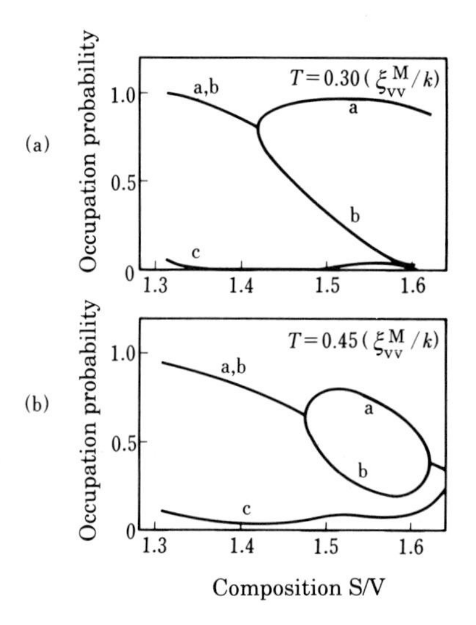

FIG. 1.78 Occupation probabilities (a, b, c) of the V–S system as a function of composition: (a) $T = 0.30(\xi^M_{vv}/k)$, (b) $T = 0.45(\xi^M_{vv}/k)$.[46] These figures are calculated results, assuming $\alpha = 0.30$ and $\xi^M_{vv} =$ 4.65 kcal mol^{-1}.

agreement between them is not satisfactory is mainly due to the assumption that in the wide composition range the parameters α and ξ^M_{vv} are constant.

Figure 1.78 shows the occupation probabilities versus composition of fixed temperature, $T = 0.30$ and 0.45 (ξ^M_{vv}/k), by using the above determined values of α and ξ^M_{vv}. At lower temperatures, the phase transition from the M_3X_4- to M_5X_8-type structure takes place, whereas at higher temperatures (below the

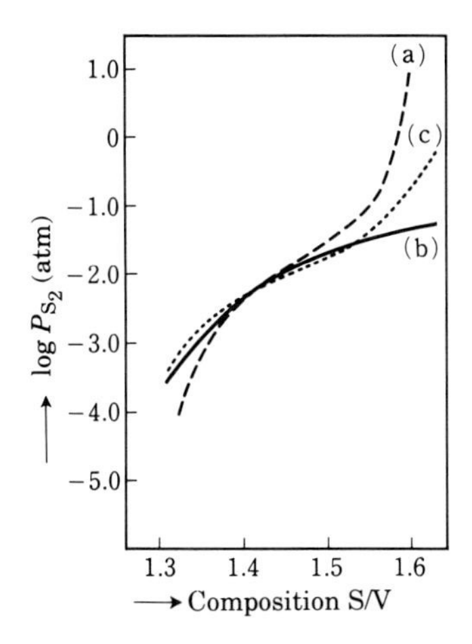

FIG. 1.79 Equilibrium sulfur pressure P_{S_2} of the V–S system as a function of composition at 800 °C.[46,47] Curve (a) is an experimental one,[47] curves (b) and (c) are calculated ones.[46]

maximum of T_{c_2}) the following successive phase transitions occur with increasing S value: M_3X_4- → M_5X_8- → M_3X_4- → CdI_2-type structure.

Figure 1.79 shows the calculated (curves (b) and (c)) equilibrium sulfur pressure, P_{S_2}, versus composition curves for the V–S system at 800 °C, together with the measured curve[47] (curve (a)). Curve (b) was calculated using eqn (1.240), with the above determined parameters. This curve is not in good agreement with the experimental curve, especially in the V_5S_8 phase region. On the other hand, curve (c) was calculated using the parameter ξ_{vv}^{M} obtained by fitting the parameter to the composition dependence of T_{c_1} as shown in Fig. 1.80, and is in rather good agreement. It is to be noted that the

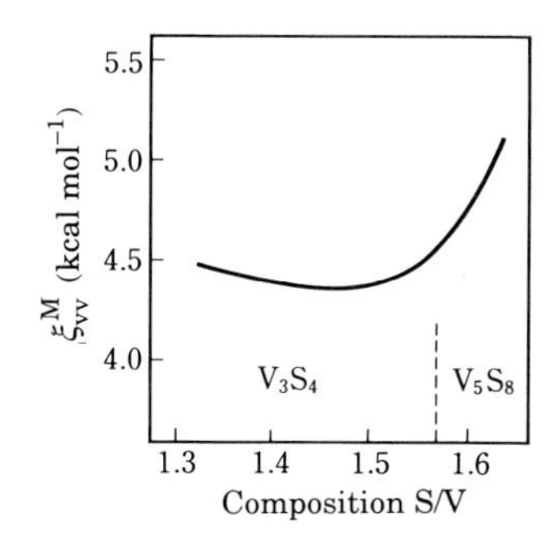

FIG. 1.80 Composition dependence of ξ_{vv}^{M} obtained by fitting the theoretical T_{c_1} to the experimental T_{c_1} as a function of composition.[46]

parameter ξ_{vv}^{M} depends significantly on the composition in the V_5S_8 phase region, but not in the V_3S_4 phase region.

The general features of the phase diagram discussed here are common to those of chalcogenides of V and Cr. More detailed experimental and theoretical studies are required to understand the nature of these substances.

1.4.12 Study of $Fe_{1-\delta}O$ (wüstite)—clustering of point defects and their ordering

As mentioned previously (see Section 1.3.5) the binary M–X system shows a phase separation phenomenon in which the phase decomposes into two phases, having lower and higher concentrations of vacancies, below the critical temperature T_c, under the condition $\varepsilon_{vv}^{M} < 0$, i.e. there is an attractive force between vacancies. In Section 1.3.5 it was not possible to refer to the details of those structures, because the model was less than simple. In any case, it can be safely said that if $\varepsilon_{vv}^{M} < 0$, vacancies cluster at low temperatures (from a thermodynamic point of view). Here let us briefly review the non-stoichiometry of 3*d* transition metal monoxides $M_{1-\delta}O$, and then discuss the $Fe_{1-\delta}O$ system as a typical example of the clustering of vacancies in detail.

As shown in Table 1.7, 3*d* transition metal monoxides with NaCl-type structure are classified into two groups: one has the narrow non-stoichiometry seen in CoO and NiO, and the other has the wide non-stoichiometry seen

Table 1.7

Non-stoichiometry of 3d transition metal monoxides

	Composition MO_x		
Oxides	Minimum x	Maximum x	Δx
TiO	0.80	1.30	0.50[a]
VO	0.80	1.30	0.50[a]
MnO	1.000	1.18	0.18[b]
FeO	1.045	1.200	0.155[b]
CoO	1.000	1.012	0.012[b]
NiO	1.000	1.001	0.001[b]

[a] The width of non-stoichiometry remarkably depends on temperature. These compounds show non-stoichiometry in both the cation-rich and cation-poor sides. At the stoichiometric composition, more than 15 per cent of the metal and oxygen lattice sites are vacant.

[b] Strictly, the expression MO_x should be changed to $M_{1-\delta}O$ for these substances, where $\delta = 1 - x^{-1}$.

in TiO, VO, MnO, and FeO. The former group, as discussed in Sections 1.4.5 and 1.4.8, can be understood as compounds with a random distribution of cation vacancies, in which the interaction between vacancies is negligible because the concentration of vacancies is very low. The compounds TiO and VO are unique in the following two points: (a) their wide non-stoichiometry for both (cation rich and anion rich) sides and (b) more than 15 per cent of their anion and cation sites are vacant at the stoichiometric composition $MO_{1.0}$. Although the large number of vacancies must show a variety of vacancy-ordered structures at low temperature, detailed phase diagrams for these systems have not been determined.

$Fe_{1-\delta}O$, called wüstite, has been studied from the viewpoints of thermodynamics and physicochemical properties. As mentioned in Section 1.1, stoichiometric FeO cannot be prepared under the usual conditions. Many investigators have studied the thermodynamic properties of wüstite by use of various kinds of techniques. Here we introduce a study carried out by Fender and Riley,[48] who used a solid electrolyte cell (see Section 1.4.8) to determine the equilibrium oxygen pressure P_{O_2}. The following cell was utilized,

$$\mathrm{Pt/(Fe + Fe_{1-\delta_1}O)\ |\ stabilized\ zirconia\ |\ Fe_{1-\delta}O/Pt} \qquad (1.241)$$

A mixture of $(Fe + Fe_{1-\delta_1}O)$ is used as a standard electrode, because the oxygen partial pressure of the mixture ($P_{O_2}^{(I)}$) is constant at fixed temperature. (The value of $P_{O_2}^{(I)}$ was determined in advance against temperature using a solid electrolyte cell.) By measuring the electromotive force E for the cell (E can be expressed as $-2FE = \Delta G_f^0(Fe_{1-\delta}O) - \frac{1}{2}\Delta\bar{G}_{O_2}$, where $\Delta G_f^0(Fe_{1-\delta}O)$ is the standard formation energy of $Fe_{1-\delta}O$ and $\Delta\bar{G}_{O_2}$ is the relative partial free energy of oxygen of $Fe_{1-\delta}O$; see below), the equilibrium oxygen pressure $P_{O_2}^{(II)}$ of $Fe_{1-\delta}O$ was determined as a function of temperature and composition δ. To change the value of δ, the coulometric titration method was applied *in situ* (see Section 1.4.8).

Figure 1.81 shows the EMF versus T curves as a function of non-stoichiometry $(1-\delta)$ in the temperature range from 600 to 1300 °C. The starting composition was $Fe_{0.886}O$, from which six different compositions were prepared by coulometric titration. This figure definitely indicates that the EMF versus T lines change their slope at fixed temperatures for a fixed composition. For example, $Fe_{0.886}O$ shows a change in slope at 850, 1030, and 1190 °C, denoted as A, B, and C in the figure. Similar behaviour is clearly observed for all the samples measured. Fender and Riley proposed a tentative phase diagram in which the wüstite region is divided into the three sub-phases, Ⓘ, Ⓘⓘ, and Ⓘⓘⓘ (dotted lines), shown in Fig. 1.82. They suggested that the origin of sub-phases Ⓘ, Ⓘⓘ, and Ⓘⓘⓘ is a kind of order–disorder of vacancies.

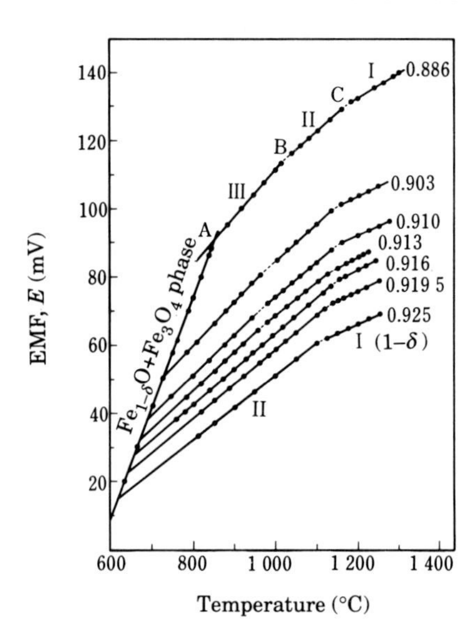

FIG. 1.81 EMF for the cell in eqn (1.241) as a function of temperature for $Fe_{1-\delta}O$.[48]

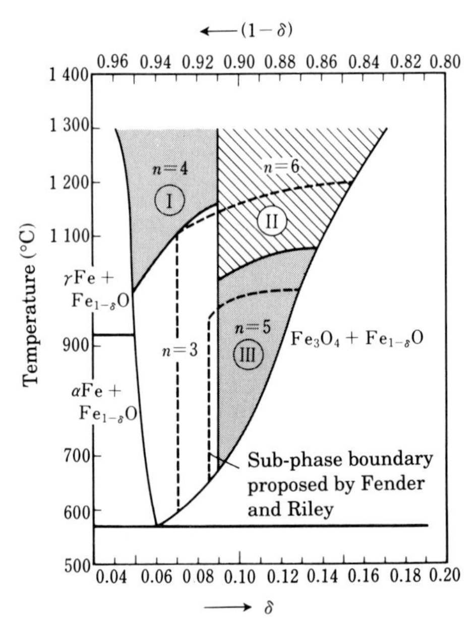

FIG. 1.82 Phase diagram of wüstite.[48,52] Sub-phases (I), (II), and (III) and $n = 3, 4, 5, 6$ were proposed by Fender and Riley[48] and Sorensen,[52] respectively (see Figs 1.81 and 1.83).

The defect structure of wüstite can be discussed from the view of a quasi-chemical equilibrium among defects, similar to the case of $Ni_{1-\delta}O$. Assuming that the predominant defects are iron ion vacancies, we obtain the following equations:

$$\tfrac{1}{2}O_2(\text{gas}) \rightleftharpoons V_M^\times + O_O^\times \qquad K_0 = [V_M^\times]/P_{O_2}^{1/2} \tag{1.242}$$

$$V_M^\times \rightleftharpoons V_M^- + h \qquad K_1 = [V_M^-]p/[V_M^\times] \tag{1.243}$$

$$V_M^- \rightleftharpoons V_M^{2-} + h \qquad K_2 = [V_M^{2-}]p/[V_M^-] \tag{1.244}$$

When the defect $V_M^\times$ is perfectly ionized to V_M^{2-} (acceptor), we obtain

$$\tfrac{1}{2}O_2(\text{gas}) \rightleftharpoons V_M^{2-} + 2h + O_O^\times \qquad K \equiv K_0K_1K_2 = [V_M^{2-}]p^2/P_{O_2}^{1/2} \tag{1.245}$$

Then we have

$$K = [V_M^{2-}]^3 P_{O_2}^{-1/2} \tag{1.246}$$

because p is proportional to $[V_M^{2-}]$. $[V_M^{2-}]$ is the concentration of cation vacancies as defined above, but by the expression of mole fraction this value is equal to δ in $Fe_{1-\delta}O$. From eqn (1.246), we have

$$\delta \propto P_{O_2}^{1/6} \tag{1.247}$$

On the other hand, the relative partial free energy of oxygen of wüstite is defined as*

$$\Delta\bar{G}_{O_2} = RT \ln P_{O_2} \tag{1.248}$$

Therefore $\Delta\bar{G}_{O_2}$ can be expressed in the general form as

$$\Delta\bar{G}_{O_2} = RT \ln P_{O_2} \propto nRT \ln \delta \tag{1.249}$$

where n shows the defect types, e.g. for the case of eqn (1.245) $n = 6$. From this equation it is clear that the value of n can be obtained by plotting $\Delta\bar{G}_{O_2}$ against $\ln \delta$, which suggests the dominant defects in the materials.

Figure 1.83 shows the $\Delta\bar{G}_{O_2}$ versus $\ln \delta$ curves measured by many workers.[48–51] The value of n varies from 3 to 6 and increases with increasing temperature and δ. In Fig. 1.82[52] the sub-phases characterized by a specific n are shown, they are closely correlated to those (Ⓘ, ⒾⒾ, ⒾⒾⒾ) obtained by Fender and Riley.

* On the right electrode of the cell expressed by eqn (1.241), the following chemical equilibrium is a reasonable assumption at the microscopic scale:

$$\frac{2}{\varepsilon}(1-\delta+\varepsilon)Fe_{1-\delta}O \rightleftharpoons \frac{2}{\varepsilon}(1-\delta)Fe_{1-\delta+\varepsilon}O + O_2 \qquad (\delta \gg \varepsilon)$$

ΔG for the chemical equilibrium corresponds to the relative partial free energy of oxygen in $Fe_{1-\delta}O$, $\Delta\bar{G}_{O_2}$ (depending on the composition and temperature). It follows that $\Delta\bar{G}_{O_2} = \bar{G}_{O_2} - G_{O_2} = RT \ln P_{O_2}$

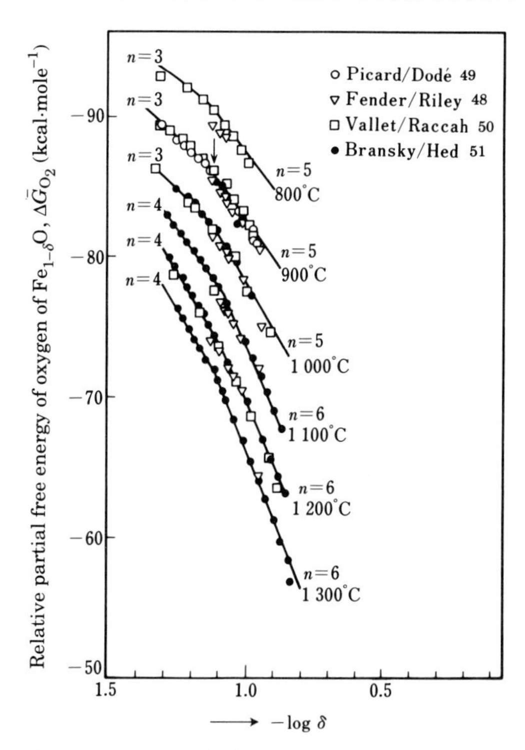

FIG. 1.83 Relative partial free energy of oxygen of wüstite as a function of δ in $Fe_{1-\delta}O$ at different temperatures.[52]

The defect structure of $Fe_{1-\delta}O$ with the NaCl-type structure had been estimated to be a random distribution of iron vacancies. In 1960, Roth[53] confirmed, by powder X-ray diffraction, that the defect structure of wüstite quenched from high temperatures consists of iron vacancies (V_{Fe}) and interstitial iron (Fe_i) (there are about half as many Fe_i as V_{Fe}). This was a remarkable discovery in the sense that it showed that different types of crystal defects with comparable concentrations are able to exist simultaneously in a substance. Roth also proposed a structure model, named a Roth cluster, shown in Fig. 1.84. Later this model (defect complex = vacancy + interstitial) was verified by X-ray diffraction on a single crystal and also by *in-situ* neutron diffraction experiments. Moreover, it has been shown[54] that the defect complex arranges regularly and results in a kind of super-structure, the model structure of which (called a Koch–Cohen model) is shown in Fig. 1.85 together with the basic structures (a) and (b).

It has also been reported[55] that the concentration ratio of metal vacancies to interstitials, R_{VI}, changes with the non-stoichiometry δ. For example, $R_{VI} = 4$ for $\delta = 0.03$ and 3 for $\delta = 0.10$.

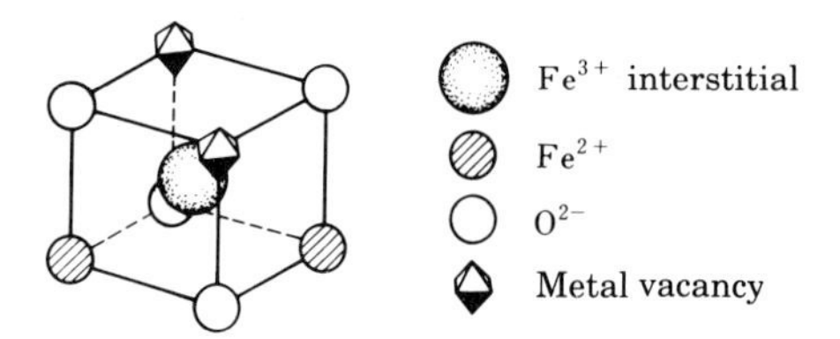

FIG. 1.84 Defect complexes of wüstite proposed by Roth (Roth cluster).[53]

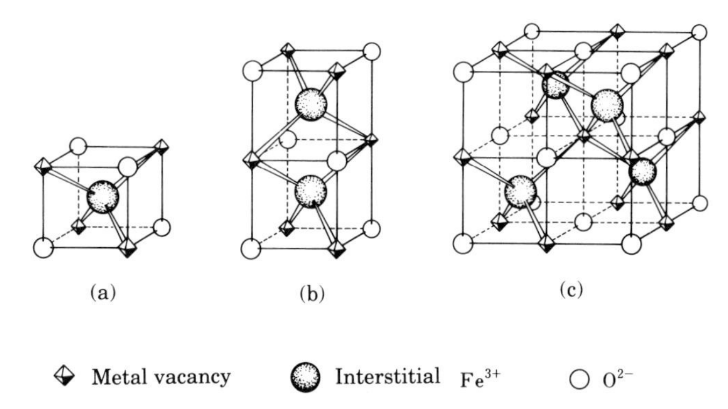

FIG. 1.85 Defect complexes of wüstite. (a) Basic structure (Roth cluster); (b) edge-sharing tetrahedra (6:2 complex); (c) corner-sharing tetrahedra (Koch–Cohen complex).

Based on these experimental results, Catlow and Fender[56] calculated the stability of various kinds of defect complexes by the use of Mott's method.[57] In principle the coulombic interaction between interstitial metal ions $(Fe_i)^{3+}$ and metal vacancies $(V_{Fe})^{2-}$ was numerically calculated. Figure 1.86 shows the model structures treated, the binding energies for these structures are tabulated in Table 1.8. In the 4:1 cluster model, for example, an interstitial iron Fe^{3+} is tetrahedrally coordinated by four iron vacancies (effective charge $= -2$) (interstitial Fe^{3+} is also tetrahedrally coordinated by four anions). The 4:1 cluster is the basic structure of the defect complexes, but to realize smaller R_{VI} it is necessary for the basic structure to be clustered. In other words, in order to reduce the R_{VI} value, it is necessary to construct a super-cell with a larger unit cell based on the 4:1 cluster (Fe^{3+}-centred tetrahedron). This can be achieved by edge- or corner-sharing of tetrahedra. The calculations showed that a small corner-shared cluster (number of units in the cluster <3) did not gain binding energy (defined as the difference in energy between the cluster and a random distribution of defects), but in a cluster of more than four units the binding energy for the corner-shared

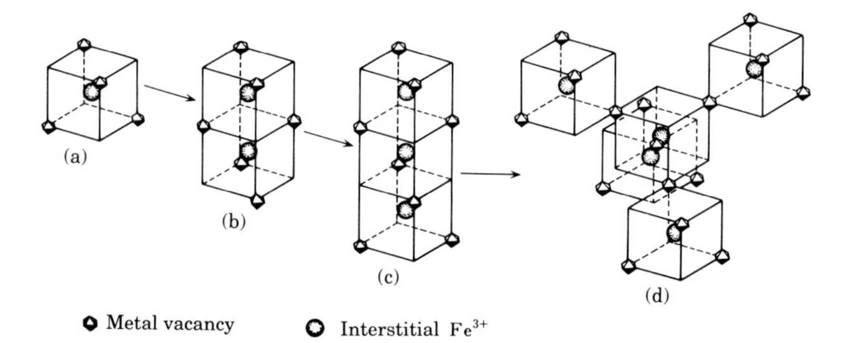

FIG. 1.86 Cluster models, for which Catlow and Fender discussed the stability from the viewpoint of binding energy.[56] (a) 4:1 cluster, (b) 6:2 cluster, (c) 8:3 cluster, (d) 16:5 cluster.

Table 1.8

Binding energy of the clusters (defect complexes) in $Fe_{1-\delta}O$[56]

Type of cluster	Binding energy (eV)
4:1	1.98
6:2	2.42
8:3	2.52
13:4	2.10
16:5	2.38

cluster is larger than that for the edge-shared cluster. The binding energy for the 13:4 cluster (Koch–Cohen model; Fig. 1.85) is about 10 per cent smaller than that for the 16:5 cluster (see Fig. 1.86 and Table 1.8). Also, from the experimental results, the basic unit in the spinell Fe_3O_4 was shown to be the 16:5 cluster, Catlow and co-workers concluded that the 16:5 cluster is more likely to be the basic unit in wüstite than the 13:4 cluster. This suggests that with increasing δ crystals with the structure of the 16:5 cluster begin to nucleate and grow. The computed result is also supported by a study on the magnetic structure of magnetite Fe_3O_4 by use of polarized neutron diffraction.[58]

The relationship between the phase diagram (Fig. 1.82) and the cluster models (Fig. 1.86) has not yet been clarified. It is important that detailed *in-situ* X-ray and neutron diffraction studies are carried out to study the structure as a function of temperature and non-stoichiometry (more detailed and recent results are reviewed in refs 52, p. 61 and 59, p. 699).

Table 1.9
Value of n in various defect complexes in $Fe_{1-\delta}O$

Defect complex	n
V_M^-, V_M^{2-}	4, 6
$(V_M V_M)^{2-}$, $(V_M V_M)^{4-}$	3, 5
V_M^{2-}–Fe_i^{3+}–V_M^{2-} (2:1)	4
$13V_M^{2-}$–$4Fe_i^{3+}$ (13:4)	3.33
$16V_M^{2-}$–$5Fe_i^{3+}$ (16:5)	3.09
$4V_M^{2-}$–Fe_i^{3+} (4:1)	4
$6V_M^{2-}$–$2Fe_i^{3+}$ (6:2)	3.5
$8V_M^{2-}$–$3Fe_i^{3+}$ (8:3)	2.4

In conclusion Table 1.9 shows the n values of eqn (1.249) for various defect complexes.

1.5 Concluding remarks

In this chapter non-stoichiometric compounds derived from point defects have been reviewed mainly from a thermodynamical point of view, and many examples have been presented for the purpose of understanding the nature of non-stoichiometry. As mentioned above it is not necessary to take the interaction between defects (ε_{vv} for vacancies) into consideration if the

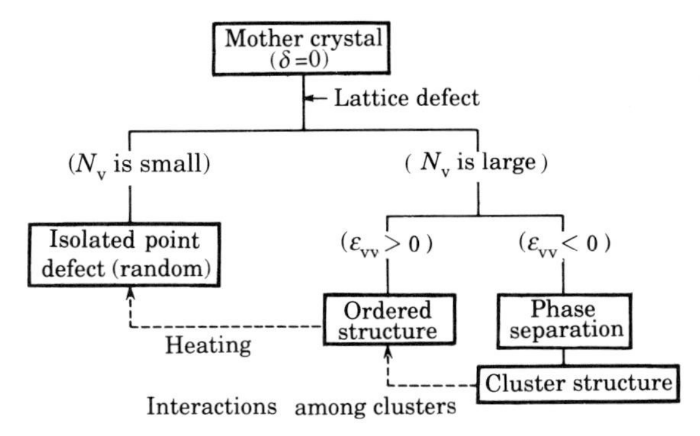

FIG. 1.87 Summary of non-stoichiometric compounds derived from point defects.

concentration of defects (N_v) is low. However, when N_v becomes higher, ε_{vv} plays an important role in the nature of the non-stoichiometric compounds. Figure 1.87 shows a classification of non-stoichiometric compounds in terms of N_v and ε_{vv}.

The precise determination of the cluster structure is indispensable in the understanding of the chemical and physical properties of these substances. The development of calculation methods for the stability of defect complexes is expected in the near future.

References

1. T. Katsura, B. Iwasaki, S. Kimura, and S. Akimoto, *J. Chem. Phys.*, 1967, **47**, 4559.
2. F. A. Kröger and H. J. Vink, *Solid State Phys.*, 1956, **3**, 310.
3. J. B. Lightstone and G. G. Libowitz, *J. Phys. Chem. Solids*, 1969, **30**, 1025.
4. G. G. Libowitz and N. B. Hannay (eds), *Treatise on solid state chemistry*, **1**, 335, Plenum Press, New York, London, 1975.
5. R. H. Fowler and E. A. Guggenheim, *Statistical thermodynamics*, Cambridge University Press, Cambridge, 1949.
6. Y. Oka, K. Kosuge, and S. Kachi, *J. Solid State Chem.*, 1978, **24**, 41.
7. J. H. Hildebrand and R. L. Scott, *The solubility of non-electrolytes*, Reinhold Pub. Co., New York, 1950.
8. G. G. Libowitz, *J. Appl. Phys.*, 1962, **33**, 399.
9. P. A. Cox, *The electronic structure and chemistry of solids*, Oxford University Press, Oxford, 1987.
10. N. B. Hannay, *Solid state chemistry*, Prentice Hall, 1967.
11. T. Katsura and M. Hasegawa, *Bull. Chem. Soc. Jpn*, 1967, **40**, 561.
12. H. Endo, M. Wakihara, M. Taniguchi, and T. Katsura, *Bull. Chem. Soc. Jpn*, 1973, **46**, 2087.
13. M. Taniguchi and M. Wakihara, *Proc. 1st US–Japan seminar on defects and diffusion of solids*, Tokyo, 1976.
14. K. Kosuge, H. Okinaka, S. Kachi, K. Nagasawa, Y. Bando, and T. Takada, *Jap. J. Appl. Phys.*, 1970, **9**, 1004.
15. Y. Goto and T. Kitamura, *J. Jpn Soc. Pow. Met.*, 1962, **9**, 109.
16. B. Kubota, *J. Am. Ceram. Soc.*, 1961, **44**, 239.
17. O. Fukunaga and S. Saito, *J. Am. Ceram. Soc.*, 1968, **51**, 362.
18. B. Kubota and E. Hirota, *Natl Tech. Rep. Jpn*, 1961, **2**, 372.
19. M. Wakihara, J. Nii, T. Uchida, and M. Taniguchi, *Chem. Lett.*, 1977, 621.
20. M. Wakihara, T. Uchida, and M. Taniguchi, *Met. Trans.*, 1978, **9B**, 29.
21. D. J. Young, W. W. Smeltzer, and J. S. Kirkaldy, *J. Electrochem. Soc.*, 1973, **120**, 1221.
22. M. Wakihara, T. Uchida, and M. Taniguchi, *Mat. Res. Bull.*, 1976, **11**, 973.
23. G. Kullerud and R. A. Yund, *J. Petrology*, 1962, **3**, 126.
24. H. Rau, *J. Phys. Chem. Solids*, 1974, **35**, 1415.
25. H. Rau, *J. Phys. Chem. Solids*, 1975, **36**, 1199.

26. T. Resenquist, *J. Iron Steel Inst.*, 1954, **176**, 37.
27. M. Laffitte, *Bull. Soc. Chim. Fr.*, 1959, 1211.
28. C. M. Osburn and R. W. Vest, *J. Phys. Chem. Solids*, 1971, **32**, 1331.
29. J. R. Hellman and V. S. Stubican, *Mat. Res. Bull.*, 1982, **17**, 459.
30. J. W. Patterson, E. C. Bogren, and R. A. Rapp, *J. Electrochem. Soc.*, 1967, **114**, 752.
31. W. D. Kingerly, J. Pappis, M. E. Doty, and D. C. Hill, *J. Am. Ceram. Soc.*, 1959, **42**, 393.
32. C. Wagner, *J. Electrochem. Soc.*, 1968, **115**, 933.
33. H. Okinaka, K. Kosuge, and S. Kachi, *Trans. JIM*, 1971, **12**, 44.
34. Y. D. Tretyakov and R. A. Rapp, *Trans. AIME*, 1969, **245**, 1235.
35. N. L. Peterson and C. L. Wiley, *J. Phys. Chem. Solids*, 1984, **45**, 281.
36. M. O'Keeffe, Y. Ebisaki, and W. J. Moore, *J. Phys. Soc. Jpn*, 1963, **18**, 131.
37. F. Perinet, S. Barbezat, and C. Monty, *J. Phys. Colloq.*, 1980, **C6**, 315.
38. J. Maruenda, R. Farhi, and G. Petat-Ervas, *J. Phys. Chem. Solids*, 1981, **42**, 911.
39. M. O'Keeffe and W. J. Moore, *J. Chem. Phys.*, 1961, **35**, 1324.
40. F. A. Kröger, *The chemistry of imperfect crystals*, North-Holland Publ. Co., Amsterdam, 1974.
41. J. Bloem, *Philip. Res. Rep.*, 1956, **11**, 273.
42. J. Bloem and F. A. Kröger, *Z. Phys. Chem.*, 1956, **7**, 1.
43. E. J. Verwey, *Bull. Soc. Chim. Fr.*, 1949, 122.
44. F. A. Kröger and H. J. Vink, *Solid State Phys.*, 1956, **3**, 307.
45. Y. Oka, K. Kosuge, and S. Kachi, *J. Solid State Chem.*, 1978, **23**, 11.
46. Y. Oka, K. Kosuge, and S. Kachi, *J. Solid State Chem.*, 1978, **24**, 41.
47. M. Wakihara, T. Uchida, and M. Taniguchi, *Mat. Res. Bull.*, 1976, **11**, 973.
48. B. E. F. Fender and F. D. Riley, *J. Phys. Chem. Solids*, 1969, **30**, 793.
49. C. Picard and M. Dode, *Bull. Soc. Chim. Fr.*, 1970, **7**, 2486.
50. P. Vallet and P. Raccah, *Mem. Sci. Rev. Metal.*, 1965, **62**, 1.
51. I. Bransky and A. Z. Hed, *J. Am. Ceram. Soc.*, 1968, **51**, 231.
52. O. Sørensen (ed.), *Nonstoichiometric oxides*, Academic Press, New York, 1981, p. 41.
53. W. L. Roth, *Acta Crystallogr.*, 1960, **13**, 146.
54. F. Koch and J. B. Cohen, *Acta Crystallogr.*, 1969, **B25**, 275.
55. A. K. Cheetham, B. E. F. Fender, and R. I. Tayler, *J. Phys. C*, 1971, **4**, 2166.
56. C. R. A. Catlow and B. E. F. Fender, *J. Phys. C*, 1975, **8**, 3267.
57. N. F. Mott and M. J. Littleton, *Trans. Faraday Soc.*, 1938, **34**, 485.
58. P. D. Battle and A. K. Cheetham, *J. Phys. C*, 1979, **12**, 337.
59. C. R. A. Catlow and W. C. Mackrodt (eds), *Nonstoichiometric compounds*, The American Ceramic Society, Inc., Westerville, Ohio, 1986.

Further reading

R. H. Fowler and E. A. Guggenheim, *Statistical thermodynamics*, Cambridge University Press, Cambridge, 1949.
R. F. Gould (ed.), *Nonstoichiometric compounds*, Adv. Chem. Ser., vol. 39, 1963.
L. Mandelcorn (ed.), *Nonstoichiometric compounds*, Academic Press, New York, 1964.

N. B. Hannay (ed.), *Treatise on solid state chemistry*, Plenum Press, New York, London, 1976.

N. B. Hannay, *Solid state chemistry*, Prentice Hall, 1967.

N. N. Greenwood, *Ionic crystals, lattice defects and nonstoichiometry*, Butterworth, London, 1968.

A. Rabeneau (ed.), *Problem of non-stoichiometry*, North-Holland, Amsterdam, 1970.

L. Eyring and M. O'Keeffe (eds), *The chemistry of extended defects in non-metallic solids*, North-Holland, Amsterdam, London, 1970.

F. A. Kröger, *The chemistry of imperfect crystals*, North-Holland, Amsterdam, 1974.

O. T. Sørensen (ed.), *Nonstoichiometric oxides*, Academic Press, New York, 1981.

A. F. Wells, *Structural inorganic chemistry*, Oxford University Press, Oxford, 1984.

C. R. A. Catlow and W. C. Mackrodt (eds), *Nonstoichiometric compounds*, The American Ceramic Society, Inc., Westerville, Ohio, 1986.

L. Smart and E. Moore, *Solid state chemistry, an introduction*, Chapman & Hall, 1992.

2

NON-STOICHIOMETRIC COMPOUNDS DERIVED FROM EXTENDED DEFECTS

2.1 Introduction

The non-stoichiometric compounds that we describe in this chapter are closely correlated with the classical non-stoichiometric compounds derived from point defects discussed in Chapter 1. For the past twenty years precise structural analyses on complex binary and ternary compounds have been carried out using X-ray and neutron diffraction techniques. Moreover, owing to the striking development of the resolving power of the electron microscope crystal structures can be seen directly as structure images. As a result, it has been shown that most complex structures can be derived by introducing extended defects regularly into a mother structure. A typical example is a 'shear structure', which is derived by introducing planar defects of anion rows into the mother lattice. A 'block structure' is derived by introducing two groups of planar defects. 'Vernier structures', 'micro-twin structures', 'intergrowth structures', and 'adaptive structures' are also described in detail in this chapter.

2.2 Shear structures

2.2.1 Shear structures with one set of planar defects—one-dimensional shear structure

At the beginning of 1950, Professor A. Magnéli's group in Sweden started a systematic study of the crystal structures of the oxides of transition metal elements such as Ti, V, Mo, and W, mainly by X-ray diffraction techniques.[1–4] As a result, they confirmed the existence of the homologous compounds expressed by V_nO_{2n-1}, Ti_nO_{2n-1} etc. ($n = 2, 3, 4, \ldots$) and also predicted that the crystal structure of these compounds could be derived from a mother structure, 'rutile'.

Figure 2.1 shows the X-ray powder diffraction patterns (CuK_α) of compounds TiO_x between Ti_2O_3 ($x = 1.5$) and TiO_2 ($x = 2.0$).[3] This clearly indicates the convergence of the diffraction patterns to that of TiO_2 (rutile) with increasing x, which is why the Magnéli school predicted the mother

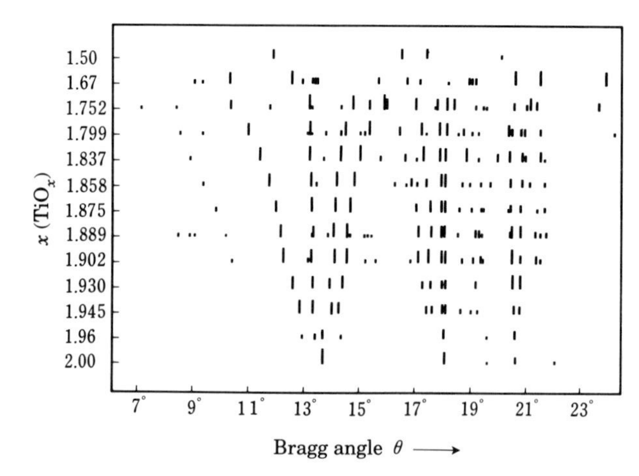

FIG. 2.1 X-ray (CuK_α) powder diffraction patterns of TiO_x ($1.5 < x < 2.0$).[4] The top of the patterns is from Ti_2O_3 (corundum) and the bottom from TiO_2 (rutile). With increasing x, the patterns converge to that of rutile.

structure to be rutile. This prediction was verified by the structure determinations of Ti_5O_9[5] and V_nO_{2n-1}.[6] These compounds are called Magnéli phases after the main investigator, and similar compounds have been discovered.

Anderson and Hyde[7] proposed the structural principle for these compounds, by which the daughter structure (they called 'shear structure'), i.e. Ti_nO_{2n-1} and V_nO_{2n-1}, etc., can be derived from the mother structure as follows:

Operation (*1*): The mother structure (for example, TiO_2 (rutile)) is divided into blocks with the dimension of $a = n \times d_{H_s}$ (n is an integer), where H_s is a shear plane with plane indices (hkl) and d_{H_s} is the plane spacing of H_s (Fig. 2.2(a)).

Operation (*2*): Material with the dimension of $t = \gamma \times d_{H_s}$ ($0 < \gamma < 1$) is cut away from the boundary region of the blocks, as shown in Fig. 2.2(b).

Operation (*3*): By operating a shear vector $\mathbf{R}$, gaps between the blocks are eliminated (Fig. 2.2(c)).

Thus, a daughter structure with a new periodicity of $d = (n - \gamma)d_{H_s}$ can be obtained. Hereafter we shall call the total operation the shear operation. Because the composition of the daughter phase generally differs from that of the mother phase, the composition of the material eliminated by operation (2) has to differ from that of the mother structure (see below). It is also to be noted that operations (1)–(3) do not represent the mechanism of

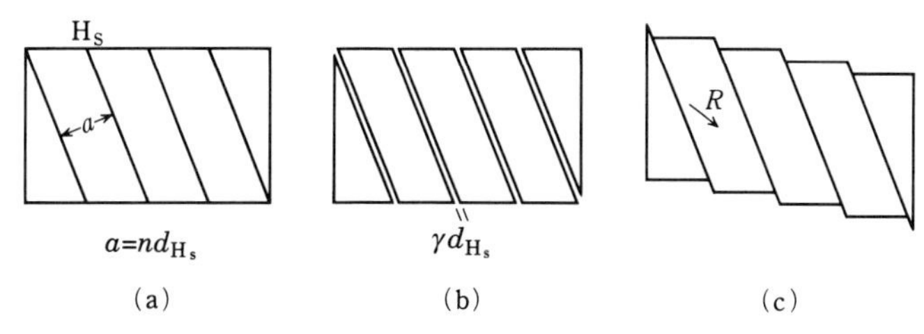

FIG. 2.2 Schematic drawing of the shear operation (see text).

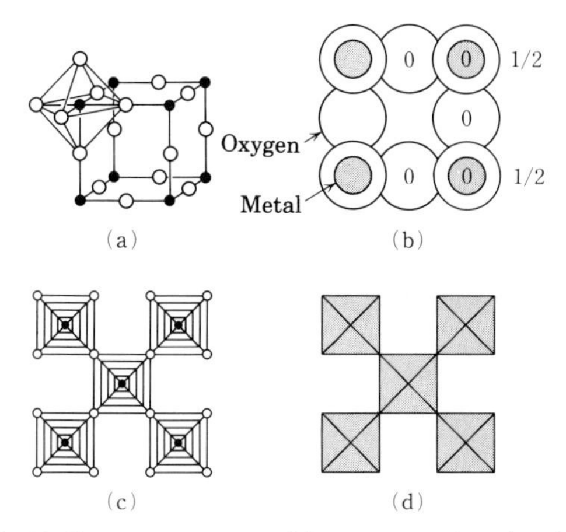

FIG. 2.3 ReO_3-type structure: (a) a perspective drawing; (b) a projection of the atomic arrangement on (001); (c) and (d) a linkage pattern of octahedra.

formation of the daughter phase from the mother phase, but only the relationship between the mother and daughter structures.

At present, only cubic ReO_3 and tetragonal TiO_2 (rutile) are known as mother compounds of shear structures. We shall, first, study the shear structure derived from ReO_3, the structure of which can be understood visually. In the ReO_3 structure, Re occupies the cube-corner and O occupies the centre of each cube-edge as shown in Figs 2.3(a) and (b). In other words the regular octahedra of oxygen, the centres of which are occupied by Re metal, are linked to each other by corner-sharing along the x-, y-, and z-directions. Figures 2.3(c) and (d) show the arrangement of the octahedra along [001]. Typical shear planes $(1k0)$, which have been observed by electron microscopy, are shown in Fig. 2.4, where $k = 1, 2, \ldots, 5$. In every shear plane, the arrangement of atomic planes, perpendicular to the shear planes, is $\ldots$**ABAB**$\ldots$, where **A** denotes an oxygen only plane and **B** denotes a plane of (oxygen + metal). For example, **A** = O (oxygen) and

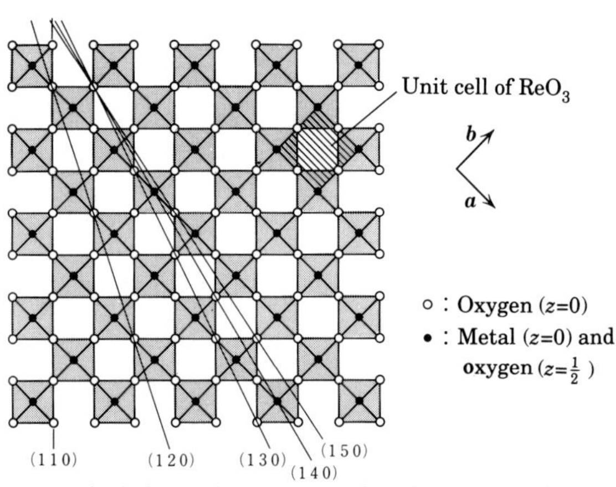

FIG. 2.4 Typical shear planes $(1k0)$ of ReO_3-type mother structure.

B = MO_2 for the (130) shear plane. In the usual shear operation, **A** planes only are periodically eliminated from the mother compounds, consequently the daughter compound has an oxygen-poor composition compared with that of the mother compound.

Figure 2.5 shows the shear operation of $(130)\frac{1}{2}[1\bar{1}0]$ on ReO_3, where (130) denotes a shear plane and $\frac{1}{2}[1\bar{1}0]$ denotes a shear vector. Figure 2.5(a) shows the oxygen polyhedra of both slabs after the elimination of one sheet of **A** plane (oxygen only plane). For visualization the two slabs are separated by vector [110], thus only the point defects of oxygen produced by operation (2) are essential in the shear operation. These point defects are annihilated by operation (3) (shear vector $\frac{1}{2}[1\bar{1}0]$), which means that edge-sharing of octahedra near the shear plane occurs, as shown in octahedron 1,1′,2,2′,3,3′ in Fig. 2.5(b) (shaded octahedra), in contrast with corner-sharing in ReO_3.

Figure 2.6 shows the structure in the vicinity of the shear plane for the series of shear operations $(1k0)\frac{1}{2}[1\bar{1}0]$, similar to Fig. 2.5(b). In these, the shear operations $(110)\frac{1}{2}[1\bar{1}0]$ and $(010)\frac{1}{2}[1\bar{1}0]$ are of special importance. For the case of $(110)\frac{1}{2}[1\bar{1}0]$, only operation (3) is performed on ReO_3, without the elimination of the oxygen only plane, therefore the composition of the daughter compound is unchanged. This structure is often called an anti-phased boundary (APB) or twin structure. The structure for the operation $(010)\frac{1}{2}[1\bar{1}0]$ is also a special one, in which a string of zigzag edge-shared octahedra runs through a crystal. These two structures are the basic ones for the shear structures $(1k0)\frac{1}{2}[1\bar{1}0]$. The shear operation $(1k0)\frac{1}{2}[1\bar{1}0]$ can be expressed as

$$\begin{aligned}(1k0)\tfrac{1}{2}[1\bar{1}0] &= p(010)\tfrac{1}{2}[1\bar{1}0] + q(110)\tfrac{1}{2}[1\bar{1}0]\\ &= (q\ \ p+q\ \ 0)\tfrac{1}{2}[1\bar{1}0] \qquad (2.1)\end{aligned}$$

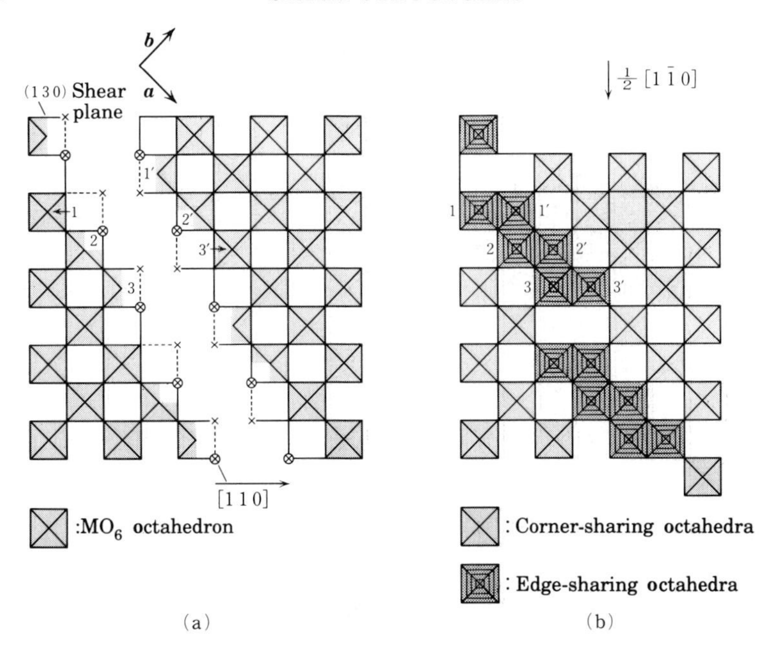

FIG. 2.5 Shear operation of $(130)\frac{1}{2}[1\bar{1}0]$ on ReO_3-type structure. (a) Arrangement of the oxygen polyhedra of both slabs after the elimination of one sheet of **A** plane (oxygen only plane). For visualization the two slabs are separated by vector [110]. The mark ⊗ denotes the point defects of oxygen on plane A, which are produced by operation (2), and the mark × denotes the point defects produced by the separation of the slabs. (b) New structure formed by shear operation (3). By this operation the point defects are annihilated, which results in the occurrence of edge-sharing octahedra near the shear plane.

which is often written as

$$(1k0) = p(010) + q(110) = (q \;\; p+q \;\; 0)$$

where $q = 1$ and $p = k - 1$ in this case. Equation (2.1) indicates that the shear operations $(1k0)\frac{1}{2}[1\bar{1}0]$ are composed of the operations $(110)\frac{1}{2}[1\bar{1}0]$ and $(010)\frac{1}{2}[1\bar{1}0]$.

Consider a crystal with a structure resulting from periodic elimination of one **A** layer (oxygen-only plane) every n **B** layers (oxygen + metal). The arrangement of the atomic planes becomes

$$\ldots \mathbf{ABAB}^* \mid \mathbf{BABAB} \ldots \mathbf{B}^* \mid \mathbf{BABA} \ldots \tag{2.2}$$

where the mark * denotes an eliminated **A** layer and $| \ldots |$ a unit cell of the

FIG. 2.6 Structure in the vicinity of the shear plane for a series of the shear operation $(1k0)\frac{1}{2}[1\bar{1}0]$ of ReO_3-type structure.[27] (a) $(110)\frac{1}{2}[1\bar{1}0]$: the structure obtained after only operation (3) on ReO_3 without the elimination of an oxygen-only plane. The structure is called APB (anti phase boundary) or twin structure. (b) $(120)\frac{1}{2}[1\bar{1}0]$; (c) $(130)\frac{1}{2}[1\bar{1}0]$; (d) $(140)\frac{1}{2}[1\bar{1}0]$; (e) $(150)\frac{1}{2}[1\bar{1}0]$; (f) $(010)\frac{1}{2}[1\bar{1}0]$, note that the string of zigzag edge-sharing octahedra is running through the crystal.

crystal. As mentioned above, the composition of the **A** and **B** layers for the $(1k0)$ shear planes of a ReO_3-type structure (mother compound) is expressed as

$$\begin{aligned} k = \text{even:} &\quad \mathbf{A} = \mathrm{O}, &\quad \mathbf{B} = \mathrm{MO_2} \\ k = \text{odd:} &\quad \mathbf{A} = \mathrm{O_2}, &\quad \mathbf{B} = \mathrm{MO} \end{aligned} \tag{2.3}$$

After closing up each block by shear vector $\frac{1}{2}[1\bar{1}0]$ to eliminate gaps, we obtain daughter crystals with the composition M_nO_{3n-1} (k = even) and M_nO_{3n-2} (k = odd) (in general, the composition of the daughter compound is $(n-1)\mathbf{A} + n\mathbf{B}$). In Fig. 2.7 an ideal structure of $W_{20}O_{58}$ derived by the shear operation of $(130)\frac{1}{2}[1\bar{1}0]$ on ReO_3 (cubic WO_3, being isomorphous

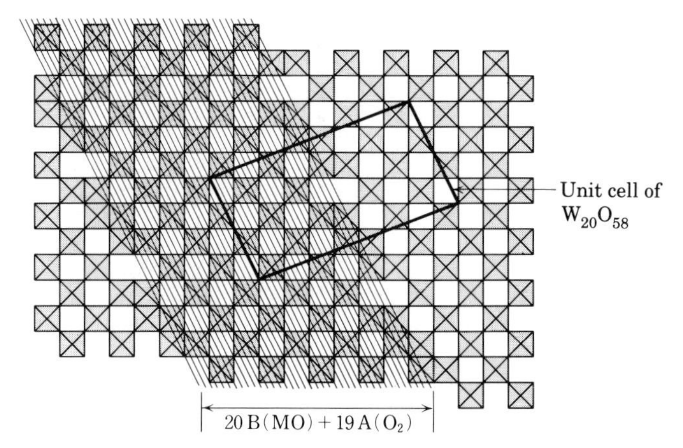

FIG. 2.7 Ideal structure of $W_{20}O_{58}$ produced by the shear operation $(130)\frac{1}{2}[1\bar{1}0]$ on cubic WO_3 structure. The unit cell is outlined.

with ReO_3) is shown as an example, which is in accord with the structure determined by X-ray diffraction.

Next, the rutile (TiO_2) based shear structure is discussed. In the rutile-type structure (tetragonal), the metal and oxygen occupy the following positions as shown in Fig. 2.8:

Metal (2a): $000;\ \frac{1}{2}\frac{1}{2}\frac{1}{2}$

Oxygen (4f): $\pm(uu0;\ u+\frac{1}{2}\ \frac{1}{2}-u\ \frac{1}{2})$

The distorted octahedra of oxygen, the centres of which are occupied by metals, are linked by sharing edges along the c-direction or by sharing corners in (001) planes. The metals have a body-centred tetragonal lattice and oxygens are pseudo-hexagonal closed packed. Figures 2.9(a) and (b)

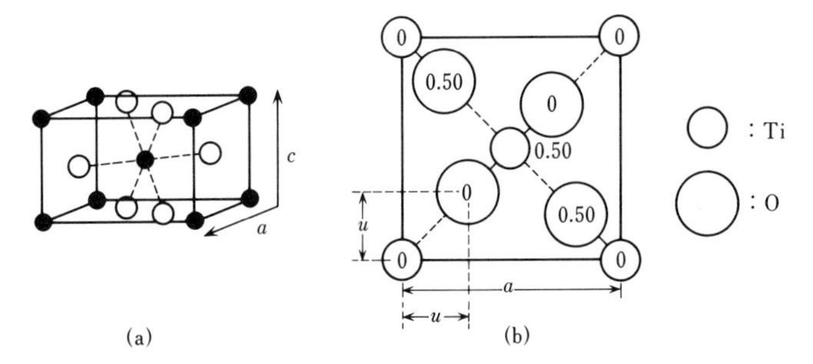

FIG. 2.8 Rutile-type structure (I): (a) a perspective drawing, (b) a projection of the atomic arrangement on (001). Parameter u is variable.

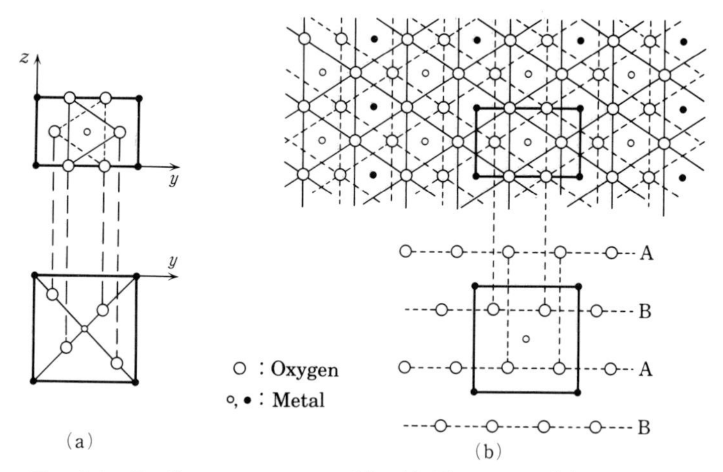

FIG. 2.9 Rutile-type structure (II): (a) The top and bottom are the projection on (100) and (001) of a real rutile structure, respectively. (b) The top and bottom are the projection on (100) and (001) of an idealized rutile structure, i.e. the oxygen packing of rutile is assumed to be HCP.

show the projection of the rutile structure and that of an idealized rutile structure (i.e. assuming that the arrangement of oxygen is undistorted hexagonal close packing, HCP) along [100] (top figures of (a) and (b)) and [001] (bottom figures), respectively. For simplicity, we assume hereafter that the arrangement of oxygen is HCP in the rutile structure.

Ti_nO_{2n-1} and V_nO_{2n-1} are well known as shear compounds derived from rutile (TiO_2, mother structure). The shear operation of these compounds is $(121)\frac{1}{2}[0\bar{1}1]$ or $(132)\frac{1}{2}[0\bar{1}1]$. The shear vector $\frac{1}{2}[0\bar{1}1]$ corresponds to the vector from an oxygen to an oxygen in the (100) plane of the rutile structure. Consequently the oxygen arrangement after the shear operation is the same as that before the operation. (This is not the case for ReO_3, because the structure has a defective face centred cubic (FCC) lattice oxygen arrangement.) This means that it is adequate to observe only the metal positions after the shear operation. However, the shear structure derived from rutile is not visible, since the shear planes are not parallel to one of crystal axes of rutile. Figure 2.10 shows the rutile structure projected along [100], together with the shear planes (121), (132), and (011). Although the arrangement of atoms on the shear planes (121) and (132) is not visible in the figure, it is possble to confirm that the arrangement of atomic planes parallel to the shear planes is . . . **ABABAB** . . . for both, where **A** = O and **B** = MO. For the case of $(011)\frac{1}{2}[0\bar{1}1]$, only the shear operation (3) is performed, without removal of

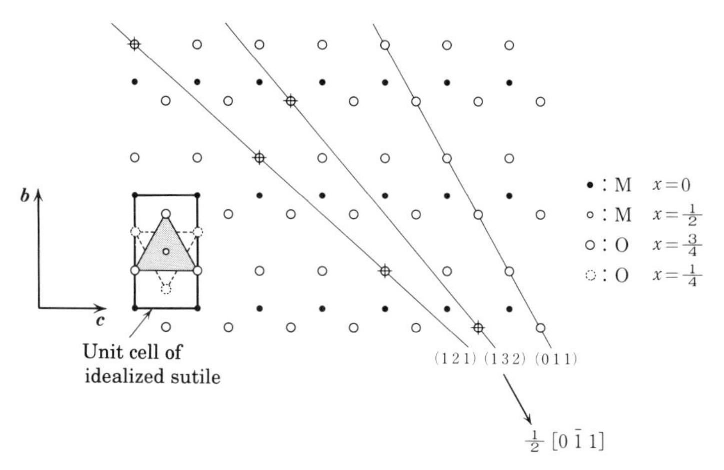

FIG. 2.10 Typical shear planes (hkl) of a TiO_2-type mother structure.

an oxygen-only plane, being similar to the shear structure $(110)\frac{1}{2}[1\bar{1}0]$ of ReO_3. This structure is a kind of APB or twin structure.

Figures 2.11(a), (b), and (c) show the structures in the vicinity of the shear planes for the shear operations $(121)\frac{1}{2}[0\bar{1}1]$, $(132)\frac{1}{2}[0\bar{1}1]$, and $(011)\frac{1}{2}[0\bar{1}1]$ (APB). The structure near the shear plane for the shear operation $(132)\frac{1}{2}[0\bar{1}1]$ can be regarded as being composed of $(121)\frac{1}{2}[0\bar{1}1]$ and $(011)\frac{1}{2}[0\bar{1}1]$ (APB), i.e. alternate stacking of (121) shear structure and (011) APB. Generally the shear operation $(hkl)\frac{1}{2}[0\bar{1}1]$ can be resolved into its components:

$$\begin{aligned}(hkl) &= p(011) + q(121)\\ &= (q \;\; p+2q \;\; p+q)\end{aligned} \tag{2.4}$$

This equation is similar to eqn (2.1) for ReO_3 families. For example, the shear structure (132) of TiO_2 (rutile) corresponds to $p = q = 1$. In Section 2.6, the structure for various (p, q) is discussed.

The homologous compounds V_nO_{2n-1} are derived by the shear operation $(121)\frac{1}{2}[0\bar{1}1]$ on rutile. As mentioned above, the arrangement of atomic planes parallel to (121) is ... **ABABAB**..., where **A** = O and **B** = MO. After the elimination of an oxygen-only plane **A** in every n **B** plane, the arrangement of the atomic planes gives compounds generally expressed as M_nO_{2n-1} $(=(n-1)\mathbf{A} + n\mathbf{B})$. As a typical example, Fig. 2.12 shows an ideal structure of V_6O_{11} $(n = 6)$. The structure has been confirmed by X-ray diffraction.[6] In Table 2.1, the shear compounds of the transition metal oxides are summarized.[8]

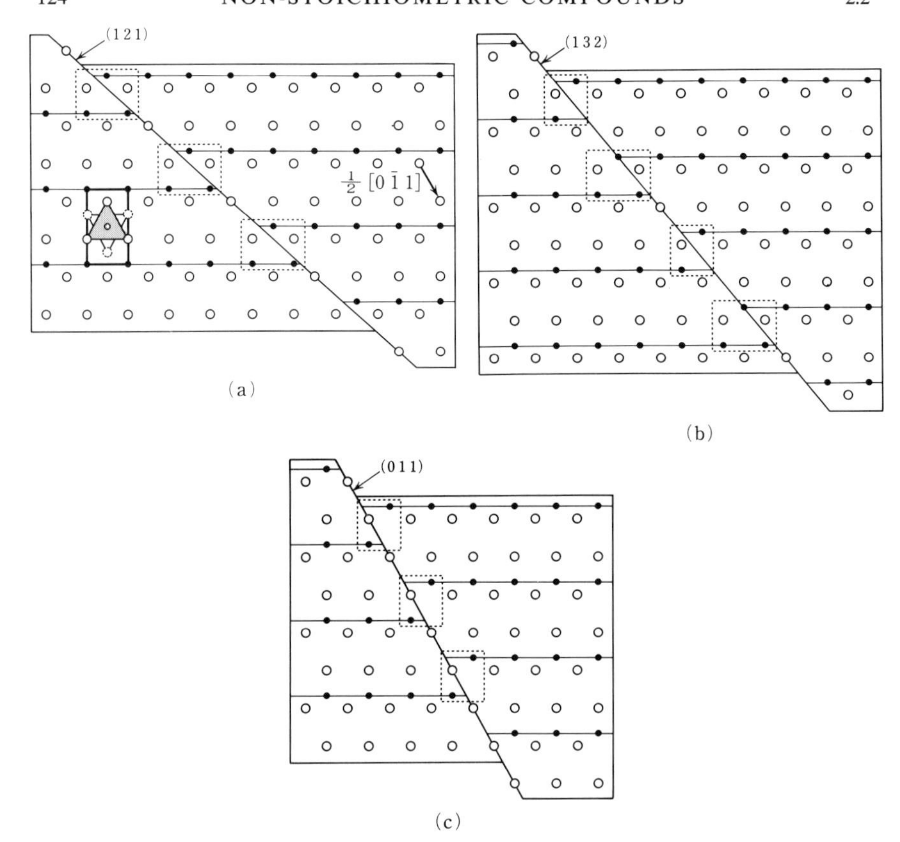

FIG. 2.11 Structure in the vicinity of the shear plane for the shear operation $(hkl)\frac{1}{2}[0\bar{1}1]$ of a TiO_2-type structure. (a) $(121)\frac{1}{2}[0\bar{1}1]$; (b) $(132)\frac{1}{2}[0\bar{1}1]$; (c) $(011)\frac{1}{2}[0\bar{1}1]$: the structure obtained after only operation (3) on TiO_2 without the elimination of an oxygen-only plane. The structure is called an APB (anti phase boundary) or twin structure, and is similar to the shear structure of $(110)\frac{1}{2}[1\bar{1}0]$ of ReO_3 (see Fig. 2.6(a)). Note the atomic arrangement in the zones framed by dotted lines (see also Fig. 2.113).

We shall now show some examples of studies on shear compounds performed using electron microscope (lattice image and structure image) and electron diffraction techniques. Figure 2.13 shows the lattice image of a slightly reduced WO_3 ($WO_{3-\delta}$),[9] the bundle of black lines (lattice image) show the kinds of planar defects. The inset shows the electron diffraction pattern, in which the strong diffraction spots (main diffraction) are from the mother structure (ReO_3-type structure) and the streaks along **g**(120) (reciprocal lattice vector) are from the randomly spaced planar defects parallel to

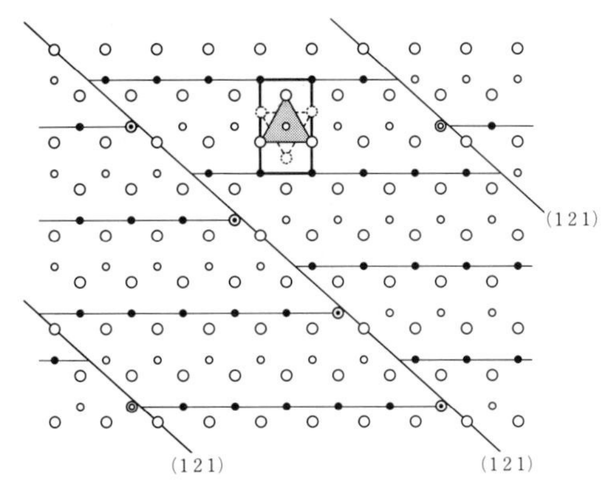

FIG. 2.12 Ideal structure of V_6O_{11} viewed along the *a*-axis of rutile, derived by the shear operation $(121)\frac{1}{2}[0\bar{1}1]$ on rutile. The marks ⊙ and ◎ show the metals at $x = 0$ and $\frac{1}{2}$, leading to face-sharing of adjacent octahedra, similar to a Corundum-type structure.

Table 2.1

Compounds with shear structure in transition metal oxides[8]

(a) Ti–O, V–O, Ti(Cr)–O

Substance	n	Shear plane	Shear vector
Ti_nO_{2n-1}	$4 \le n \le 10$	{121}	$\frac{1}{2}\langle 011\rangle$
	$16 \le n < 36$	{132}	$\frac{1}{2}\langle 011\rangle, \frac{1}{6}\langle 121\rangle, \frac{1}{10}\langle 155\rangle$
V_nO_{2n-1}	$4 \le n \le 9$	{121}	$\frac{1}{2}\langle 011\rangle$
V_nO_{2n+1}	$2 < n < ?$	Block structure (see Section 2.2.2)	
$Ti_{n-2}Cr_2O_{2n-1}$	$6 \le n \le 9$	{121}	$\frac{1}{6}\langle 123\rangle$

(b) Mo–O system

x in MoO_x	Chemical composition
2.00	MoO_2
2.75	Mo_4O_{11}
2.765	$Mo_{17}O_{47}$
2.80	Mo_5O_{14}
2.875	Mo_8O_{23}
2.887	$Mo_{26}O_{75}$
2.889	$Mo_{18}O_{52}$
2.894	$Mo_{19}O_{55}$
2.923	$Mo_{13}O_{38}$
3.00	MoO_3

(c) W–O system

x in WO_x	Chemical composition
2.00	WO_2
2.72	$W_{18}O_{49}$
2.75	W_4O_{11}
2.90	$W_{20}O_{58}$
2.916	$W_{24}O_{70}$
2.95	$W_{40}O_{118}$
2.96	$W_{25}O_{74}$
3.00	WO_3

(d) Mo(W)–O system

x in MO_x	Chemical composition	Ratio of Mo to W
2.735	Mo_8O_{23}	—
2.889	Mo_9O_{26}	—
2.90	$(Mo, W)_{10}O_{29}$	1:4
2.909	$(Mo, W)_{11}O_{32}$	—
2.917	$(Mo, W)_{12}O_{35}$	1:2
2.929	$(Mo, W)_{14}O_{41}$	1:1
3.00	WO_3	—

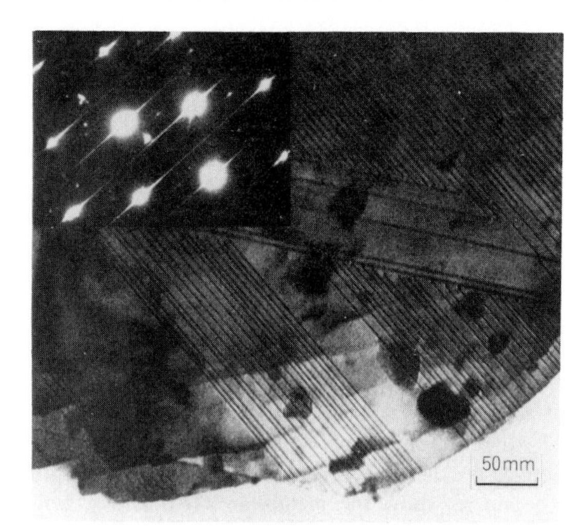

FIG. 2.13 Lattice image of slightly reduced WO_3.[9] The black lines show a kind of planar defect, which corresponds to the shear plane {120}. The inset shows a streaked diffraction pattern for the crystal parallel to **g**(120), indicating the random distribution of shear planes (see also Fig. 2.16).

(120) of ReO_3. These pictures give us information on the shear structure formation. The sample is a slightly reduced one and the lattice image shows the early stages of the shear structure formation or the 'embryo' of the shear structure. The planar defects (shear planes) parallel to (120) (it is noted that there are eight equivalent planes to (120) in the cubic system, denoted as {120}) introduced into the mother structure are randomly arranged, which results in the diffuse streaks along **g**(120) in the electron diffraction pattern. In other words, many kinds of shear compounds W_nO_{3n-1} are mixed. The ordered shear planes (*n*: fixed) gives a group of super spots between the main spots, as shown in Fig. 2.16.

Figure 2.14 shows the structure image of the shear compound $W_{18}O_{52}$ derived by the shear operation $(130)\frac{1}{2}[1\bar{1}0]$ on ReO_3.[10] The image of the shear plane shows the six edge-sharing octahedra as dark blocks and the empty tunnels as pale strips, the mother structure (ReO_3-type) between the shear planes is seen as white (tunnel) and black (W centred octahedron of oxygen) dots (see Figs 2.5 and 2.7). The shear plane spacing is variable. A high-resolution image of an isolated single shear plane (120) of reduced WO_3 is shown in Fig. 2.15(a).[11] The black dots forming a square array can be regarded as the W atom position of WO_3 projected on to the (100) plane. The dark regions along the shear plane correspond to the four edge-sharing octahedra and the large white dots elongated along [110] to the doubled

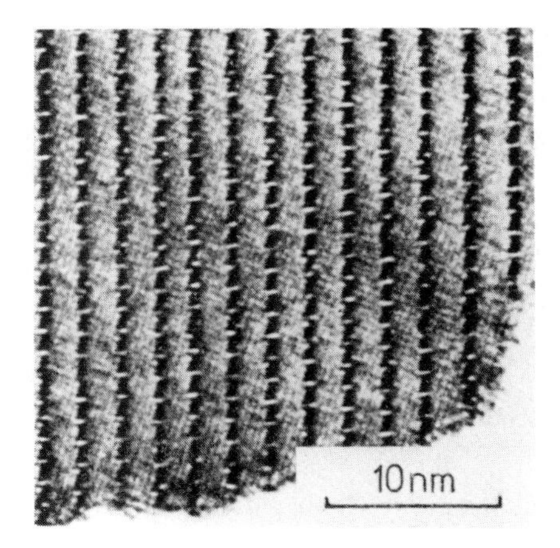

FIG. 2.14 Structure image of $W_{18}O_{52}$ in a series of W_nO_{3n-2}, derived by the shear operation $(130)\frac{1}{2}[1\bar{1}0]$.[10] For the interpretation of the electron micrograph, see text.

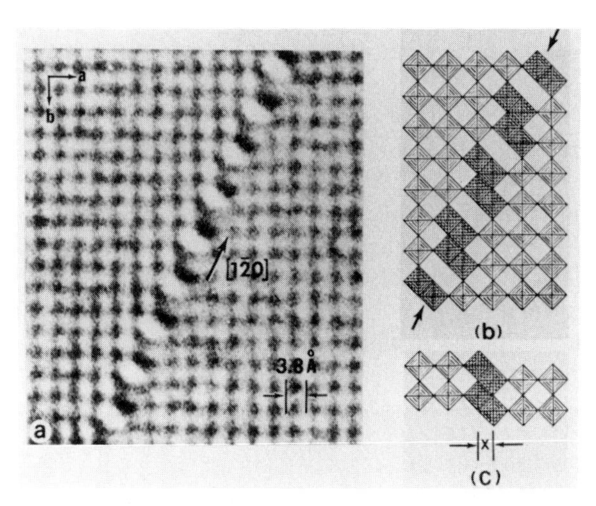

FIG. 2.15 Structure image of an isolated single shear plane of (120) from a fragment of slightly reduced WO_3 (a) and structure models used for interpretation (b) and (c),[11] (see text).

tunnels (compare with the ideal structure depicted in (b)). On close observation of the picture it was concluded that the edge-sharing octahedra are considerably distorted, which results in a larger distance x in Fig. 2.15(c) than in the ideal structure. Thus, the high-resolution structure image has become a useful technique for the study of the microscopic, local structure of crystals.

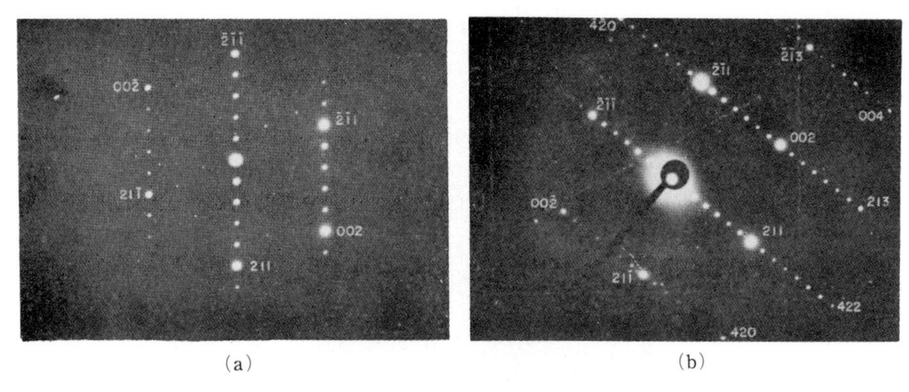

FIG. 2.16 Electron diffraction patterns of V_nO_{2n-1}, derived by shear operation $(121)\frac{1}{2}[0\bar{1}1]$ on rutile.[12] (a) $n = 5$ (V_5O_9), (b) $n = 7$ (V_7O_{13}).

Next we show the shear structure derived from rutile. Figure 2.16 shows the electron diffraction patterns of V_nO_{2n-1} ($n = 3, 4, \ldots, 9$),[12] which are derived by the shear operation $(121)\frac{1}{2}[0\bar{1}1]$ on rutile. For example, the diffraction pattern for V_5O_9 ($n = 5$ in the general formula M_nO_{2n-1}) shows a series of super spots along **g**(121), these are equally spaced between the main diffraction spots (four super spots between the main spots). Figure 2.17 shows lattice images of Ti_nO_{2n-1}, which was prepared by *in-situ* beam heating of rutile thin films in an electron microscope. Assuming the shear operation is known, the composition can be estimated by measuring the shear plane spacing. A high-resolution structure image of V_8O_{15} of V_nO_{2n-1} families is shown in Fig. 2.18.[14] For the case of a rutile-based shear structure, the interpretation of the structure image is not easy, because the structure

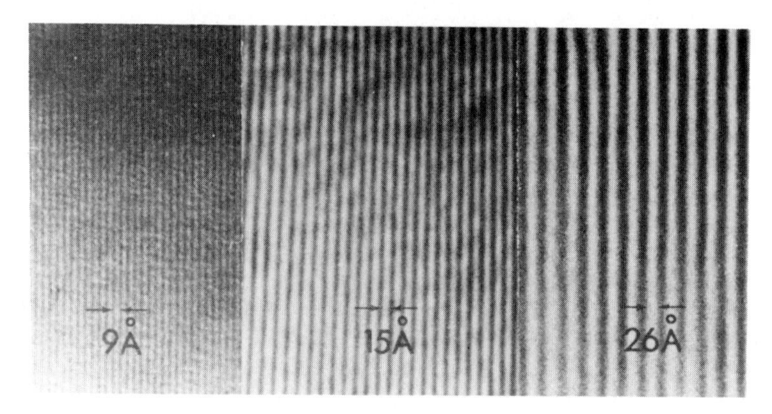

FIG. 2.17 Lattice image of Ti_nO_{2n-1}[13] (see text).

FIG. 2.18 High resolution electron micrograph of V_nO_{2n-1} ($n = 8$).[14] Note that the regular intergrowth of $2V_8O_{15}$ and V_7O_{13} is seen in a rather wide region.

does not include the empty tunnels as seen in ReO_3-based shear structures. It is to be noted that the intergrowth structure consisting of $2V_8O_{15}$ and V_7O_{13} units is observed in a rather wide region, which indicates the possibility of the adaptive structure in the V–O system (see Section 2.6).

2.2.2 *Shear structure with two sets of planar defects—block structure*

The binary or ternary oxides of Nb often crystallize in the block structure, which is closely related to the shear structure mentioned above. The simple block structure can be derived from the ReO_3-type structure by a successive shear operation: $(100)\frac{1}{2}[\bar{1}0\bar{1}] \rightarrow (010)\frac{1}{2}[0\bar{1}\bar{1}]$. Figure 2.19(a) shows the ReO_3 structure projected along [001]. The stacking of the atomic planes parallel to the shear plane of (100) is . . . **ABAB** . . . , where **A** = oxygen-only plane and **B** = (oxygen + metal) plane. On removing an **A** plane in every n (=3 in this figure) **B** planes and closing the blocks by the shear vector of $\frac{1}{2}[\bar{1}0\bar{1}]$ to eliminate the gaps, we obtain the shear structure . . . ***BABAB*** . . . , shown in Fig. 2.19(b). The successive shear operation $(010)\frac{1}{2}[0\bar{1}\bar{1}]$ ($n = 4$ in this figure) is performed on the structure (b) to give the block structure shown in Fig. 2.19(c). The structure consists of L and U sheets (L and U mean the lower and upper sheets in the unit cell). The U sheet is composed of the unit block of the [3 × 4] octahedra (in general [$m \times n$] octahedra; the dimension of the unit block clearly depends on the distance between the eliminated shear planes by the shear operation) at $z = 1$, four of which at the corner of the blocks share the edges of the adjacent blocks, forming a check pattern. The L sheet at $z = 0.5$ has the same structure as the U sheet (generally, the L

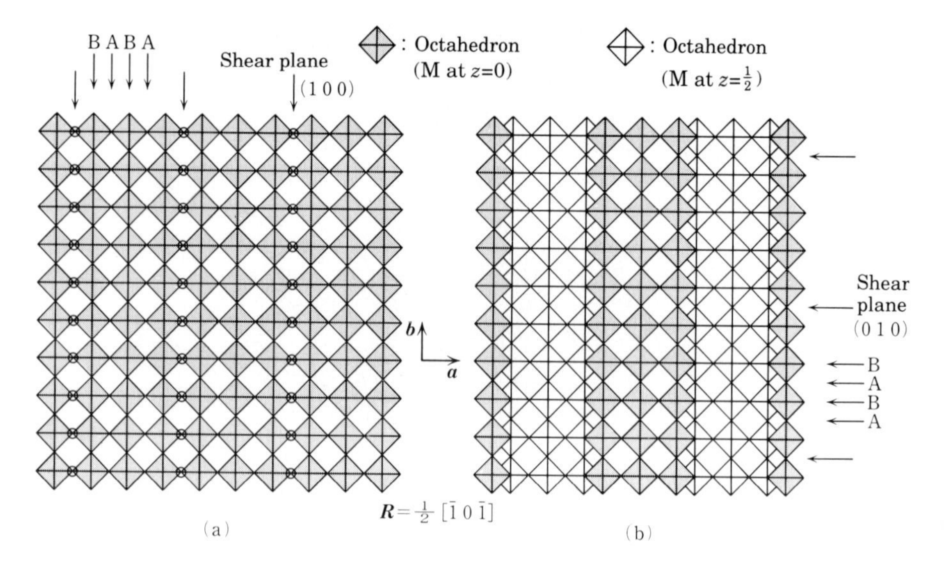

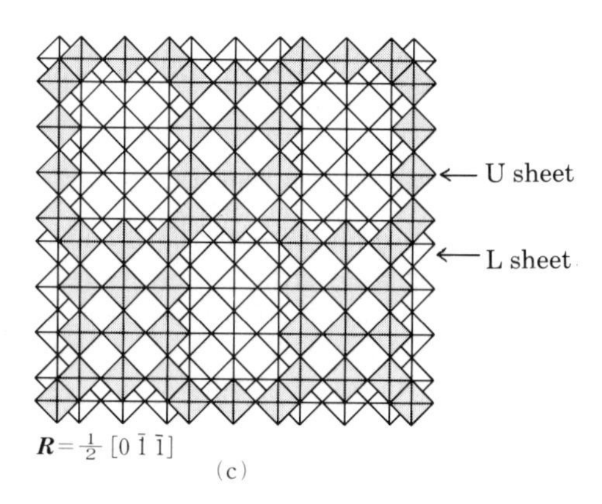

FIG. 2.19 Example of a process for deriving a block structure by a successive shear operation on ReO_3-type structure: $(100)\frac{1}{2}[\bar{1}0\bar{1}] \rightarrow (010)\frac{1}{2}[0\bar{1}\bar{1}]$. (a) Shear plane of (100) of ReO_3-type. The mark ○ denotes the oxygens to be removed. (b) Derivation of a one-dimensional block (shear) structure by the shear operation $(100)\frac{1}{2}[\bar{1}0\bar{1}]$. (c) Block structure [3 × 4] obtained by a successive shear operation of $(010)\frac{1}{2}[0\bar{1}\bar{1}]$ on structure (b). This type of block structure is characterized by edge-sharing of all octahedra at the corners of the blocks with four adjacent blocks.

FIG. 2.20 Ideal structure of M–Nb_2O_5 $[4 \times 4]_{\infty}$, belonging to the type of block structure shown in Fig. 2.19(c). The unit cell is outlined. For the notation, see text.

and U sheets differ in structure and composition as mentioned later). Both sheets are mutually connected by edge-sharing as shown in Fig. 2.19(c). The structure of M–Nb_2O_5, shown in Fig. 2.20, has a similar structure to Fig. 2.19(c), in which the dimension of the unit block is $[4 \times 4]$.

By the second shear operation of $(010)\frac{1}{2}[\bar{2}\bar{1}1]$ on the structure shown in Fig. 2.19(b), we obtain a new structure shown in Fig. 2.21. On each sheet each unit block ($[3 \times 4]$ octahedron) shares two opposite corners with those of two adjacent blocks, forming one-dimensional chains of blocks. N–Nb_2O_5

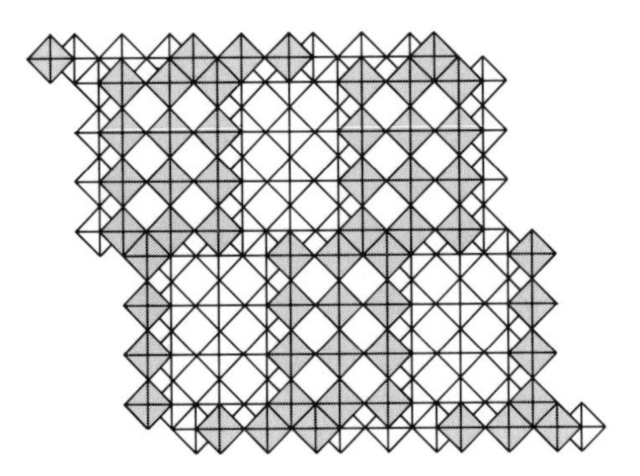

FIG. 2.21 Another example of a block structure derived by the shear operation of $(010)[\bar{2}\bar{1}1]$ on structure (b) of Fig. 2.19. Each block $[3 \times 4]$ shares only two opposite corners with two adjacent blocks.

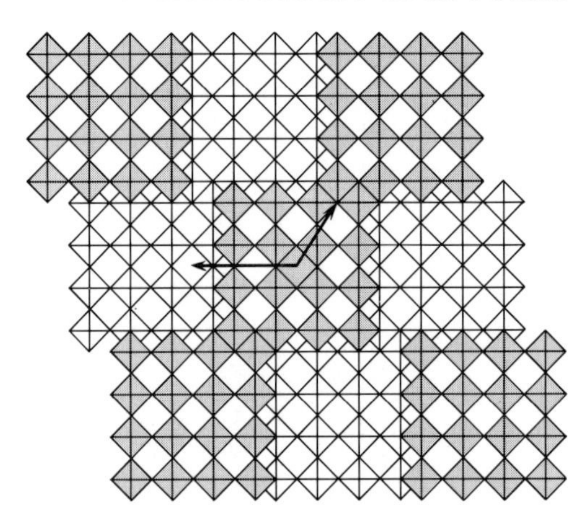

FIG. 2.22 Ideal structure of N–Nb_2O_5 $[4 \times 4]_\infty$, belonging to the type of block structure shown in Fig. 2.21.

has a similar structure as shown in Fig. 2.22. Thus the polymorphism of Nb_2O_5 (M- and N-) is easy to understand from consideration of the block structure (as for the other polymorph H–Nb_2O_5, see Fig. 2.27).

Let us consider the structure derived by a second shear operation of $(\bar{2}50)\frac{1}{2}[\bar{2}\bar{1}1]$ on the structure shown in Fig. 2.19(b). On periodically eliminating the oxygen-only planes parallel to $(\bar{2}50)$ as shown in Fig. 2.23(a), we get a crystal with the gaps between the slabs, as shown in Fig. 2.23(b). In this figure the slabs are separated by the vector [010] for visualization, similar to Fig. 2.5(a), and special consideration for combining the octahedra in each slab was taken in order to form the block structure. On closing up the slabs by the shear vector $\frac{1}{2}[\bar{2}\bar{1}1]$, a new type of block structure is obtained. The mark (°) in Fig. 2.23(c) denotes oxygen vacancies. By supplying oxygen to the vacancy sites, tetrahedral sites for metals are obtained. Usually the metal sited at the tetrahedral positions is different from that sited in the matrix (octahedral sites). In this structure the blocks in each sheet are connected by sharing the corners of tetrahedra. A typical compound having this type of structure is $WNb_{12}O_{33}$, shown in Fig. 2.24. W atoms are sited at tetrahedral positions.

Block structures can therefore, in principle, be derived from the mother structure by the double shear operation. The process, however, is not always as straightforward as that shown in Fig. 2.23. Here, we discuss the block structure from another point of view. The block structure can be interpreted in the following two ways: how to link the blocks $[m \times n]$ in each sheet, and how to connect the sheets U and L. In Fig. 2.25 the basic units, [A], [B], and [C], for the block structure are shown, reduced from the above

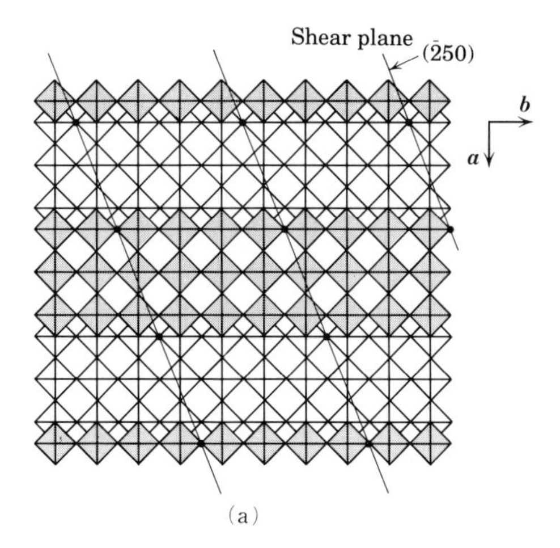

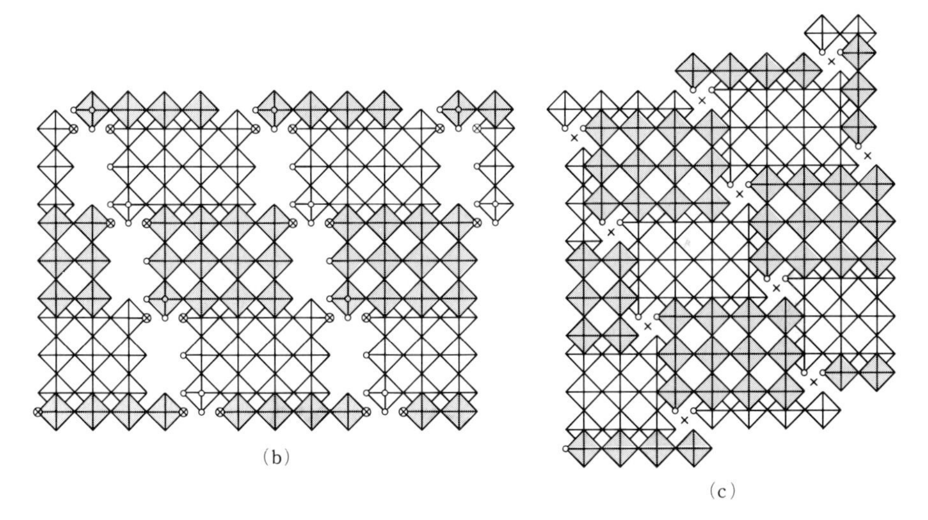

FIG. 2.23 Shear operation $(\bar{2}50)\frac{1}{2}[\bar{2}\bar{1}1]$ on structure (b) of Fig. 2.19. (a) The oxygen arrangement on the shear plane $(\bar{2}50)$. The oxygens marked ● are removed. (b) After removing oxygens on the shear plane, each slab is separated by the vector [010] of ReO_3 for visualization. Note that special consideration is taken in order to form the block structure. The marks ⊗ and ○ denote the oxygen vacancies derived from the shear operation and from the separation of the slabs for visualization, respectively. (c) On closing up the slabs by the shear vector $\frac{1}{2}[\bar{2}\bar{1}1]$, a new type of block structure is obtained. By supplying oxygen to oxygen vacancies (○), tetrahedral sites for metals (×) are possible.

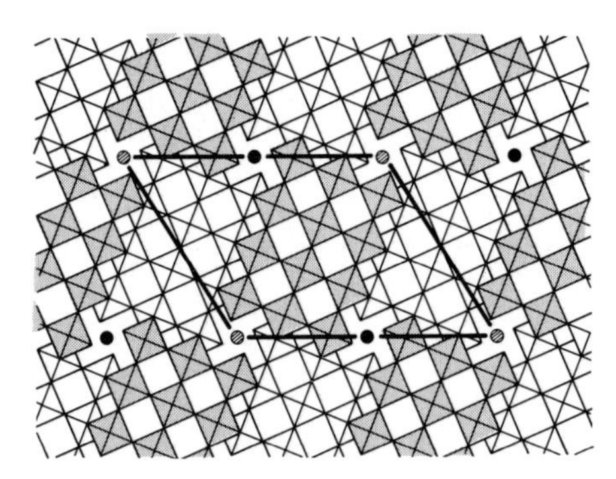

FIG. 2.24 Ideal structure of $WNb_{12}O_{33}$ $[3 \times 4]_1$, belonging to the type of block structure shown in Fig. 2.23(c). W atoms (●, ◍) are sited at tetrahedral positions.

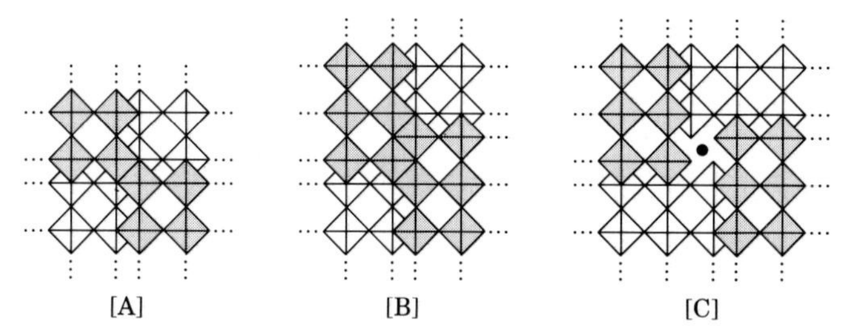

FIG. 2.25 Basic units of block structure: [A] each block shares one edge of one octahedron at the corner of the block (see Fig. 2.19(c)); [B] each block shares three edges of two octahedra at the corner of the block (see Fig. 2.21); [C] each block shares one corner of one tetrahedron (see Fig. 2.23(c)).

mentioned examples (see, Fig. 2.19(c), Fig. 2.21, and Fig. 2.23(c)). On combining these basic units, various kinds of block structures can be derived. Figure 2.26 shows the structure of $TiNb_{24}O_{62}$, in which the basic units [B] and [C] are combined. In this structure each block $[3 \times 4]$ in each sheet is linked by only one adjacent block. How many blocks are connected to each block unit is very important in describing the block structure, which is done by using suffix p as $[m \times n]_p$. The isolated block is expressed by $p = 1$, for example, the structure shown in Fig. 2.23(c) is expressed as $[3 \times 4]_1$. The paired block in the same sheet is expressed by $p = 2$. For example, the structure shown in Fig. 2.26 is expressed as $[3 \times 4]_2$. A block linked to more than two adjacent blocks is expressed by $p = \infty$. Both of the structures shown in Fig. 2.19(c) and Fig. 2.21 are expressed as $[3 \times 4]_\infty$.

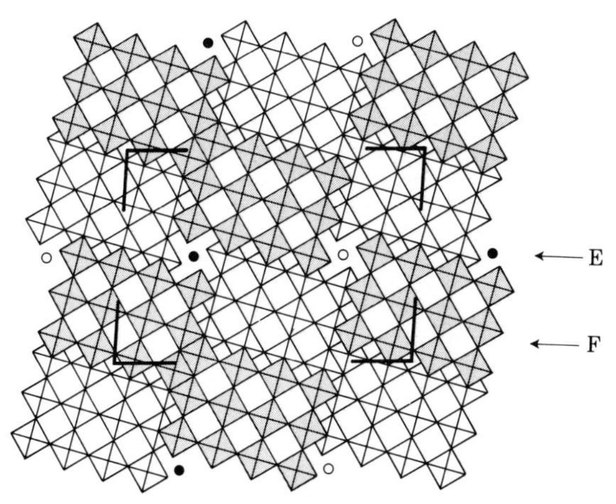

FIG. 2.26 Ideal structure of $TiNb_{24}O_{62}$ $[3 \times 4]_2$. Ti atoms are sited at tetrahedral positions. This structure can be obtained by combining the basic units [B] and [C]. The marks E → and F → denote the direction of a row of edge-sharing octahedra between the adjacent sheets and tetrahedral metals and the direction of a row of edge-sharing octahedra between the blocks both in the sheets and in the adjacent sheets, respectively.

The structure shown in Fig. 2.27 (H–Nb_2O_5) is based on the units of [B] and [C] as a whole. The U and L sheets have different block units, i.e. the structure of the U sheet is $[3 \times 4]_1$ with a composition of $Nb_{13}O_{33}$ and that of the L sheet is $[3 \times 5]_\infty$ with a composition of $Nb_{15}O_{37}$. The structure $W_4Nb_{26}O_{77}$, shown in Fig. 2.28, which is based on the basic unit [C], is composed of $[4 \times 3]_1$ and $[4 \times 4]_1$ in each sheet. This structure can be regarded as an intergrowth structure ($WNb_{12}O_{33} + W_3Nb_{14}O_{44}$) (see Section 2.5).

The composition of block structures $[m \times n]_p$ can be generally calculated by[15]

$$M_{mnp+1}O_{3mnp-p(m+n)+4} \tag{2.5}$$

For example, the composition of the structure $[3 \times 4]_1$, shown in Fig. 2.24, is calculated to be $M_{13}O_{33}$. When different block structures are mixed (see Figs 2.27 and 2.28), the composition of the crystal is expressed as

$$\sum_i [M_{mnp+1}O_{3mnp-p(m+n)+4}]_i \tag{2.6}$$

where i denotes the number of different types of the block unit that are mixed. For example, the composition of the structure $\{[3 \times 4]_1 + [3 \times 5]_\infty\}$ (Fig. 2.27) is $([3 \times 4]_1 + [3 \times 5])_\infty) = M_{13}O_{33} + M_{15}O_{37} = 14M_2O_5$ (for the case of $p = \infty$, the composition is expressed as $M_{mn}O_{3mn-(m+n)}$).

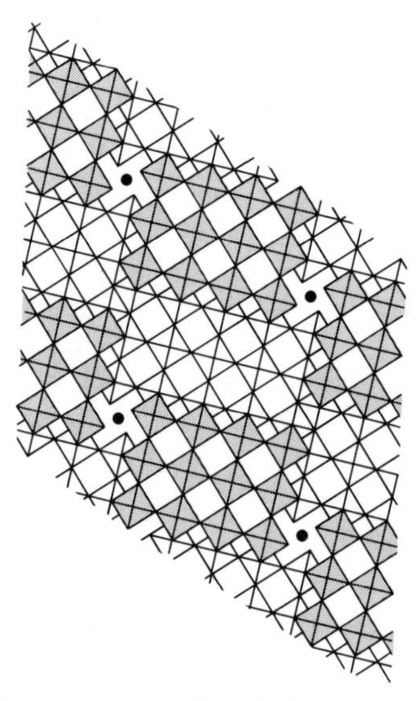

FIG. 2.27 Ideal structure of H–Nb_2O_5 ($[3 \times 4]_1 + [3 \times 5]_\infty$), based on the basic units [B] and [C]. Note that the U and L sheets differ in both composition and the dimension of the blocks. Therefore this structure can be regarded as a kind of intergrowth structure.

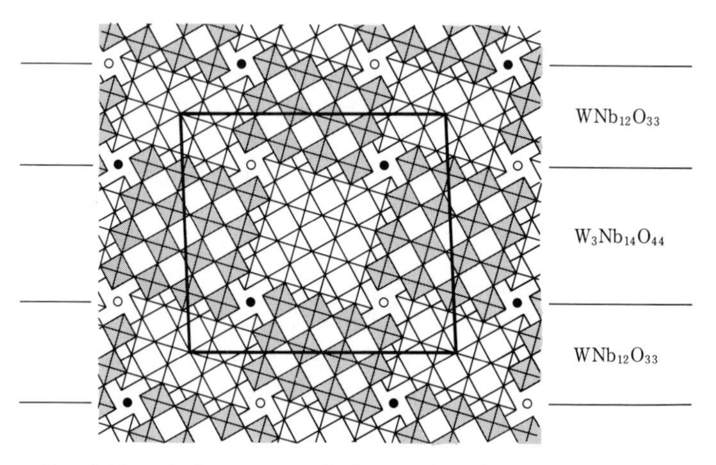

FIG. 2.28 Ideal structure of $W_4Nb_{26}O_{77}$ ($[3 \times 4]_1 + [4 \times 4]_1$). There are two kinds of block structures, $[3 \times 4]_1$ and $[4 \times 4]_1$, in the same sheet. The structure can also be regarded as a kind of intergrowth structure as shown in the figure.

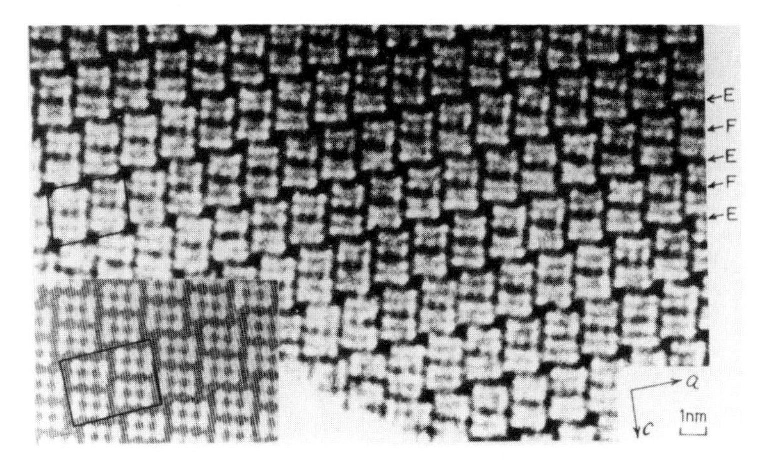

FIG. 2.29 Structure image of $TiNb_{24}O_{62}$.[16] The inset shows the calculated image, which is in good accordance with the observed one. The marks E → and F → are referred to in Fig. 2.26. For the interpretation of the image, see text.

Here let us show some examples of structure images of block structures studied by high-resolution electron microscopy. Figure 2.29 shows a structure image of $TiNb_{24}O_{62}$ $[3 \times 4]_2$ (see Fig. 2.26).[16] Regions of high electron density, such as Nb and Ti, show up as dark spots, regions of low electron density, such as O, appear as grey spots, and tunnels characteristic of the block structure appear as white spots. For the $[3 \times 4]_2$ structure, the $[3 \times 4]$ dark spots enclosing $[2 \times 3]$ white spots (corresponding to tunnels of the unit block) are clearly observed. The marks E → and F → (see Fig. 2.26) denote the direction of a row of edge-sharing octahedra between the adjacent sheets and tetrahedral metals, and the direction of a row of edge-sharing octahedra between the blocks both in the sheets and in the adjacent sheet, respectively. In this figure, structural defects are not observed.

It is well known that five compounds, i.e. $Nb_{12}O_{29}$ ($NbO_{2.416}$), $Nb_{22}O_{54}$ ($NbO_{2.4545}$), $Nb_{47}O_{116}$ ($NbO_{2.4681}$), $Nb_{25}O_{62}$ ($NbO_{2.480}$), and $Nb_{53}O_{132}$ ($NbO_{2.4905}$), exist in the composition range between NbO_2 and Nb_2O_5 in the Nb–O system, and that the structures of these phases are closely related. It was also found that the $Nb_{12}O_{29}$ phase always deviates slightly from stoichiometry towards a composition with excess oxygen.[17] In order to identify the origin of this non-stoichiometry, structure images of $Nb_{12}O_{29}$[18] were taken using a high-resolution microscope, one such image is shown in Fig. 2.30(a). The structure of stoichiometric $Nb_{12}O_{29}$ is $[3 \times 4]_\infty$ with the orthorhombic system ($a = 28.90$, $b = 3.835$, $c = 20.72$ Å), as shown in Fig. 2.30(c). In the centre of the structure image, a larger black square-spot is

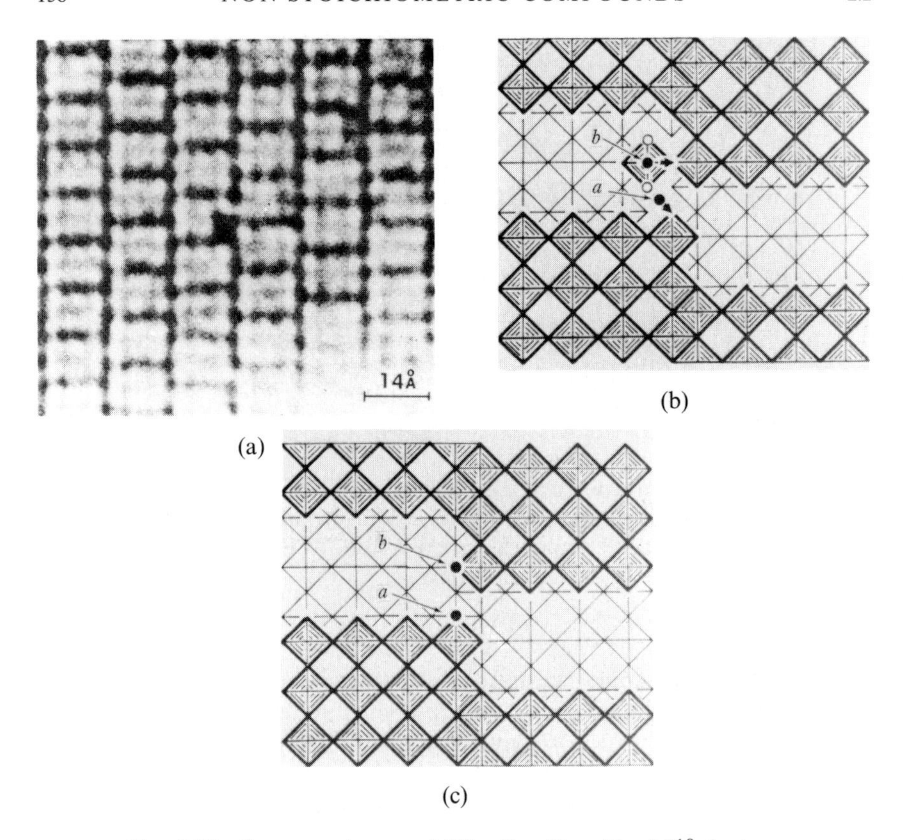

FIG. 2.30 Structure image of $Nb_{12}O_{29}$ $[3 \times 4]_\infty$ (a)[18] (note a defect in the centre of the image; see text) and corresponding structure model (b), which has two extra metals (●) and two extra oxygens (○), and the perfect crystal (c), which is derived from the defect model (b) by removing two oxygens and transferring two metals to regular positions.

seen, which indicates the existence of a structural defect composed of heavy atoms. On close examination, it was concluded that the defect is isolated and is composed of two extra oxygen (white circles) and two metal atoms (Nb; black circles *a* and *b*), as shown in Fig. 2.30(b). Metal atom *b* is octahedrally coordinated by oxygen and *a* is tetrahedrally coordinated by oxygen. This type of defect is randomly distributed in the crystal, which explains why $Nb_{12}O_{29}$ always shows oxygen-rich non-stoichiometry. It is to be noted that, after elimination of the extra oxygen and rearrangement of metals *a* and *b* along the arrows, we have a perfect structure of $Nb_{12}O_{29}$ (Fig. 2.30(c)).

Figure 2.31 shows a structure image of $9Nb_2O_5 \cdot 8WO_3$ ($W_8Nb_{18}O_{69}$, $[5 \times 5]_1$), which was taken at 1000 kV with a point-to-point resolution of

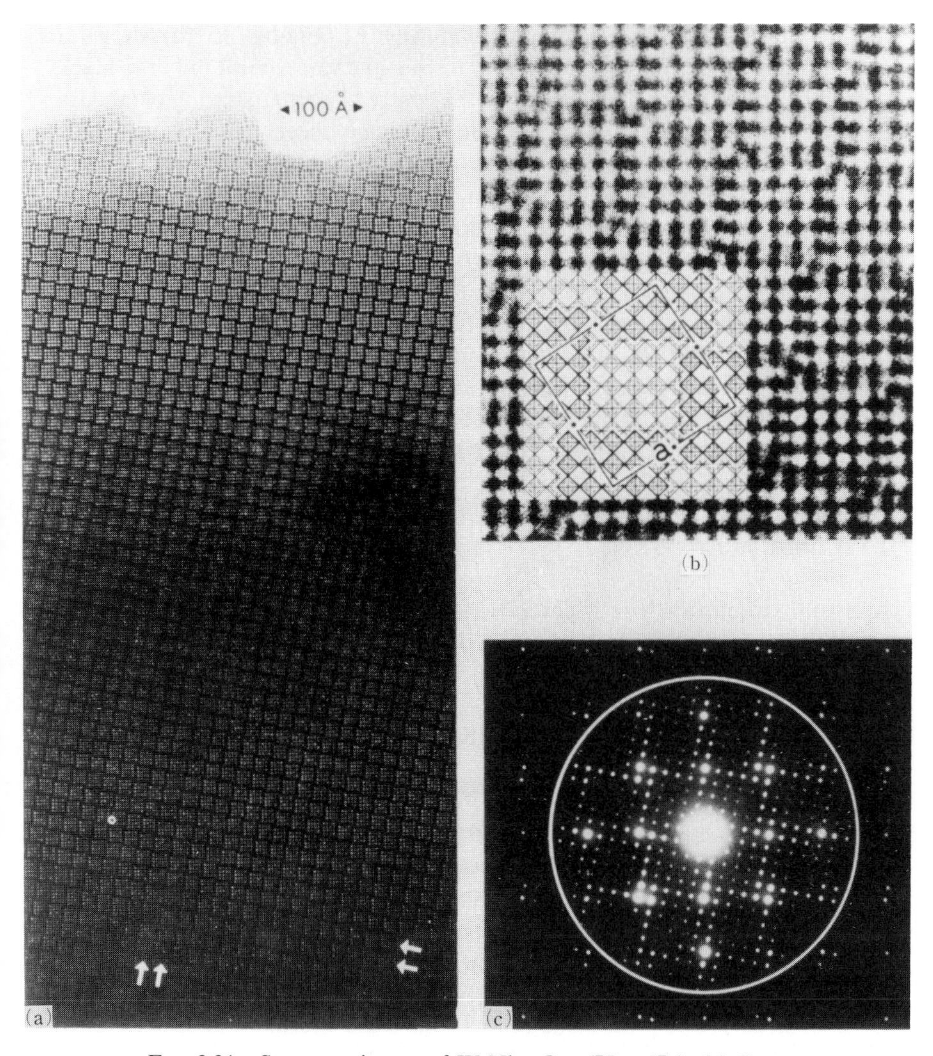

FIG. 2.31 Structure image of $W_8Nb_{18}O_{69}$ ($[5 \times 5]_1$). (a) Structure image at low magnification. (b) Structure image at high magnification. The inset shows a structure model for this compound. (c) Corresponding diffraction pattern. The diffraction spots in the circle were used for structure imaging.[19]

2 Å.[19] Figure 2.31(c) is a corresponding diffraction pattern, the diffraction spots in the circle were used for taking the structure image. The structure is a block structure of $[5 \times 5]_1$ shown in the inset in Fig. 2.31(b). Figure 2.31(a) shows a structure image at low magnification, indicating the $[5 \times 5]_1$ block structure, i.e. $[5 \times 5]$ black spots enclosing $[4 \times 4]$ white dots. The arrows

in the figure show the defect block of [5 × 6]. Figure 2.31(b) shows a structure image at high magnification. In this one can see not only the black dots corresponding to metals sited at the centre of corner-sharing octahedra (a dot-to-dot distance of 3.8 Å) within the same sheet, but also the black spots corresponding to metals at the centre of edge-sharing octahedra (a dot-to-dot distance of 1.9 Å) between the adjacent sheets. Also, the tetrahedral sites, i.e. four small white dots enclosing a black dot, are visible. A detailed comparison between the observed and calculated structure image is given in Ref. 20.

Though many kinds of block structures have been recognized so far, we have described only the structural principle and typical examples of block structures. The mechanism of the shear structure formation, and the redox processes of these oxides are of interest in solid state chemistry, and detailed and comprehensive descriptions can be found in the literature.[21–25]

2.3 Vernier structures

A crystal structure which is composed of more than two sublattices, which are independent, though modulated, is named a vernier or chimney and ladder structure. If the dimension of each sublattice (named A, B, C, ...) along the c-axis is supposed to be $c_A, c_B, c_C, \ldots$, the unit length of the whole structure (supercell structure) along the c-axis is expressed as

$$c = lc_A = mc_B = nc_C = \cdots \tag{2.7}$$

where $l, m, n, \ldots$ are integers. Suppose that the number of sublattices is two (c_A and c_B) for simplicity. In this case two types of vernier structure are considered as shown in Fig. 2.32. Structure [a] is composed of A- and B-planes, which are stacked alternatively along the b-axis as ... ABAB ... Due to the difference between the dimensions of the unit cells of each plane, the unit length of the supercell structure along the c-axis is expressed by eqn (2.7). In structure [b], the structures A and B are mutually interpenetrated. The dimension of the super structure along the c-axis is also expressed by eqn (2.7). In each case, the relation between structures A and B is very similar to that between the main scale and vernier scale in vernier calipers.

The compounds $Y_nO_{n-1}F_{n+2}$ (n = integer) in the Y_2O_3–YF_3 system are typical examples of structure [a]. In the expression $Y(O,F)_x$, the composition range of these compounds is $2.13 < x < 2.22$. The basic structure (mother structure) of these compounds is a Fluorite (CaF_2) type structure (see Fig. 1.42). In this structure, F ions can be regarded as sited at all of the tetrahedral holes of the FCC packing of Ca, as shown in Fig. 2.33(a). This is, however, not realistic, taking the difference of ionic radii (Ca^{2+}: 0.99 Å, F^-: 1.36 Å) into consideration, the structure has to be regarded as a kind of defect CsCl

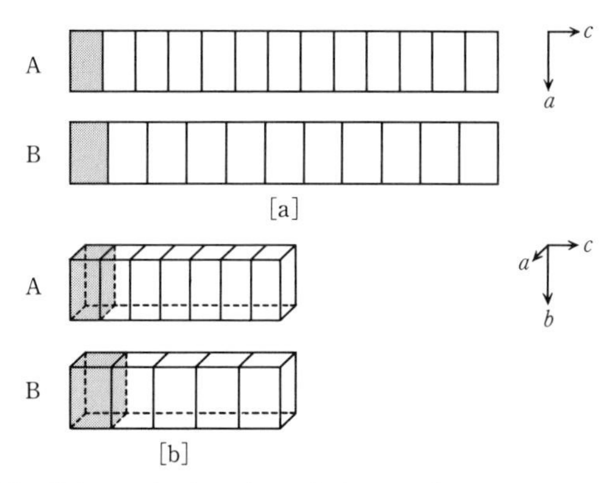

FIG. 2.32 Schematic drawing of the vernier structure. [a] Sub-lattices A and B are stacked as ... ABAB ... along the b-axis. [b] Sub-lattices A and B are mutually interpenetrated.

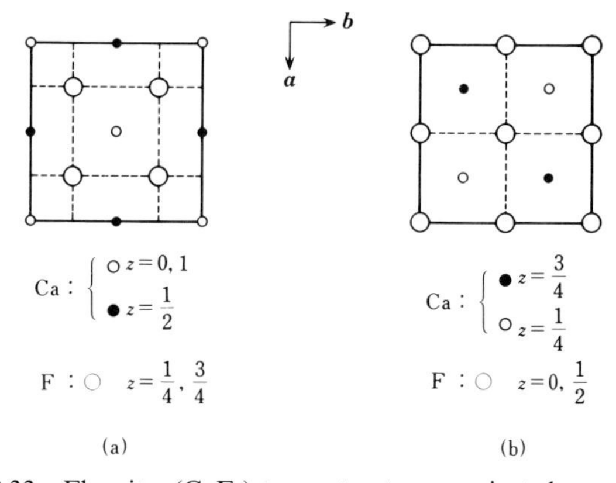

FIG. 2.33 Fluorite (CaF_2)-type structure projected on (001). (a) F ions are sited at all of the tetrahedral holes of CCP of Ca. (b) Fluorite structure as a defect CsCl-type structure.

structure, as shown in Fig. 2.33(b). In the figure, half of the body-centre sites for cations are empty. In the system $Y(O,F)_x$ ($x > 2$), excess anions have to be incorporated into the defect CsCl structure. Possible positions for excess anions are ($\frac{1}{4}, \frac{1}{4}, \frac{1}{4}$) and equivalent sites in Fig. 2.33(b).

A structure model for $Y_7O_6F_9$ ($n = 7$ in $Y_nO_{n-1}F_{n+2}$) is depicted in Fig. 2.34.[26] In the unit cell of a Fluorite-type structure (Fig. 2.33(b)), there are four molecules of MX_2 (M = cation, X = anion). In order to describe the structure of $Y_7O_6F_9$ based on a Fluorite-type structure, it is necessary to consider at least seven times the super structure of Fluorite, i.e. $M_{28}X_{56}$,

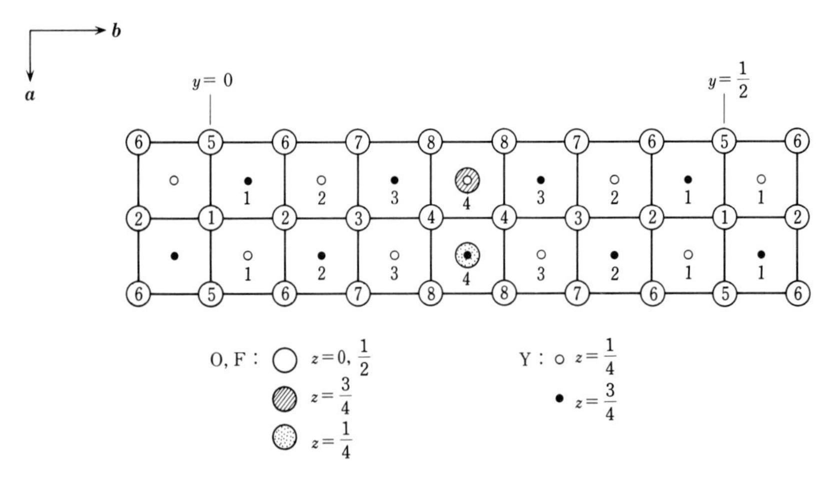

FIG. 2.34 Structure model of $Y_7O_6F_9$ (I) based on a Fluorite-type structure, projected on (001)[26] (half of the unit cell is drawn).

and also to discuss where the four excess anions are to be sited. In Fig. 2.34 excess anions are sited at vacant cation sites marked by ◍ and ◎. In other words extra anions are incorporated as interstitial atoms, at the centre of a simple cube formed by anions.

Figure 2.35 shows the structure of $Y_7O_6F_9$ determined by single crystal

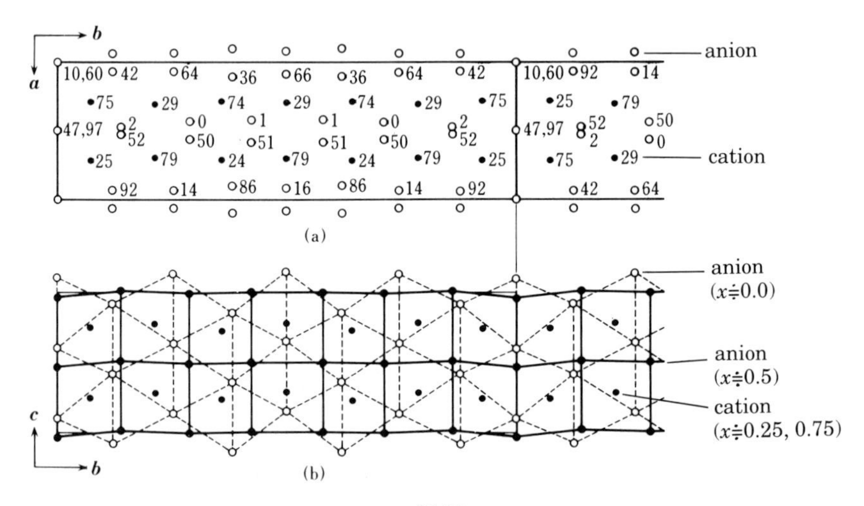

FIG. 2.35 Structure of $Y_7O_6F_9$[26,27] projected on (001) (a) and on (100) (b). Heights in (a) are in units of $c/100$. (Compare with Fig. 2.34.)

X-ray analysis,[26,27] in which figure (a) is the projection on (001), corresponding to Fig. 2.34. It is to be noted that in the real crystal two excess anions in half of a unit cell are placed at $x = 0$ and 1, instead of at $x = \frac{1}{4}$ and $\frac{3}{4}$ as in the ideal structure model. This causes a vernier structure, i.e. the anion along [010] at $x = \frac{1}{2}$ is seven times the unit length, while that at $x = 0$ is eight times the length. To confirm this relation, the projection on (100) is shown in Fig. 2.35(b), where the dotted net plane A_1 (anion plane, 3^6 net) is at $x = 0$ and the solid net plane A_2 (anion plane, 4^4 net) is at $x = \frac{1}{2}$. The figure clearly shows that the A_1 plane with a dimension of eight times that of the subcell along [010] and the A_2 plane with a dimension of seven times that of the subcell are alternately stacked along [100], which exactly corresponds to the vernier structure [a] in Fig. 2.32.

Therefore the ideal structure of $Y_7O_6F_9$ can be depicted as a stacking of $\ldots A_1B_1A_2B_2 \ldots$ along [100], as shown in Fig. 2.36. In the real crystal, shown in Fig. 2.35, mutual correlations among these atomic planes exist. These result in structural modulation, i.e. the movement of not only anions but also cations from the ideal positions, although the dimensions of the subcells of each plane (A_1, B_1, A_2, B_2) are assumed to be constant in the ideal structure, and, moreover, the site occupancy of anions is differentiated.

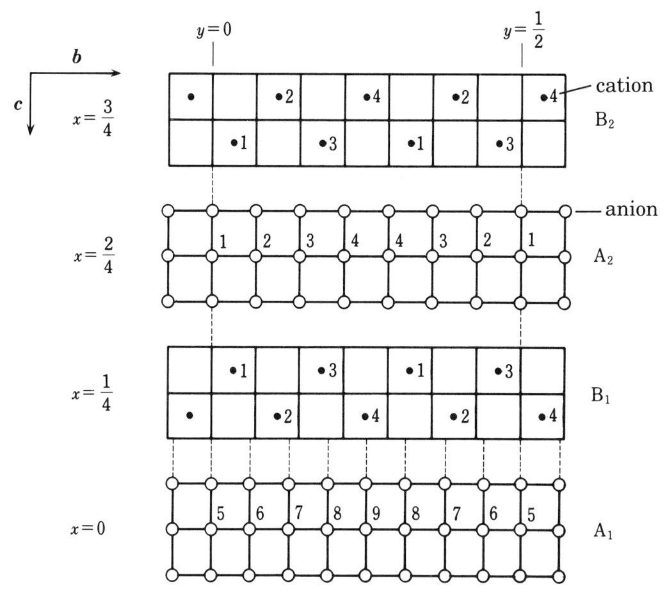

FIG. 2.36 Structure model of $Y_7O_6F_9$ (II). The layer stacking is $\ldots A_1B_1A_2B_2 \ldots$ along the *a*-axis. The unit cell of each layer is assumed to be constant.

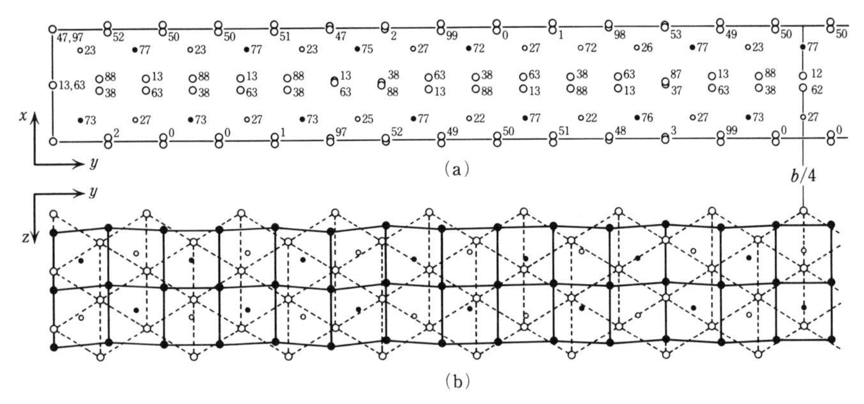

FIG. 2.37 Structure of $Zr_{108}N_{98}F_{138}$, projected on (001) (a) and on (010) (b).[29] The figure is drawn in the same manner as Fig. 2.35.

Oxygen ions occupy the sites labelled 1,2,3,5 and fluorine ions the sites labelled 4,6,7,8,9 on the A_1 and A_2 planes, which indicates clustering of anions.

In the $Y_nO_{n-1}F_{n+2}$ system, compounds $n = 4,5,6,7,8$ have been confirmed, the structures of which can be understood in terms of the vernier structure, similar to structure model (II) (Fig. 2.36). Closely related compounds have also been observed in the Nb–Zr–O[28] and Zr–N–F[29] systems. Figure 2.37 shows the structure of $Zr_{108}N_{98}F_{138}$ drawn in the same way as $Y_7O_6F_9$ in Fig. 2.35. This structure is also an example of a vernier structure based on a Fluorite-type structure. A vernier structure with layer type [a] (as in Fig. 2.32) has been critically and systematically reviewed by Makovicky and Hyde.[30]

Next we discuss the type [b] vernier structure. One example is non-stoichiometric $Ba_{1+x}Fe_2S_4$ ($1.06 < x < 1.15$), in which many phases have been confirmed to exist as one phase. The basic structure type of these phases is a stoichiometric β-$BaFe_2S_4$, shown in Fig. 2.38. This structure is composed of two sublattices, one-dimensional chains of FeS_4 tetrahedron and one-dimensional chains of Ba along [001], which are mutually interpenetrated. The FeS_4-chain sublattice forms a simple tetragonal lattice ($a_F = b_F$, c_F) and the Ba-chain sublattice a body-centred tetragonal lattice ($a_B = b_B$, c_B), where $a_B = a_F$ and $c_B = c_F$. Non-stoichiometric $Ba_{1+x}Fe_2S_4$ has a supercell of $BaFe_2S_4$ along [001], because $c_B \neq c_F$ ($a_B = a_F$). The length along the c-axis is expressed as

$$c = pc_B = qc_F \tag{2.8}$$

where p and q are integers and $1 + x = p/q$. Figure 2.39 shows the structure of $Ba_5Fe_9S_{18}$ ($p = 10$, $q = 9$) and $Ba_9Fe_{16}S_{32}$ ($p = 9$, $q = 8$).[31,32]

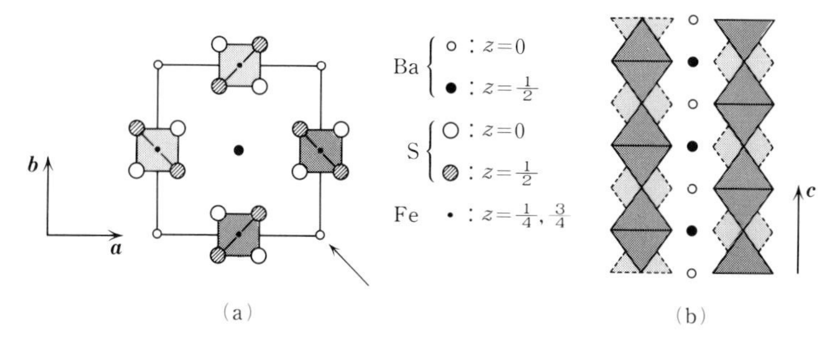

FIG. 2.38 Structure of β-$BaFe_2S_4$. (a) Projection on (001). Tetrahedral FeS_4 is outlined. An arrow is normal to the projection plane of (b). (b) Projection on (110). FeS_4- and Ba-chain subcells along [001] are seen.

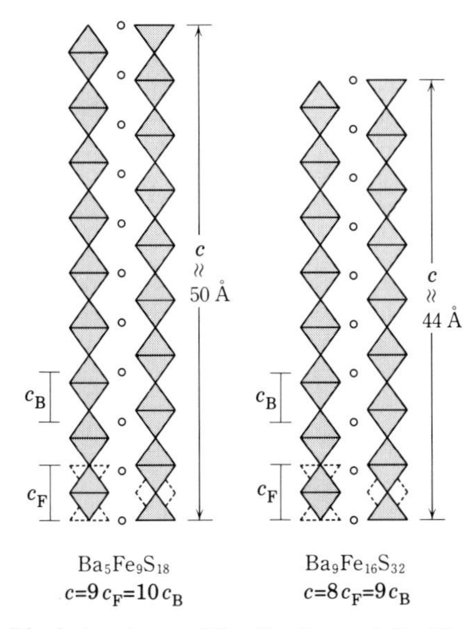

FIG. 2.39 Ideal structures of $Ba_5Fe_9S_{18}$ and $Ba_9Fe_{16}S_{32}$. These figures are drawn in the same way as Fig. 2.38(b).

We shall now review experimental work on non-stoichiometric $Ba_{1+x}Fe_2S_4$.[33,34] Non-stoichiometric $Ba_{1+x}Fe_2S_4$ was prepared by heating weighed mixtures of BaS, Fe, and S in evacuated silica tubes. Figure 2.40 shows the relation between the ratio of Ba to Fe ($\alpha = 2Ba/Fe$), annealing temperature (T_a) (the samples were quenched from T_a), and lattice parameters (a, c_B, c_F),[33] measured by X-ray powder diffraction. The values of c_B and c_F are closely correlated to T_a, almost independent of nominal composition.

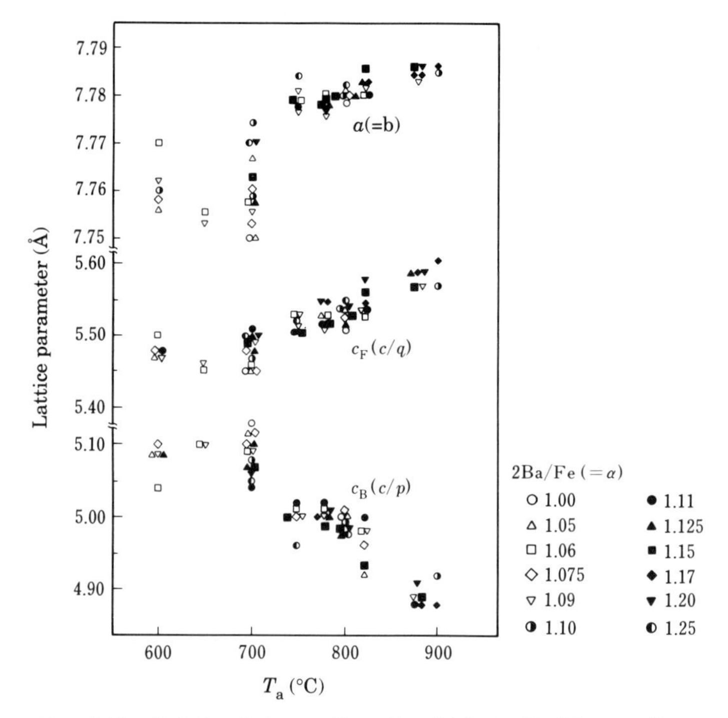

FIG. 2.40 Relation between the ratio of 2Ba to Fe (α), annealing temperature (T_a), and lattice parameters (a, c_B, c_F).[33] The samples were prepared by quenching from T_a.

This fact is very interesting when considering the origin of the complexity of this type of the compound (see ref. 33). In Fig. 2.41 is shown the relation between the lattice parameters and p/q ($=1 + x$, composition of Ba). This figures indicates that non-stoichiometric $Ba_{1+x}Fe_2S_4$ is stable in the composition range $0.06 < x < 0.15$, and that extrapolation of the lattice parameters to $x = 0$ gives good agreement with those of β-$BaFe_2S_4$. The lattice parameter changes continuously with Ba composition, x, this bears a close resemblance to the relation of lattice parameter versus composition of typical, classical non-stoichiometric compounds, e.g. VO_{1+x} ($-0.20 < x < 0.30$) with an NaCl-type structure. However, this is incorrect, as shown later each value of p/q, which can be continuously changed, forms a dependent structure (though they are mutually related), and therefore the compound differs in nature from classical non-stoichiometric compounds.

Figure 2.42 shows the electron diffraction patterns (EDP) of $Ba_{1+x}Fe_2S_4$, $x \neq 0$ and $x = 0$, with $[100]_\beta$ and $[310]_\beta$ zone axes.[34] The subscript means that the direction refers to the β-$BaFe_2S_4$ structure. The pattern for $x \neq 0$

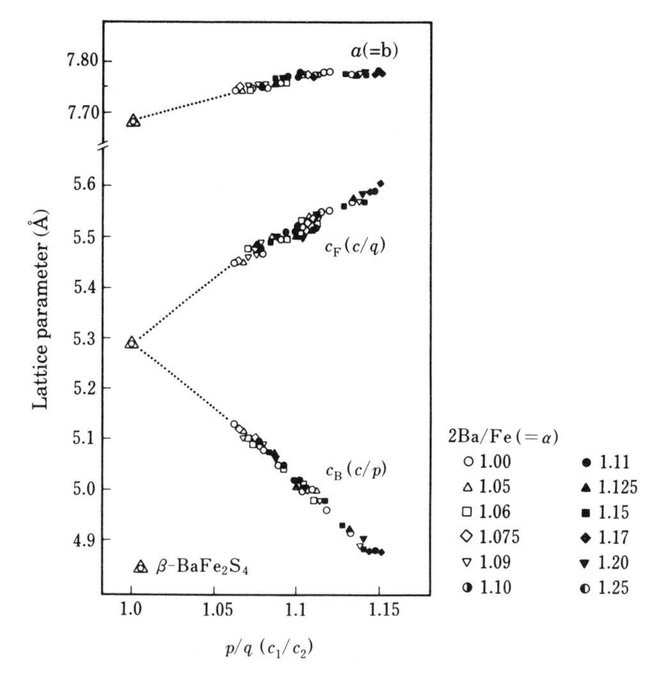

FIG. 2.41 Relation between p/q and the lattice parameters.[33]

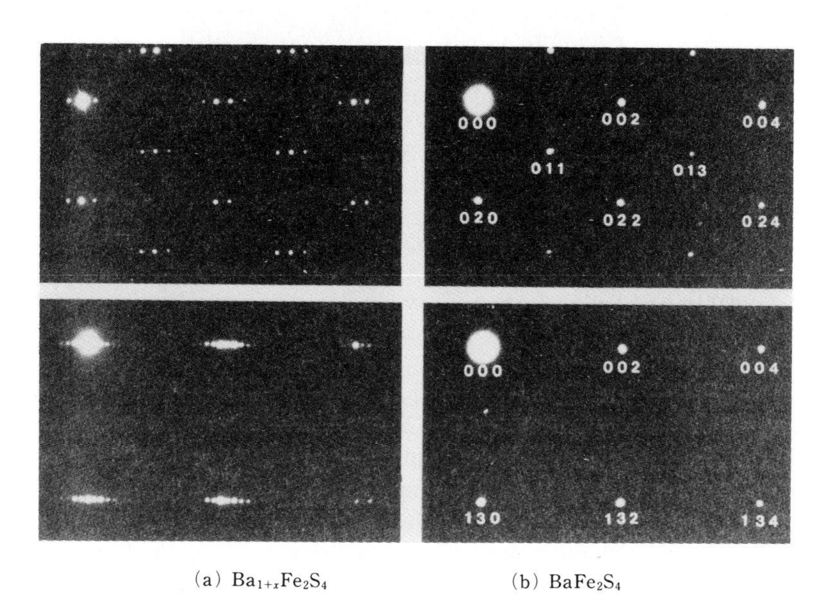

(a) $Ba_{1+x}Fe_2S_4$ (b) $BaFe_2S_4$

FIG. 2.42 Electron diffraction patterns (EDP) of $Ba_{1+x}Fe_2S_4$ (a) and β-$BaFe_2S_4$ (b), with $[100]_\beta$ and $[3\bar{1}0]_\beta$ zone axes.[34]

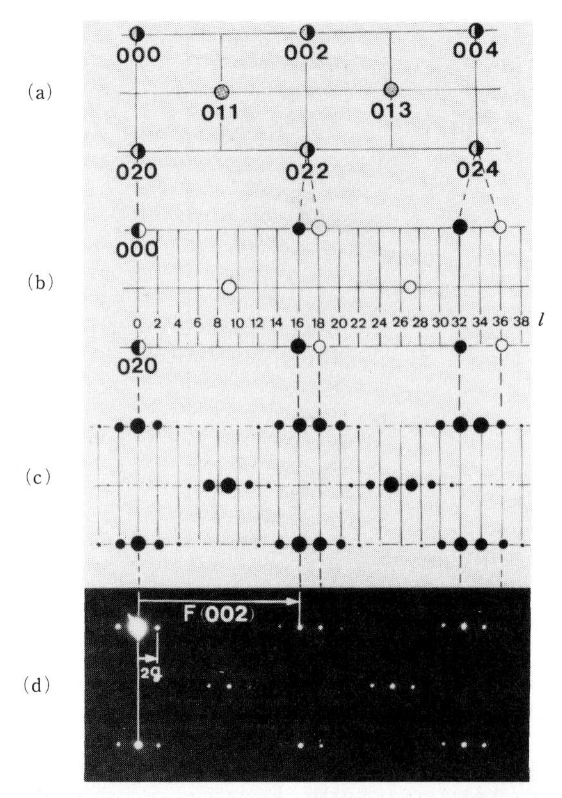

FIG. 2.43 Calculated and observed EDPs of $Ba_9(Fe_2S_4)_8$ with [100] zone axis:[34] (a) calculated EDP of ideal β-$BaFe_2S_4$; (b) calculated EDP of ideal $Ba_9(Fe_2S_4)_8$; (c) calculated EDP of real $Ba_9(Fe_2S_4)_8$, i.e. the lattice modulation is taken into consideration; (d) observed EDP of $Ba_9(Fe_2S_4)_8$. In the calculated EDPs, the diffraction spots marked as open circles are from the Ba subcell, closed ones from the FeS_2 subcell, and half closed ones from both the Ba and FeS_2 subcells.

shows super-reflections along the [001] zone axis. For reference the calculated EDPs with $[100]_\beta$ are shown in Fig. 2.43. Figure 2.43(a) shows an EDP of β-$BaFe_2S_4$, in which the diffraction spots marked as open circles are from the Ba subcell, and those marked as half-closed ones are from both the Ba and FeS_2 subcells. Figure 2.43(b) shows EDP for an ideal structure of $Ba_9(Fe_2S_4)_8$, in which those marked as closed ones are from the FeS_2 subcell (see Fig. 2.39). By ideal structure we mean that both the Ba subcell periodicity c_B and the Fe_2S_4 framework subcell periodicity c_F along the c-axis are constant, i.e. the structure does not have any lattice modulation. In this case, the diffraction patterns can be interpreted as the superposition of EDPs for each subcell, as shown in this figure. In a real crystal, however, each subcell mutually interacts and all of the atomic positions are displaced from their ideal positions as seen in the Y–O–F system. This structure is called a displacive modulated structure. Figure 2.43(c) shows the calculated EDP of $Ba_9(Fe_2S_4)_8$ with displacive modulation, which is qualitatively in agreement with the observed EDP shown in Fig. 2.43(d). In these figures, we can see the satellite diffraction along $[001]_\beta$. These EDPs are characterized by vectors **F**(002) and 2**g**, shown in Fig. 2.43(d). The vector **F**(002) shows the FeS_2 subcell and the vector 2**g** the superlattice cell.

Observed EDPs[34] for $Ba_{1+x}Fe_2S_4$ are classified into the following two groups:

(**1**) Commensurate structure: In this case, the vector **F**(002) is parallel to the vector 2**g** and moreover these vectors are related as $\mathbf{F}(002) = n \times 2\mathbf{g}$ ($n =$ positive integer); this equation leads to $n(c_F - c_B) = c_B$ or $nc_F = (n+1)c_B$. The supercell with this relation is generally called a commensurate structure. It has been concluded from the structural principle mentioned above that these have composition $Ba_{i+1}(Fe_2S_4)_i$ ($i =$ integer). Figure 2.44 shows examples of EDPs for these compounds. In these, only the structures with $i = 8$ and 9 have been determined by X-ray single crystal analyses.

(**2**) Incommensurate structure: This type can be grouped into the following two sub-classes.

[1] The vector **F**(002) is parallel to the vector 2**g**, and these vectors generally show the relation of $\mathbf{F}(002) \neq n \times 2\mathbf{g}$, in contrast to the commensurate structure. This relation is called a spacing anomaly and a structure showing a spacing anomaly is a kind of incommensurate structure.

We consider the following equation $\mathbf{F}(002) = (n - \frac{1}{2}) \times 2\mathbf{g}$ as the special case. From the equation we get

$$(n - \tfrac{1}{2})/(c_F - c_B) = c_B \qquad \text{or} \qquad (2n-1)/c_F = (2n+1)/c_B$$

Therefore the composition of compounds with this relation is $Ba_{2j+1}(Fe_2S_4)_{2j-1}$. Figure 2.45 shows EDPs for the compounds with $j = 8,9,10,11$.

Generally the vector **g** can change continuously. For example, the structure with $\mathbf{F}(002)/2\mathbf{g} = 14.60$ ($p/q = 1.068$), which has a composition of $Ba_{78}(Fe_2S_4)_{73}$, shows a rather longer c-axis of about 400 Å. It is to be noted that the structural principle of the incommensurate structure type is the same as that of the commensurate structure type, although the former type shows incommensurate behaviour on EDPs.

[2] The vector **F**(002) is not parallel to the vector 2**g**, and these vectors generally show the relation $|\mathbf{F}(002)| \neq n \times |2\mathbf{g}\cos\beta|$, where β ($0 < \beta < 5°$)

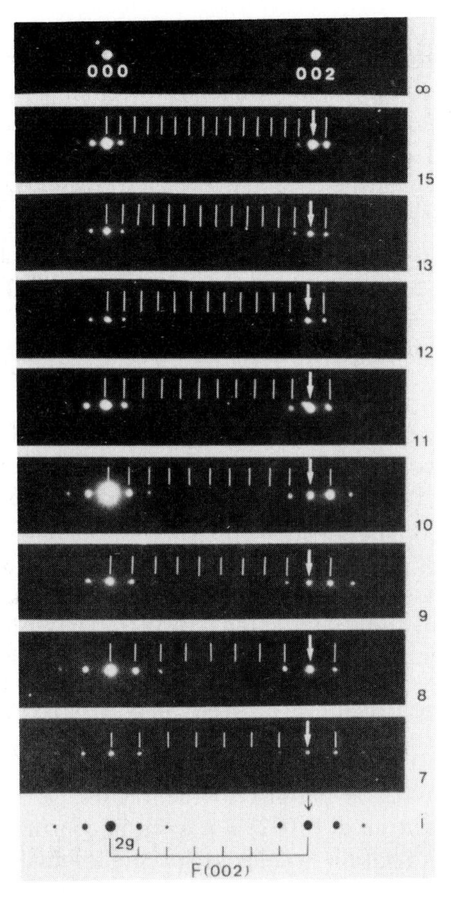

FIG. 2.44 EDPs from the commensurate structures $Ba_{i+1}(Fe_2S_3)_i$ ($i = 7, 8, \ldots$) with $[100]_\beta$ zone axis.[34]

denotes the angle between the vectors **F**(002) and 2**g**. This type of EDP is called an orientation anomaly.

Typical examples of types [1] and [2] are shown in Figs 2.46(a) and (b), respectively. The proposed structure model of a compound having a type [2] EDP is shown in Fig. 2.47.[33] Details of the structural principle showing spacing or orientation anomalies in the EDP are given in ref. 35.

It could be concluded[34] that the $Ba_{1+x}Fe_2S_4$ system generally shows a (**2**)-[2] type of EDP, and that the types (**1**) and (**2**)-[1] are special cases of (**2**)-[2]. The satellite spacings (**g**) of observed EDPs are plotted as a histogram in Fig. 2.48.[34] According to the ideal vernier structure model, the ratio $|2\mathbf{g}/\mathbf{F}(002)|$ relates to the composition as $x = |2\mathbf{g}/\mathbf{F}(002)|$ in $Ba_{1+x}Fe_2S_4$.

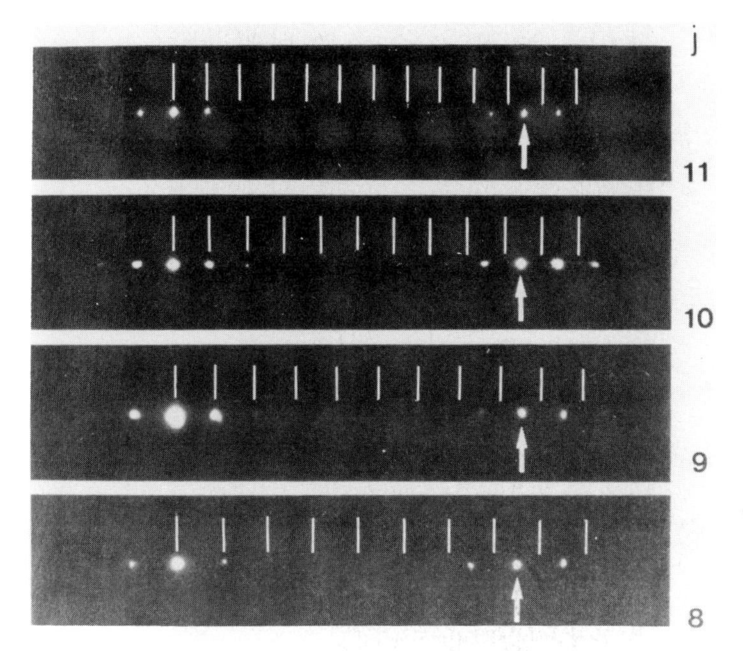

FIG. 2.45 EDPs from the incommensurate structures $Ba_{2j+1}(Fe_2S_4)_{2j-1}$ ($j = 8, 9, \ldots$) with $[100]_\beta$ zone axis[34]. These are a special case of incommensurate structure and satisfy the equation $\mathbf{F}(002) = (n - \frac{1}{2}) \times 2\mathbf{g}$.

The satellite spacings are nearly continuously distributed in the composition range $0.07 < x < 0.14$ (see also Fig. 2.41). The histogram seems to peak around the integral and half-integral of the value of $|\mathbf{F}(002)/2\mathbf{g}|$.

Figures 2.49 and 2.50 show the high resolution electron micrographs, corresponding to the case of (**2**)-[1] and (**2**)-[2], respectively. Although intuitive interpretations for these pictures are not simple, comparing them with those for the shear or block structures, the calculated images are qualitatively in agreement with the observed ones.[36, 37]

Here let us consider the structure of $Ba_{1+x}Fe_2S_4$ from the viewpoint of displacive modulated structures. As clearly seen from Fig. 2.38, Ba ions in the basic compound $BaFe_2S_4$ occupy regularly half of the face-capped tetragonal prisms (FCTP: twelve coordinations) formed by S ions. The FCTP sites are at $z = 0$ and $\frac{1}{2}$ along the c-axis. The other sites for Ba ions are at $z = \frac{1}{4}$ and $\frac{3}{4}$, where the Ba ions are eight coordinated by S ions (square anti-prism, SAP). The former sites are more stable than the latter ones, however, the occupation of all of the former sites seems to be impossible, due to the repulsive force between Ba ions.

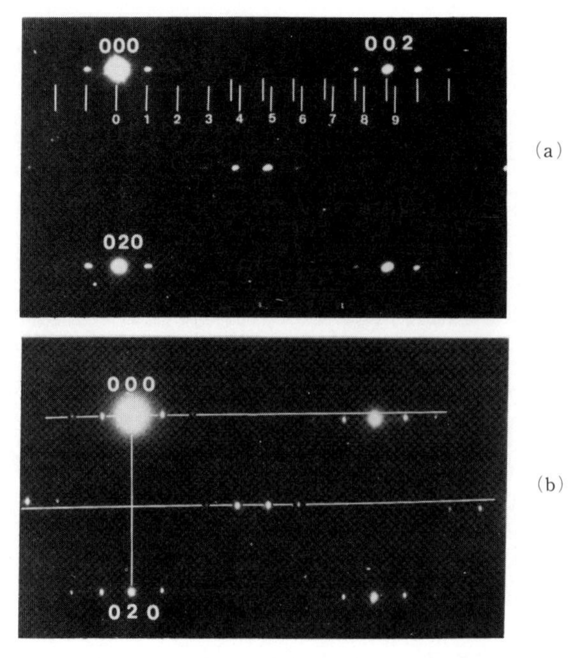

FIG. 2.46 EDPs from the incommensurate structures with $[100]_\beta$ zone axis[34]. (a) EDP shows spacing anomaly. (b) EDP shows orientation anomaly. (See text.)

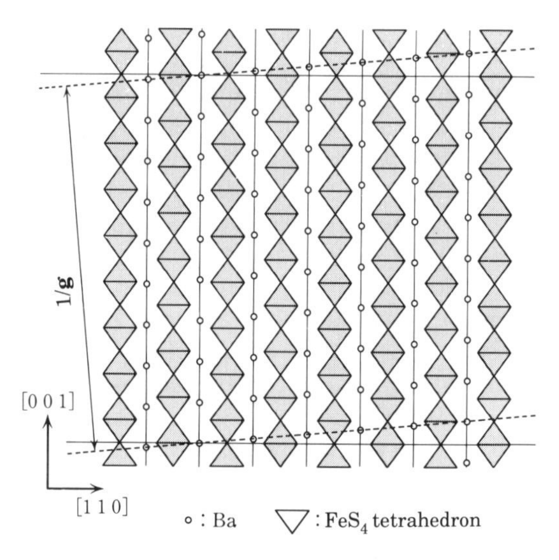

FIG. 2.47 Structure model for the crystal showing the EDP in Fig. 2.46(b).[33]

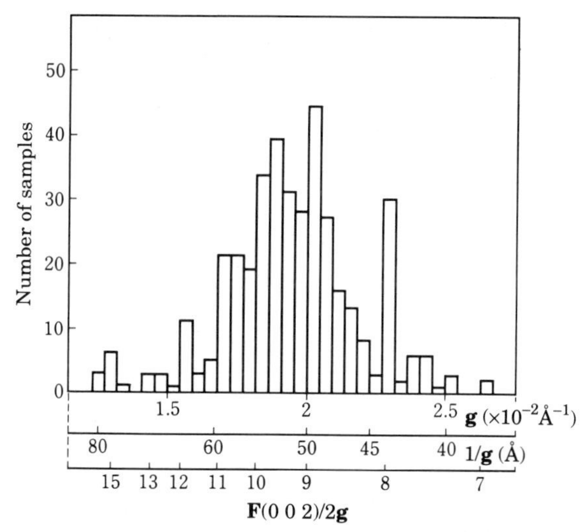

FIG. 2.48 Histogram of |**g**| value.[34]

Figure 2.51(a) shows the two sites (■: FCTP, □: SAP) for Ba ions for the ideal structure, similar to the description of Fig. 2.38(b). In Fig. 2.51(b), the simplified figure corresponding to (a) is shown. In the vernier structure Ba ions are not always able to occupy the FCTP sites. The mismatch distance Δz is defined as the distance between the real position of Ba ions and the nearest position of FCTP, as shown in Fig. 2.51(b). Figure 2.52(a) shows part of an ideal structure of $Ba_{300}(Fe_2S_4)_{271}$ ($p/q = 1.10701$) and in (b) the Δz versus z curve is shown. It is noted that the curve shows a periodicity of $c_F c_B/2(c_F - c_B)$ ($=2\mathbf{g}$). In the real vernier structure, Ba ions near to the FCTP sites would shift to the stable FCTP sites, and the shift would be accompanied by a slight contraction of the sulfur framework along the c-axis. On the other hand, Ba ions far from the FCTP sites would shift to the SAP sites, which results in a slight expansion of the sulfur framework along the c-axis. Thus, it is expected that in a real crystal the periodic structure of both subcells are mutually modulated along the c-axis.

The deviations from the ideal positions ($\Delta x, \Delta y, \Delta z$) for each ion in $Ba_9(Fe_2S_4)_8$ have been precisely determined by X-ray single crystal analyses.[32] The results show that the deviation along the x- and y-axes for both Ba and Fe is not significant and that that along the z-axis is important, while the deviation for S ions is significant along the x- and y-axis. Figure 2.53 shows the lattice modulation of the ideal and real structure of $Ba_9(Fe_2S_4)_8$. In (a) the ideal structure and the Δz versus z curve are shown. In (b), the distances between Ba–Ba, Fe–Fe, and S–S ions along the c-axis are plotted against z

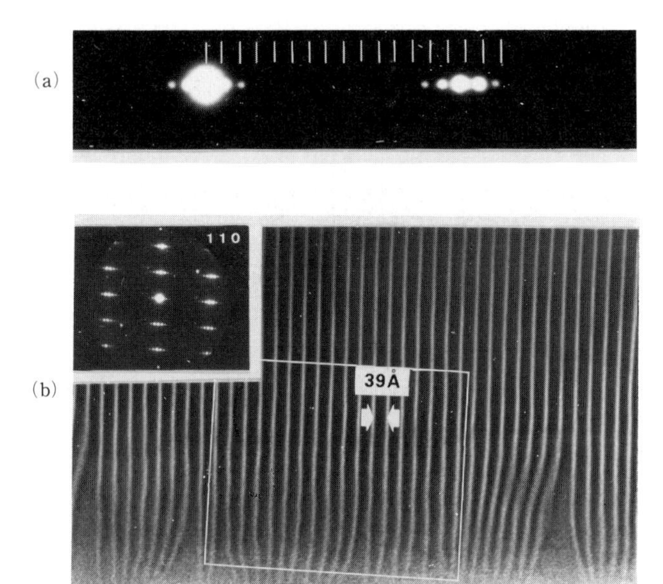

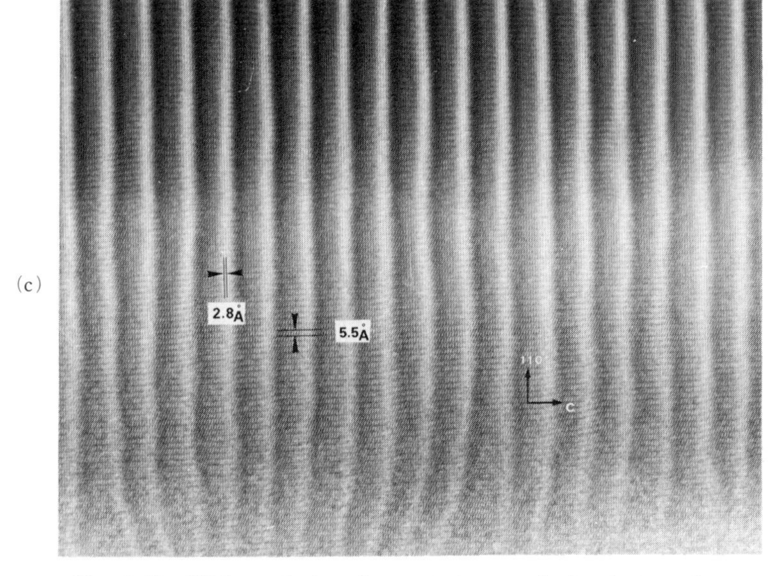

FIG. 2.49 High-resolution electron micrograph showing spacing anomaly:[34] (a) EDP showing spacing anomaly; (b) electron micrograph at low magnification; (c) electron micrograph of the enclosed area of (b) at high magnification.

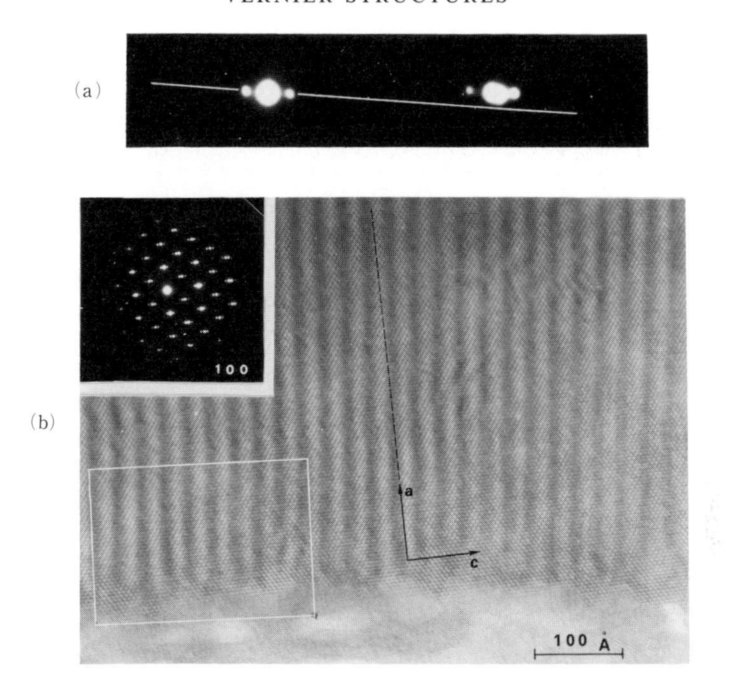

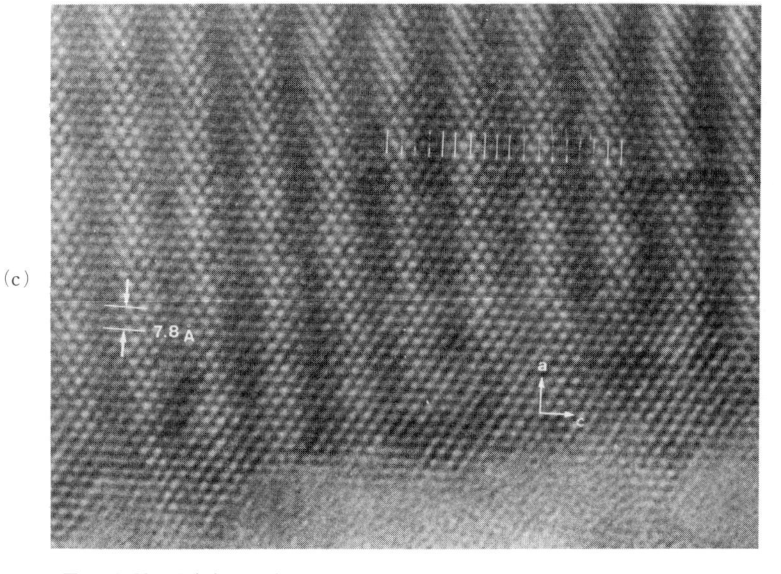

FIG. 2.50 High-resolution electron micrograph showing orientation anomaly:[34] (a) EDP showing orientation anomaly; (b) electron micrograph at low magnification; (c) electron micrograph of the enclosed area of (b) at high magnification.

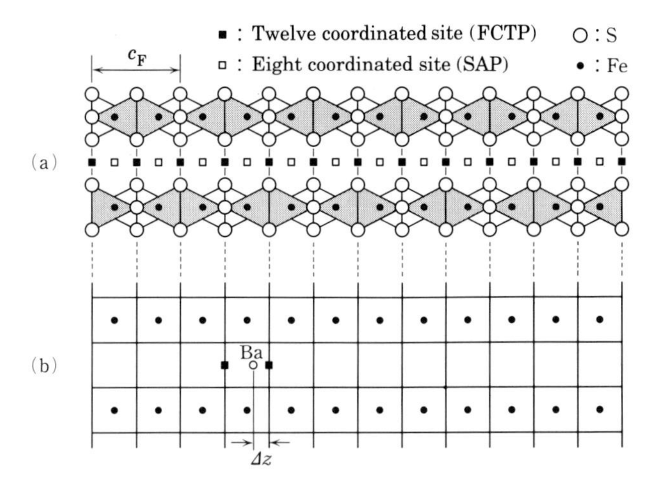

FIG. 2.51 Possible Ba sites in $BaFe_2S_4$.[34] (a) Sites for Ba, twelve coordinated by S (face-capped tetragonal prism: ■) and eight coordinated by S (square anti-prism: □). (b) Abbreviated drawing of (a). The mismatch distance Δz is defined as the distance between the real position of Ba ions and the nearest position of FCTP.

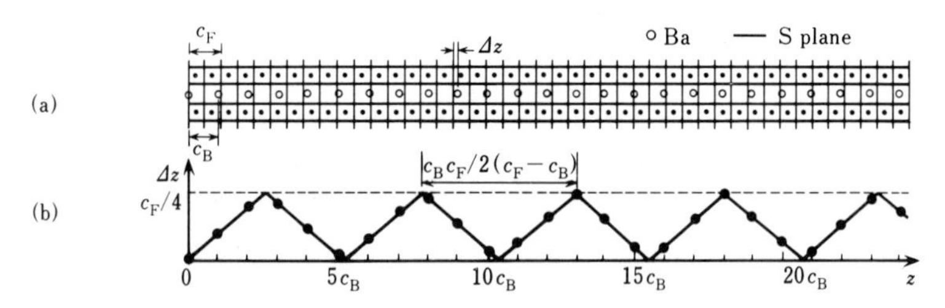

FIG. 2.52 (a) Ideal structure of $Ba_{300}(Fe_2S_4)_{271}$ without the displacive modulation. (b) Mismatch distance Δz versus z curve, indicating a periodicity of $c_F c_B/[2(c_F - c_B)] = 2\mathbf{g}$.[34]

for the real structure.[32] The figure clearly shows the displacive lattice modulation, especially for Ba ions.

Thus we have studied the structure of the $Y_nO_{n-1}F_{n+2}$ and Ba–Fe–S systems as typical examples of the vernier structure. In the vernier structure, the deviation from stoichiometry originates from the coexistence of different unit dimensions of the subcells, i.e. A_1 and A_2 sheets for the Y–O–F system, and Fe_2S_4 and Ba subcells for the Ba–Fe–S system. The structural principle of the vernier structure can be extended to a more general one, called 'adaptive structure', in which independent, though mutually related, structures for all the compositions can exist in a limited composition range[38] (see Section 2.6).

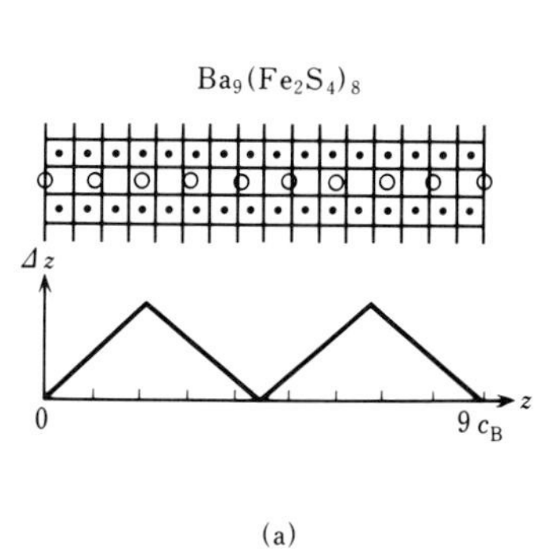

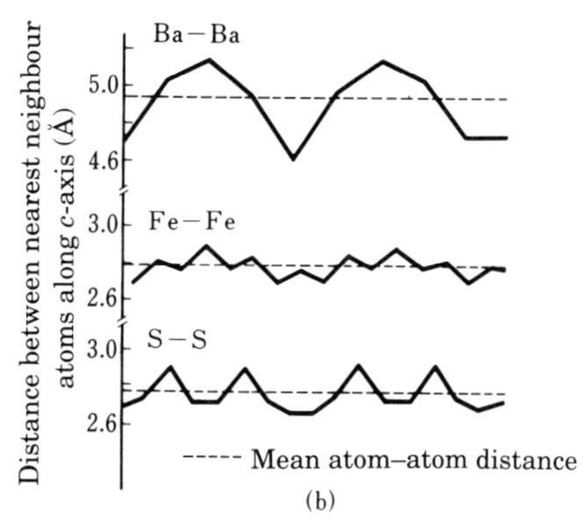

FIG. 2.53 Ideal and real structures of $Ba_9(Fe_2S_4)_8$.[32,34] (a) Ideal vernier structure and its Δz versus z relation for $Ba_9(Fe_2S_4)_8$. (b) Distances Ba–Ba, Fe–Fe, and S–S along the c-axis of a real crystal of $Ba_9(Fe_2S_4)_8$.[32].

2.4 Micro-twin structures

Crystal structures of various compounds can be generally understood from the view that the atoms with the larger radius (this means various kinds of radius such as atomic, ionic, metallic, and covalent radii, depending on chemical bonding) have the closest packing, i.e. CCP (FCC) or HCP, and those with the smaller radius occupy the tetrahedral or octahedral holes formed by the larger atoms. In some cases, however, the atoms with smaller radius can be regarded as the framework of the structure, leading to large deviations from the ideal positions of closest packing for the larger atoms.

For some combinations of elements it is more appropriate to regard the trigonal prism formed by the larger atoms, the centre of which is occupied by the smaller atom, as the structural unit, rather than the closest packing of the larger atoms. The structure based on trigonal prisms (micro-twin structure) is obtained by the operation of unit cell level twinning to the mother structure with the closest packing. This structural principle was firstly proposed by Andersson and Hyde[39] in 1974, later they published review papers on the structure,[27,40] showing many examples.

We shall discuss first the micro-twin structure derived from an HCP lattice. Figure 2.54(a) shows a projection of HCP on $(001)_{HCP}$. In Fig. 2.54(b) the ortho-hexagonal axes (a_{oh}, b_{oh}, c_{oh}) are shown as well as the usual hexagonal axes (a_h, b_h, c_h). Hereafter we use the former axes. Figure 2.55(a) shows a projection of HCP on $(010)_{oh}$, in which the stacking of ... ABAB ... along the c_{oh}-axis is seen. For simplicity, the atoms at $y = 0$ and $\frac{1}{6}$ are marked by ○ and those at $y = \frac{1}{2}$ and $\frac{4}{6}$ by ● in Fig. 2.55(b). Figure 2.55(c) shows part of the arrangement of octahedra along the b_{oh}-axis.

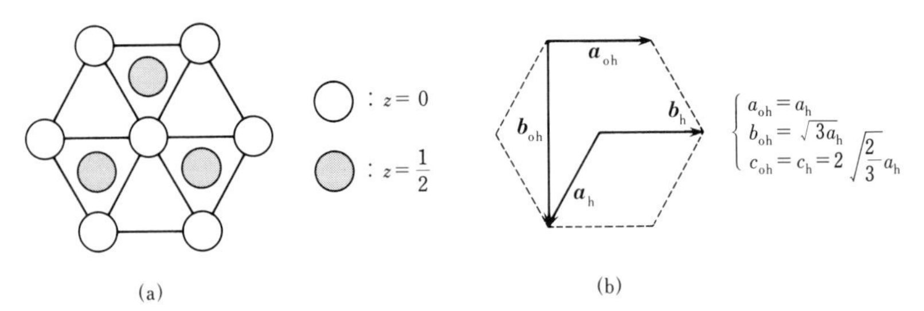

FIG. 2.54 Hexagonal close packing—relation between the hexagonal and ortho-hexagonal axes. (a) Hexagonal close packing projected on $(001)_{HCP}$. (b) Relationship between the hexagonal (a_h, b_h, c_h) and ortho-hexagonal (a_{oh}, b_{oh}, c_{oh}) axes.

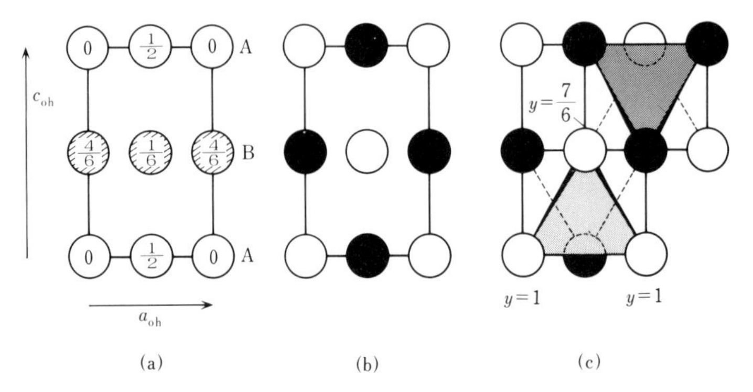

FIG. 2.55 (a) Projection of hexagonal close packing on $(010)_{oh}$. (b) For simplicity, the atoms at $y = 0$ and $\frac{1}{6}$ are marked by ○, and those at $y = \frac{1}{2}$ and $\frac{4}{6}$ by ●. (c) Octahedron stacking along the b_{oh}-axis.

Figure 2.56 shows a twin structure derived from HCP, projected on $(010)_{oh}$, in which the twin plane is $(101)_{oh}$. A new type of interstice (trigonal prisms) appears on the twin plane. As shown in Fig. 2.57(a), the trigonal prism edge across the twin plane is shorter than the other edges, which are all equal: the short edge is calculated to be $\sqrt{\frac{8}{11}}a$ or $0.85a$, where a is the length of the other edges ($a = a_h$). Figure 2.57(b) shows an arrangement of trigonal prisms, where the projection axis is normal to the twin plane. The trigonal prisms share the common edges of the nearest neighbour prisms, forming zigzag chains which are connected by sharing corners as shown in Fig. 2.57(c). In the region of the twin plane there remain other interstices A (square pyramid) and B (tetrahedra).

Thus, the regularly spaced twinning operation to the HCP lattice produces a new type of structure having trigonal prism interstices in the region of the

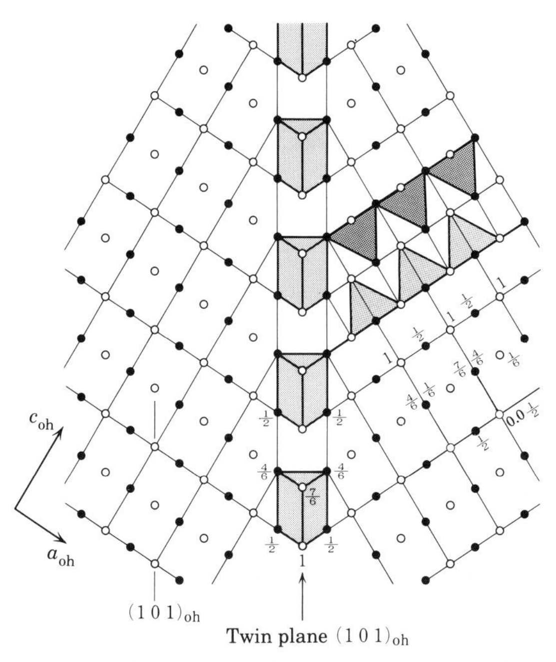

FIG. 2.56 Twin structure of hexagonal close packing projected on $(010)_{oh}$. Twin plane is $(101)_{oh}$. Note that a new type of interstice (trigonal prism) appears on the twin plane.

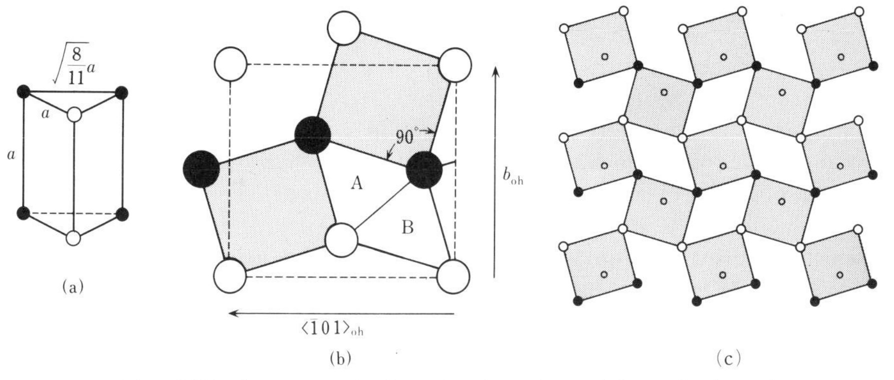

FIG. 2.57 Structure of trigonal prisms formed by twinning of HCP. (a) Dimension of trigonal prism. The edge of the trigonal prism across the twin plane is shorter than the other edges. (b), (c) Arrangement of trigonal prisms, normal to the twin plane. Additional interstices A (pyramids) and B (tetrahedron) appear.

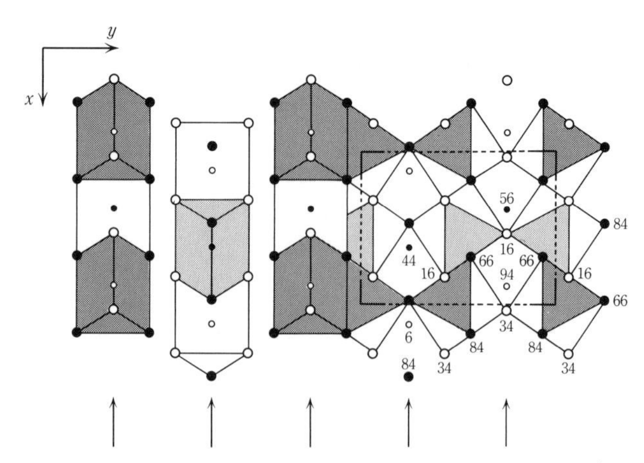

FIG. 2.58 Structure of Fe_3C (cementite)-type, $(3, 3)_{HCP}$ $Fe_3^{PS}C^P$, projected on (001).[40] Large circles are metal atoms and small ones are carbon atoms. Heights are in units of $c/100$. On the left are drawn the M_6C prisms and on the right the empty M_6 octahedra. The arrows indicate the twin planes. The unit cell is outlined.

twin plane. The distance between twin planes is denoted by $D = n \times [d_{101}/2]$, where n is positive integer. As an example, the structure of Fe_3C (cementite) is shown in Fig. 2.58.[40] The structure can be described by the micro-twinning of HCP, i.e. a twinning operation of HCP Fe with $n = 3$ and C occupying the centres of the trigonal prisms. In the right half of the figure, the arrangement of octahedra is shown, the centre of which is empty. The micro-twin structure derived from the HCP lattice is generally expressed by $(m, n, \ldots)_{HCP}$, where $m, n, \ldots$ denotes the interplanar spacing of successive twinning planes measured using a unit of $[d_{101}/2]$. For the case of Fe_3C, the structure is expressed by $(3)_{HCP}$ or $(3, 3)_{HCP}$. Taking also the (partial) occupancy of interstices of the octahedra as well as of the trigonal prism into consideration, the structure $(m, n, \ldots)_{HCP}$ can give rise to various kinds of compounds, i.e. homologous and non-stoichiometric compounds.

As mentioned above, the structure of Fe_3C is expressed by $(3, 3)_{HCP}$, in which Fe forms a trigonal prism (C in the centre) and the octahedral holes are empty. We denote this structure as $(3, 3)_{HCP}Fe_3^{PS}C^P$. Superscript PS and P stand for the skeleton of trigonal prism and the centre of prism. As shown in Table 2.2, there are many compounds with a structure similar to Fe_3C or anti-Fe_3C. The compound YF_3 is an anti-type, denoted $(3, 3)_{HCP}Y^PF_3^{PS}$. The compound $LuFeO_3$ is also an anti-type and in addition the octahedral interstices are occupied by Fe ($Lu^PFe^OO_3^{PS}$). The structure of compound Pd_5B_2 is $(3,2)_{HCP}Pd_5^{PS}B_2^P$, as shown in Fig. 2.59. The compound U_2FeS_5 is an anti-Pd_5B_2 type, $U_2^PFe^OS_5^{PS}$ (Fig. 2.60). It is to be noted that the

Table 2.2

Compounds with micro-twin structure derived from HCP[40,66]

$(m, n)_{HCP}$	Structure type	Examples (1)	Examples (2)[a]
(3, 3)	$Fe_3^{PS}C^P$ (2.58)[b]	$Fe_3B_xC_{1-x}$, Pd_3B, Er_3Rh	
	anti-Fe_3C	$Y^PF_3^{PS}$, $BiCl_3$, PBr_3, XeO_3	$Lu^PFe^OO_3^{PS}$, $UCrS_3$, $CaZrS_3$, $YScS_3$
(3, 2)	$Pd_5^{PS}B_2^P$ (2.59)	Fe_5C_2	
	anti-Pd_2B_2	Mn_5C_2, R_5Co_2 (R = Pr, Nd, Sm)	$U_2^PFe^OS_5^{PS}$ (2.60)
(2, 2)	$Yb^{PS}O^P(OH)^P$ (2.62)	ZrO_2, ScOF, TaON, SrI_2	
(1, 1)	Fe^PB^P (2.63)	LnSi (Ln = La–Er)	

[a] Compounds with (partially) filled octahedra.
[b] Figures in parentheses show the number of the figure in the text.

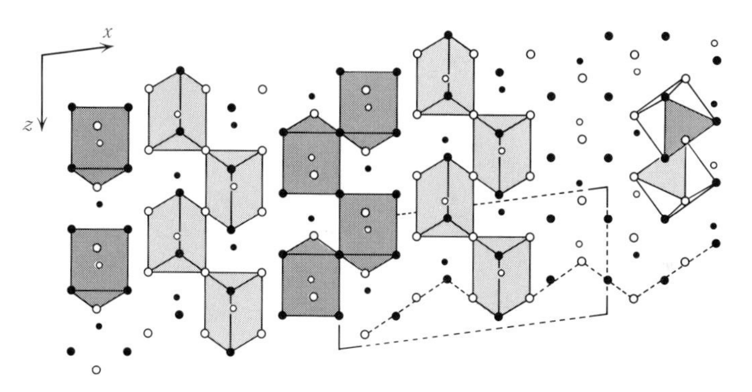

FIG. 2.59 Structure of Pd_5B_2-type, $(3, 2)_{HCP}Pd_5^{PS}B_2^P$, projected on (010).[40]

compound Fe_5C_2, having a $(3, 2)_{HCP}Pd_5^{PS}B_2^P$-type structure, shown in Table 2.2, can also be derived from $(3, 3)_{HCP}Fe_3^{PS}C^P$ by the following shear operation: Regular elimination of metal (Fe)-only planes from alternate twin individuals, parallel to the twin plane, and closing up neighbouring blocks by a shear vector so as to share the edges of trigonal prisms, shown in Fig. 2.59.

The compound having the ideal $(2, 2)_{HCP}$ structure shown in Fig. 2.61 has never been found, but similar structures showing a significant distortion of trigonal prisms have been known in a few compounds. Figure 2.62 shows the structure of the high-pressure form of YbO(OH), in which the peaks of the trigonal prisms are distorted anti-clockwise, leading to seven coordination

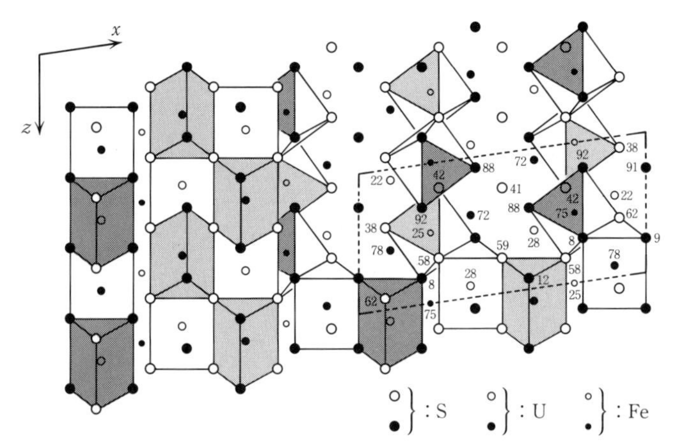

FIG. 2.60 Structure of U_2FeS_5, $(3, 2)_{HCP}S_5^{PS}U_2^{P}Fe^{O}$, projected on (010).[40] This structure is an anti-Pd_5B_2-type, additionally, the octahedral interstices of S atoms are occupied by Fe.

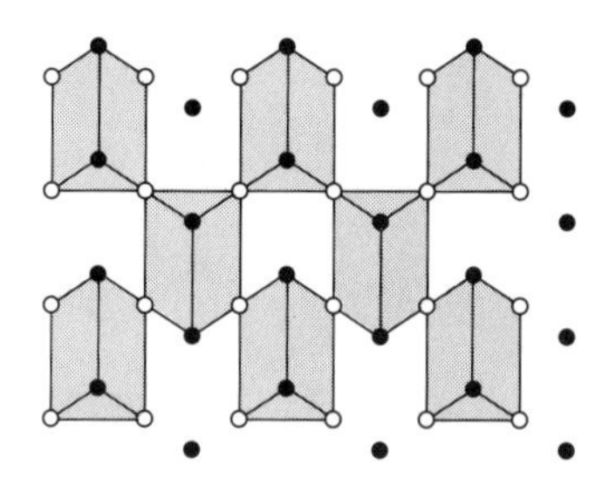

FIG. 2.61 Ideal structure of $(2, 2)_{HCP}$.

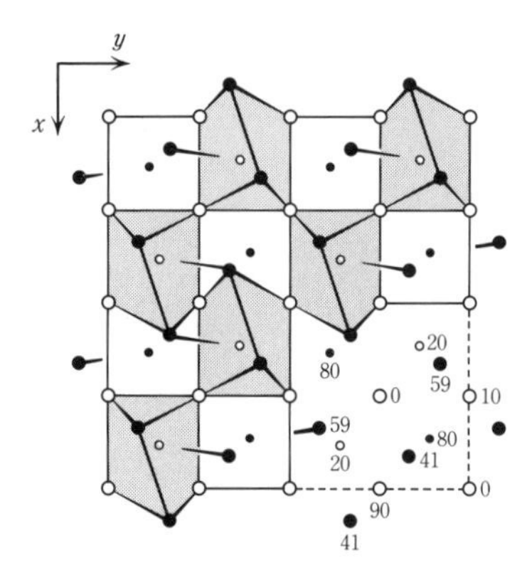

FIG. 2.62 Structure of YbO(OH), $(2, 2)_{HCP}Yb^{PS}O^{P}(OH)^{P}$.[40] This is the high-pressure form of YbO(OH). The peaks of trigonal prisms are distorted anti-clockwise, and thereby the coordination number of O or (OH) by Yb becomes seven.

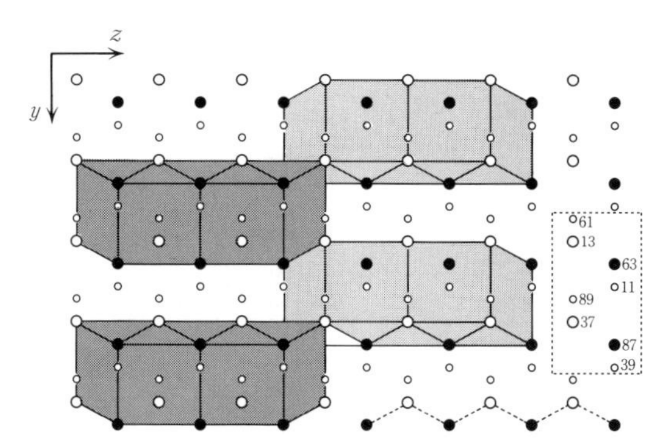

FIG. 2.63 Structure of FeB-type, $(1, 1)_{HCP}Fe^{PS}B^{P}$.[40]

of O by Yb. In the monoclinic form of ZrO_2, trigonal prisms are alternately distorted clockwise and anti-clockwise and thereby the coordination of O by Zr is also seven.

Figure 2.63 shows the structure of FeB, $(1, 1)_{HCP}Fe^{PS}B^{P}$. Many compounds of LnSi (Ln = La–Er) have this structure.

Next we discuss the micro-twin structure derived from a CCP lattice. Using the hexagonal expression, the CCP structure is composed of layer stacking ... ABCABC ... along the c_{hex}-axis $(=3\sqrt{\frac{2}{3}}a)$. The ortho-hexagonal axes (a_{oh}, b_{oh}, c_{oh}) are related to the hexagonal axes $(a_{hex}, b_{hex}, c_{hex})$ by the relation $a_{oh} = a_{hex}$, $b_{oh} = \sqrt{3}a_{hex}$, and $c_{oh} = c_{hex} = 2\sqrt{\frac{2}{3}}a_{hex}$. Figure 2.64(a)

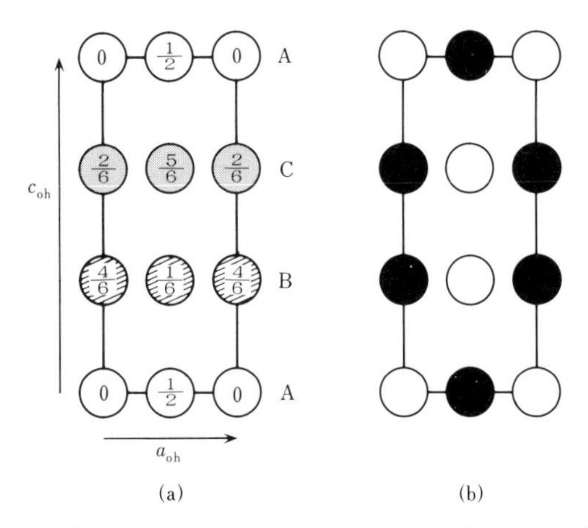

FIG. 2.64 (a) Projection of cubic close packing on $(010)_{oh}$; (b) for simplicity, the atoms at $y = 0, \frac{1}{6}$, and $\frac{5}{6}$ $(= -\frac{1}{6})$ are marked by ○ and those at $y = \frac{1}{2}, \frac{2}{6}$, and $\frac{4}{6}$ by ●.

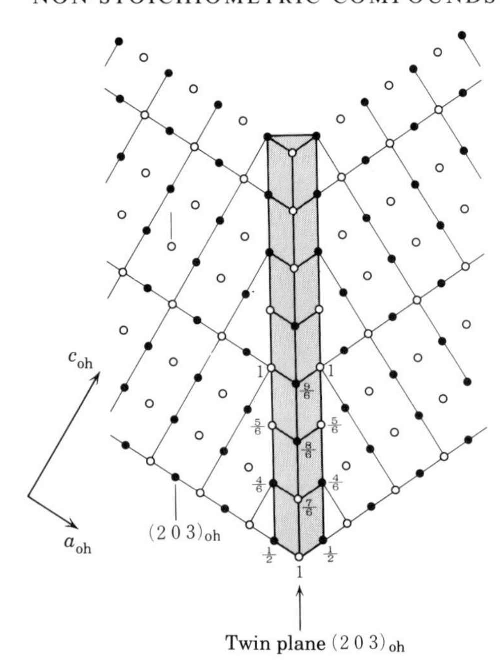

FIG. 2.65 Twin structure of cubic close packing projected on $(010)_{oh}$. The twin plane is $(203)_{oh}$. Note that a new type of interstice (trigonal prism) appears on the twin plane, being the same as that of HCP (see Fig. 2.56).

shows a projection of CCP on $(010)_{oh}$, in which the stacking . . . ABCABC . . . along c_{oh}-axis is also seen (compare with Fig. 2.55). For simplicity, the atoms at $y = 0$ and $\pm\frac{1}{6}$ are marked by ○ and those at $y = \frac{1}{2}$ and $y = \frac{3}{6} \pm \frac{1}{6}$ by ● in Fig. 2.64(b).

Figure 2.65 shows a twin structure derived from CCP, projected on $(010)_{oh}$, where the twin plane is $(203)_{oh}$. In the region of the twin plane there are trigonal prisms, which are the same size as that of the HCP. Each of the prisms shares the trigonal planes of the nearest neighbour trigonal prisms, forming pillars of trigonal prisms shown in Fig. 2.66.

It is convenient to draw micro-twin structures derived from the micro-twinning of a CCP lattice projected on the normal plane to the pillars of the trigonal prisms. Let us consider a twin structure derived from a NaCl-type structure, in which both anions and cations have a CCP arrangement.[41] Figure 2.67 shows the structure of a NaCl-type: (a) is projected on $(001)_{NaCl}$ and (b) on $(\bar{1}10)_{NaCl}$. It can be easily shown that the twin plane $(203)_{oh}$ in the ortho-hexagonal expression corresponds to the (113) plane in the cubic one. Figure 2.68 shows a twin structure of NaCl-type,

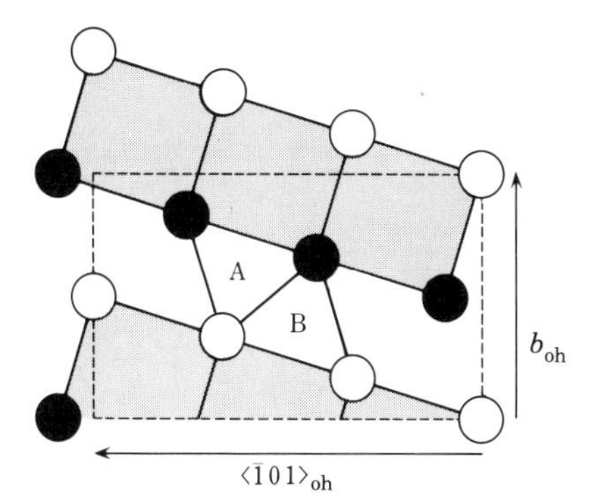

FIG. 2.66 Structure of trigonal prisms formed by twinning of CCP. Each of the trigonal prisms shares the trigonal planes of the nearest neighbour trigonal prisms, forming pillars of trigonal prisms. Additional interstices A (pyramids) and B (tetrahedron) appear.

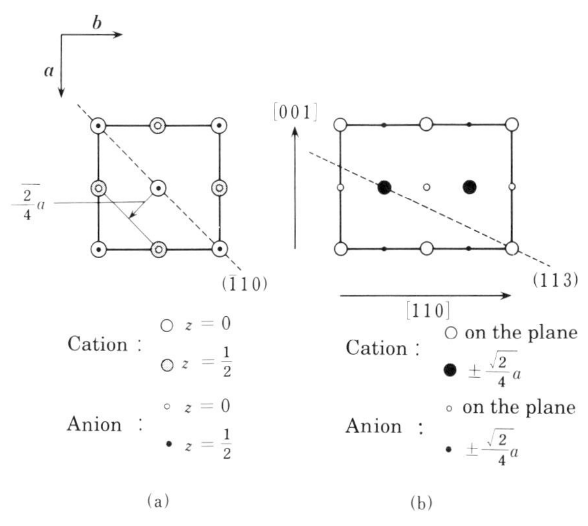

FIG. 2.67 Structure of NaCl-type projected on (001) (a) and on $(\bar{1}10)$ (b). It is noted that NaCl-type structure composed of interpenetrated subcells of CCP Na and Cl and the twin plane $(203)_{oh}$ mentioned above corresponds to the plane $(113)_{NaCl}$.

having the twin plane $(113)_{NaCl}$. In Fig. 2.69(a) is shown the structure in the region of twinning, where the anions (larger circles) form the pillars of trigonal prisms, perpendicular to the plane and the cation pairs *a*–*a*, *b*–*b*, . . . are on the faces (cubic plane) of the trigonal prisms. In Fig. 2.69(b), one cation is eliminated from each cation pair and the remaining cation is sited

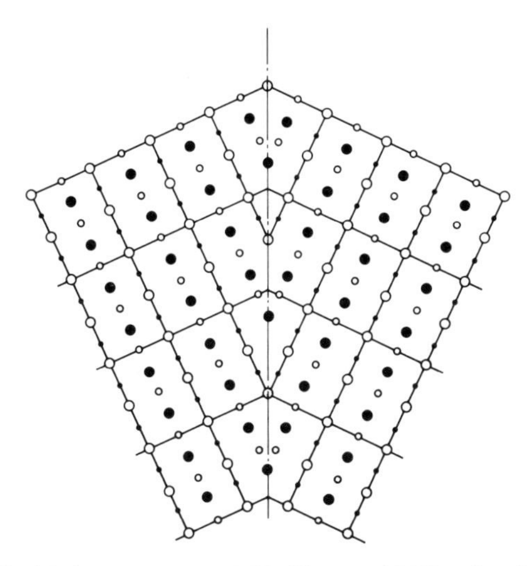

FIG. 2.68 Twin structure of NaCl-type (CCP of anions and cations) projected on $(\bar{1}10)_{NaCl}$. Twin plane is $(113)_{NaCl}$, corresponding to $(203)_{oh}$.

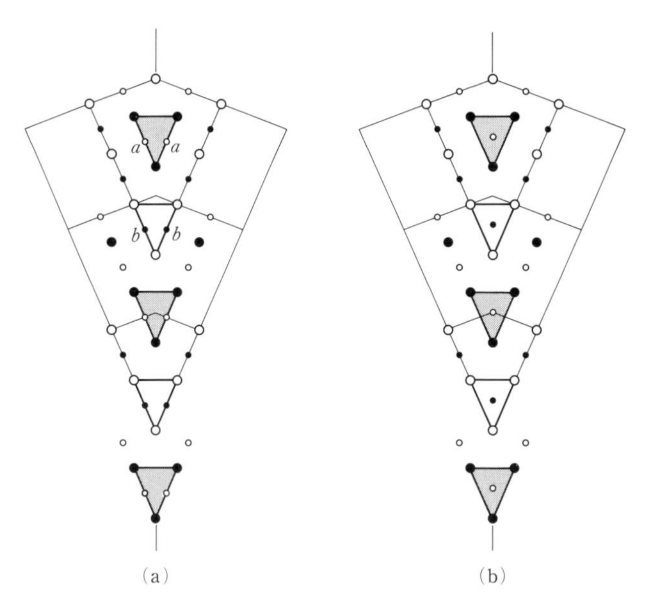

FIG. 2.69 Reconstruction of twinning structure of $(113)_{NaCl}$, projected on $(\bar{1}10)_{NaCl}$. (a) Structure of $(113)_{NaCl}$ twinning. Note that *a–a*, *b–b*, . . . cation pairs are on the cubic face of trigonal prisms. (b) Reconstruction of $(113)_{NaCl}$ twinning structure by removing one cation from each pair of cations and moving the other cations to the centre of the trigonal prisms.

on the centre of the trigonal prism. This is the twin structure derived from a NaCl-type structure. The anion arrangement in this figure is the same as the one in Fig. 2.65, thus we can obtain a basic drawing of the twin structure derived from the CCP lattice, normal to the pillars of trigonal prisms.

Taking the (partial) occupancy of interstices of octahedra and tetrahedra as well as of trigonal prisms into consideration, the structure $(m, n, \ldots)_{\mathrm{CCP}}$ can give rise to various kinds of compounds. Figure 2.70 shows the structure Re_3B, where the dotted lines on the left indicate the repeated twinning (3, 3), together with the arrangement of octahedra (in the centre) and trigonal prisms (on the right). The centres of the trigonal prisms (Re_6) are occupied by B, though the centres of the octahedra are empty. The structure can be expressed as $(3, 3)_{\mathrm{CCP}}\mathrm{Re}_3^{\mathrm{PS}}\mathrm{B}^{\mathrm{P}}$. It is noted that B atoms are coordinated by six Re atoms in the first near neighbour and by three face-capped Re atoms in the second near neighbour. (Strictly, one of the three capped Re atoms is more distant than the other two, due to the distortion of the trigonal prism. The three distances d(B–Re) in Re_3B are 2.33, 2.54, and 2.95 Å.) The structure appears in most of the tri-ionides of the lanthanide and actinide metals (anti-Re_3B type). The compound Cr_3GeC is isostructural to the Re_3B-type, with C filling the octahedral interstices, as $\mathrm{Cr}_3^{\mathrm{PS}}\mathrm{Ge}^{\mathrm{P}}\mathrm{C}^{\mathrm{O}}$. Figure 2.71 shows the structure of $CaIrO_3$. Oxygen atoms form $(3, 3)_{\mathrm{CCP}}$, Ca atoms occupy the interstices of trigonal prisms and Ir atoms the interstices of octahedra, expressed as $(3, 3)_{\mathrm{CCP}}\mathrm{O}_3^{\mathrm{PS}}\mathrm{Ca}^{\mathrm{P}}\mathrm{Ir}^{\mathrm{O}}$, being anti-$Cr_3GeC$.

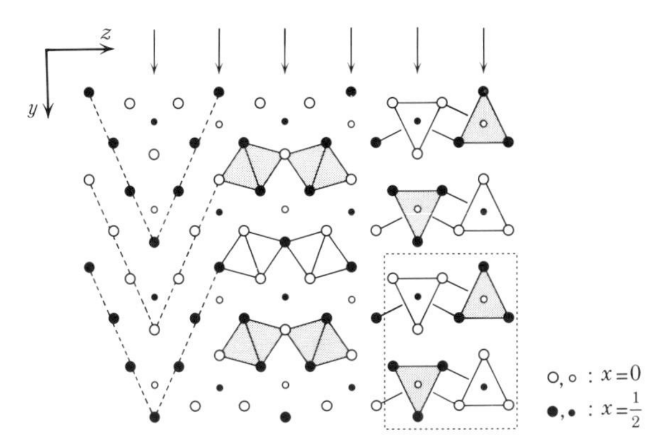

FIG. 2.70 Structure of Re_3B-type, $(3, 3)_{\mathrm{CCP}}\mathrm{Re}_3^{\mathrm{PS}}\mathrm{B}^{\mathrm{P}}$, projected on (100).[40] Large circles are metal atoms and small ones are boron atoms. The figure shows: on the left, the herring-bone pattern of $\langle 110\rangle_{\mathrm{NaCl}}$; in the centre, the Re_6 empty octahedra; and on the right, the Re_6B trigonal prisms. The arrows indicate the twin planes and the unit cell is indicated.

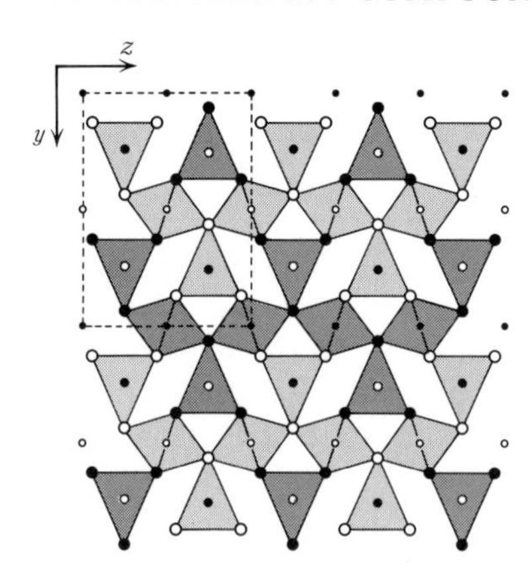

FIG. 2.71 Structure of $CaIrO_3$, $(3, 3)_{CCP}O_3^{PS}Ca^{P}Ir^{O}$, projected on (100).[40] This structure is an anti-Re_3B-type, and in addition the octahedral interstices of O atoms are occupied by Ir.

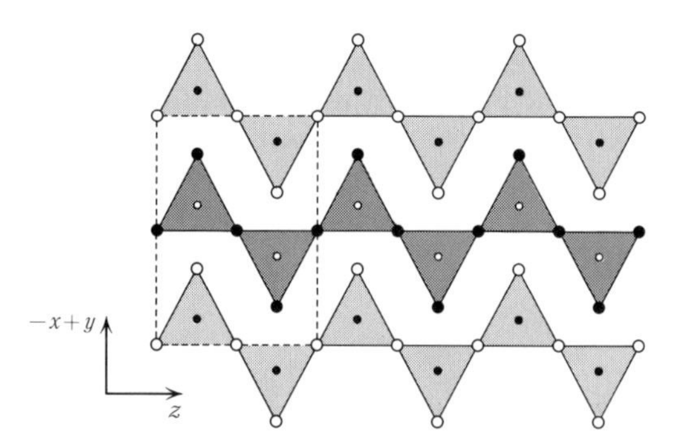

FIG. 2.72 Structure of Ni_2In-type, $(2, 2)_{CCP}Ni_2^{PS}In^{P}$.[40]

Figure 2.72 shows the structure of Ni_2In, which can be expressed as $(2, 2)_{CCP}Ni_2^{PS}In^{P}$. This structure has been generally described as HCP of In with Ni in all the octahedral and trigonal bi-pyramidal interstices. This is not rigorous, because the value of the axial ratio of c to a in the hexagonal lattice is about 1.23, which is significantly shorter than the ideal value of 1.663. The structure of Sr_2GeS_4 is also expressed as $(2, 2)_{CCP}Sr_2^{P}Ge^{T}S_4^{PS}$, in which Ge occupy half of the tetrahedral sites. This structure is an anti-Ni_2In type.

In Fig. 2.73 is shown the structure of CrB, which is a typical example of $(1, 1)_{CCP}(Cr^{PS}B^{P})$. The compound YOOD has an anti-CrB type structure as expressed by $(1, 1)_{CCP}Y^{P}O^{PS}O^{PS}D$, where half of the prismatic sites are occupied by Y. Table 2.3 shows some examples of micro-twin structure related to CCP.

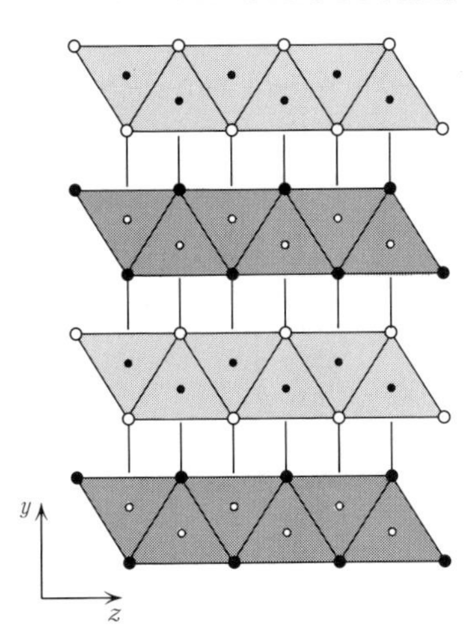

FIG. 2.73 Structure of CrB-type, $(1, 1)_{CCP}Cr^{PS}B^{P}$, projected on (100).[40]

Table 2.3

Compounds with micro-twin structure derived from CCP[40,66]

$(m, n)_{CCP}$	Structure type	Examples (1)	Examples (2)[a]
(3, 3)	$Re_3^{PS}B^{P}$ (2.70)[b]	Zr_3Co, Pd_6P	$Cr^{PS}Ge^{P}C^{O}$, $Cr_3As(C, N)$, Cr_2VC_2
	anti-Re_3B	$PuBr_3$	$Ca^{P}Ir^{O}O_3^{PS}$ (2.71), $UFeS_3$, $TlPbI_3$
(2, 2)	$Ni_2^{PS}In^{P}$ (2.72)	ACuB (A = Ca, Sr, B = Bi, Sb)	
	anti-Ni_2In		$Sr_2^{P}Ge^{T}S_4^{PS}$, K_2CoBr_4, Pb_2SiSe_4
(1, 1)	$Cr^{PS}B^{P}$ (2.73)	αNaOH, HoO(OH)	$Zr^{PS}Si_2^{TO}$
	anti-CrB	$Y^{P}O^{PS}O^{PS}D$, TlI	

[a] Compounds with (partially) filled octahedra and tetrahedra.
[b] Figure in the parentheses show the number of the figure in the text.

We shall now present studies relating to the micro-twin structure performed by B. G. Hyde, who is a main proposer of the concepts of shear and micro-twin structures. One study is on the mineral 'humite' family,[42] the composition of which is generally expressed by $n Mg_2SiO_4 \cdot Mg(OH)_2$ (this has to be altered to $n Mg_2SiO_4 \cdot Mg(OH, F)_2$, because $(OH)^-$ is often replaced by F^- in the minerals). In Table 2.4, the mineral name, together with the compositions and micro-twin structures proposed by the authors, are collected, the space group is *Pmcn* (orthorhombic) for n = odd and

Table 2.4

Minerals of humite family with micro-twin structure[42]

n	Name of mineral	Composition	Micro-twin structure
1	Norbergite	$Mg_2SiO_4 \cdot Mg(OH)_2$	$(3, 3)_{CCP}$
2	Chondrodite	$2Mg_2SiO_4 \cdot Mg(OH)_2$	$(3, 2)_{CCP}$
3	Humite	$3Mg_2SiO_4 \cdot Mg(OH)_2$	$(3, 2^2)^3_{CCP}$ [a]
4	Clinohumite	$4Mg_2SiO_4 \cdot Mg(OH)_2$	$(3, 2^3)_{CCP}$
	Forsterite	Mg_2SiO_4	$(2, 2)_{CCP}$

[a] The structure $(3, 2^2)^3_{CCP}$ denotes the structure $(3, 2, 2, 3, 2, 2, 3, 2, 2)_{CCP}$.

$P2_1/c$ (monoclinic system) for n = even. The structure of these compounds had been recognized as follows: the framework of the structure is close packing of O, which is significantly distorted from the ideal arrangement, and Mg atoms are sited at the centre of the octahedron and Si at the centre of the tetrahedron of oxygen arrays. White and Hyde noticed that the arrangement of Mg in the humite family is nearly CCP, and they proposed the micro-twin structures for the humite family, as shown in Table 2.4. They examined artificial and natural minerals of humite by electron microscopy.

Figure 2.74 shows the structure of forsterite (Mg_2SiO_4) projected on (100).[42] To the left of the figure is shown the unit cell in which atom parameters ($100x/a$) are given, and above this the arrangement of tetrahedral SiO_4. In the centre the trigonal prism of Mg_6Si, and to the right the arrangement of $(2, 2)_{CCP}$ of Mg are shown. Considering the arrangement of the trigonal prisms, this structure belongs to the Ni_2In-type structure

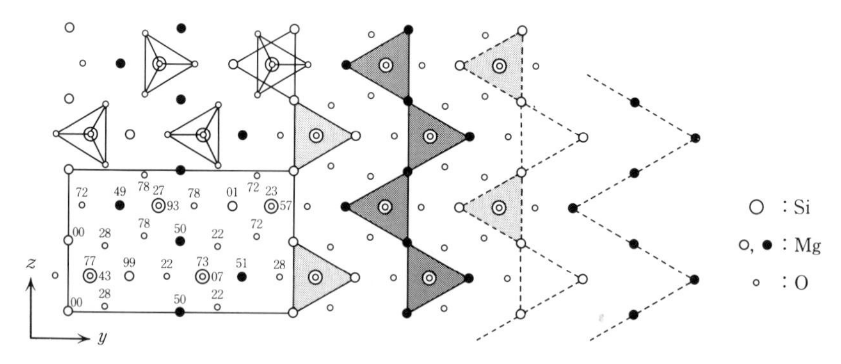

FIG. 2.74 Structure of forsterite (Mg_2SiO_4), $(2, 2)_{CCP}$ of Mg.[42] In the top left is drawn the arrangement of tetrahedral SiO_4, in the bottom left the unit cell, in the centre the arrangement of $(2, 2)_{CCP}$ of Mg. This is a Ni_2In-type structure.

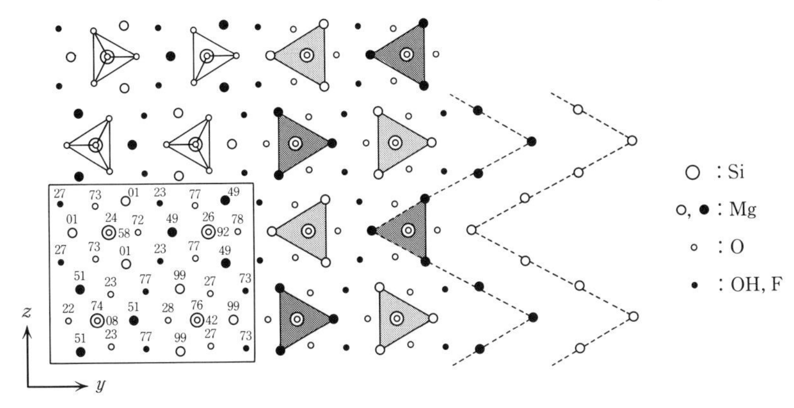

FIG. 2.75 Structure of norbergite ($Mg_2SiO_4 \cdot Mg(OH)_2$), $(3, 3)_{CCP}$ of Mg.[42] This is a Re_3B-type structure.

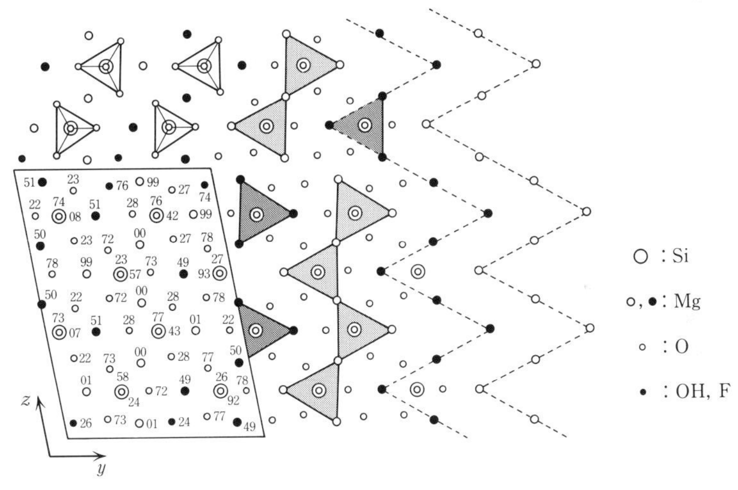

FIG. 2.76 Structure of clinohumite ($4Mg_2SiO_4 \cdot Mg(OH)_2$), $(3, 2^3)_{CCP}$ of Mg.[42] This structure can be interpreted as the intergrowth of norbergite and forsterite.

(see Fig. 2.72; $(2, 2)_{CCP}$). Figure 2.75[42] shows the structure of norbergite ($Mg_2SiO_4 \cdot Mg(OH)_2$) (in a similar manner to forsterite in Fig. 2.74), which belongs to the Re_3B-type micro-twin structure (see Fig. 2.70; $(3, 3)_{CCP}$). The structure of humite stones with $n = 2, 3, 4$ can be interpreted as the regular intergrowth of Ni_2In- and Re_3B-type structures. As an example, the structure of clinohumite with $n = 4$ is shown in Fig. 2.76, in which the arrangement of Mg is $(3, 2, 2, 2)_{CCP}$ (abbreviated to $(3, 2^3)_{CCP}$ hereafter), i.e. the intergrowth of Ni_2In- and Re_3B-type structures or of norbergite and forsterite.

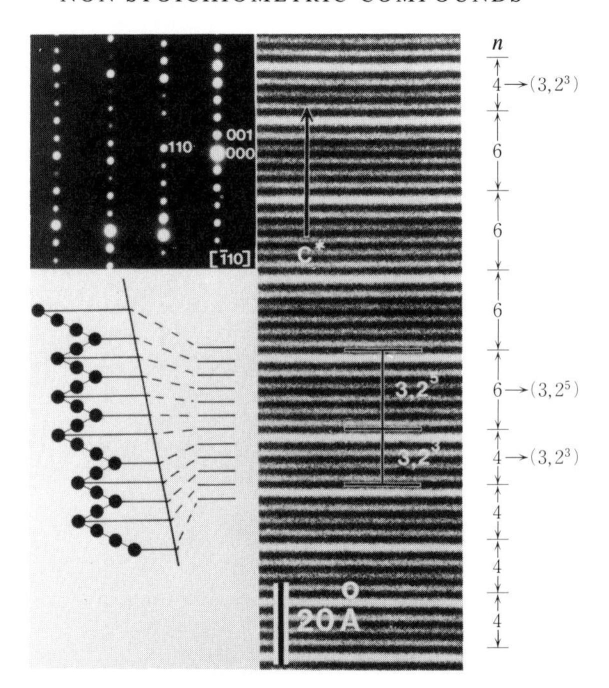

FIG. 2.77 Lattice image of clinohumite $(3, 2^3)_{CCP}$.[42] On the right is shown the lattice image of clinohumite, in the top left the diffraction pattern and in the bottom left the corresponding structure model (micro-twin model of Mg lattice). In the middle of the lattice image, the $(3, 2^5)$ structure intergrows in the clinohumite $(3, 2^3)$ matrix.

By measuring the spacing of lattice fringes of electronmicrographs, the micro-twin structures can be estimated (the spacings between lattice fringes along the c^* direction are 3 and 4.4 Å for norbergite and forsterite, respectively). Figure 2.77 shows the lattice image of clinohumite.[42] It is clearly seen that the new phase with the structure $(3, 2^5)_{CCP}$ intergrows in the matrix of clinohumite with the structure $(3, 2^3)_{CCP}$ (to identify the phases, the broad white fringes are conveniently used as a guide for the eyes). This picture definitely shows that this new phase, corresponding to $n = 6$, does exist microscopically, though not macroscopically as a mineral. Moreover, new phases with the structure of regular intergrowth of $(3, 2^5)$ and $(3, 2^3)$, generally expressed as $[(3, 2^5), (3, 2^3)^p]$, were also observed, where $p = 4, 7, 11$, corresponding to $n = 22, 34, 50$, respectively. Figure 2.78 shows the lattice image of $[(3, 2^5), (3, 2^3)^4]$ with 74 Å of the c-axis. Figure 2.79 shows the random intergrowth of materials with larger n values in the clinohumite matrix.

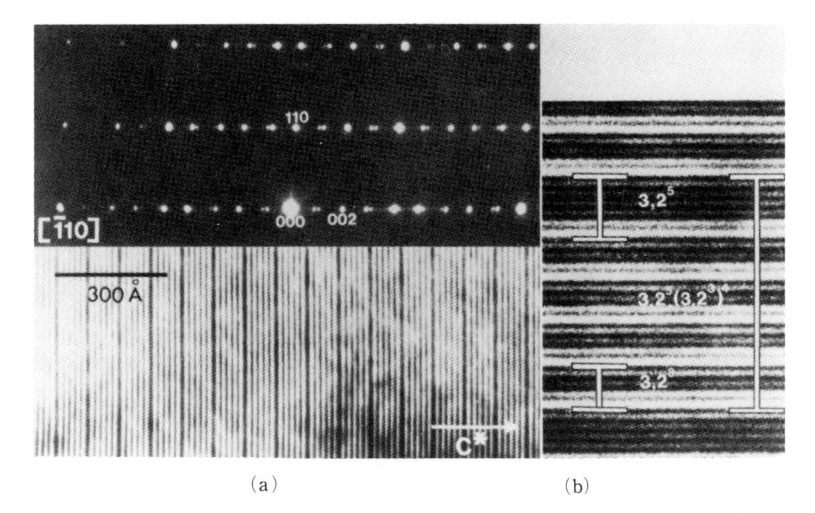

FIG. 2.78 Lattice image of an intergrowth structure $[(3, 2^5), (3, 2^3)^4]$ in clinohumite $(3, 2^3)_{CCP}$.[42] On the right (b) is shown the high-resolution lattice image of clinohumite, in the top left (a) the diffraction pattern and in the bottom left (a) the lattice image at low magnification. An ordered intergrowth of $(3, 2^5)$ in every four clinohumite $(3, 2^3)$ is seen, i.e. $[(3, 2^5), (3, 2^3)^4]$.

White and Hyde extended their observations to the manganohumite family[43] (Mg is replaced by Mn in humite) and leucophoenicite $3Mn_2SiO_4 \cdot Mn(OH)_2$[44] and obtained fruitful results from the viewpoint of the micro-twin structure.

2.5 Intergrowth structures

An intergrowth structure is a structure composed of more than two kinds of different, though related, structures, which are stacked up regularly along the superlattice axis, shown schematically in Fig. 2.80. We have already studied many examples having an intergrowth structure, e.g. the block structure of H–Nb_2O_5 (Fig. 2.27) is composed of U-sheet $[3 \times 4]_1$ and L-sheet $[3 \times 5]_\infty$. $W_4Nb_{26}O_{77}$ (Fig. 2.28) is also an intergrowth structure in a somewhat different sense from that of H–Nb_2O_5. The Vernier structure of the $Y_nO_{n-1}F_{n+2}$ system (Figs. 2.35 and 2.36) can be regarded as an intergrowth structure in the sense of being composed of A_1 and A_2 sheets, whose structure and composition are different. And the structure of humite (Figs. 2.74–2.76) could also be interpreted as the intergrowth of Re_3B- and Ni_2In-type micro-twin structures. Thus, intergrowth structures are often

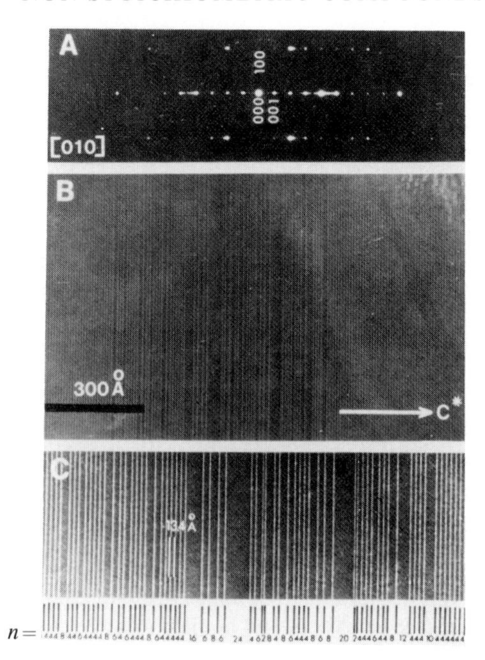

FIG. 2.79 Lattice image of random intergrowth of various n in clinohumite:[42] A: streaked diffraction pattern; B: lattice image at low magnification; C: lattice image at high magnification. $n = 4$ corresponds to clinohumite.

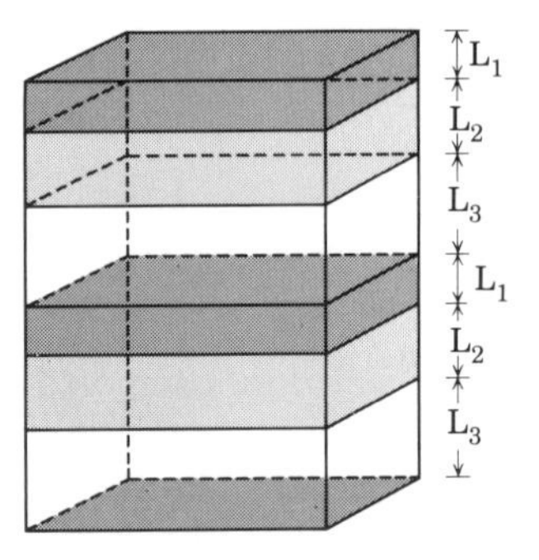

FIG. 2.80 Schematic drawing of the intergrowth structure. More than two kinds of different, though related, structures are stacked as $\ldots L_1L_2L_3L_1L_2L_3 \ldots$ along the superlattice axis.

observed and are important in structural inorganic chemistry. Here we show other typical examples of the intergrowth structure.

The structure of so-called hexagonal ferrite is based on the spinel ($MgAl_2O_4$) structure. Arrangement of oxygen in spinel is CCP, and Mg and

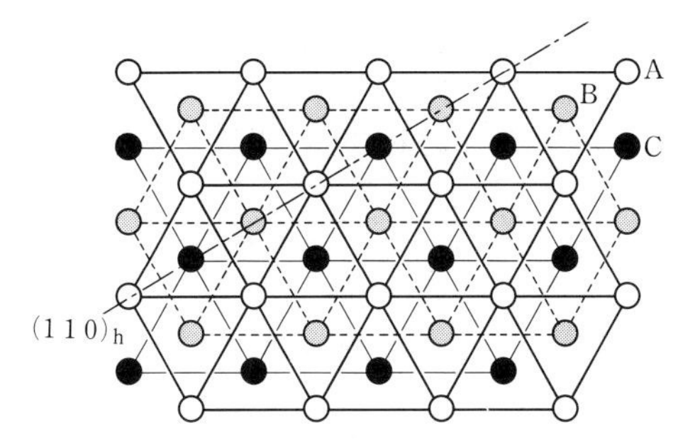

FIG. 2.81 Cubic close packing of oxygen projected on $(111)_{CCP}$ or $(001)_h$. Layer stacking along $[001]_h$ is . . . ABCABC . . . .

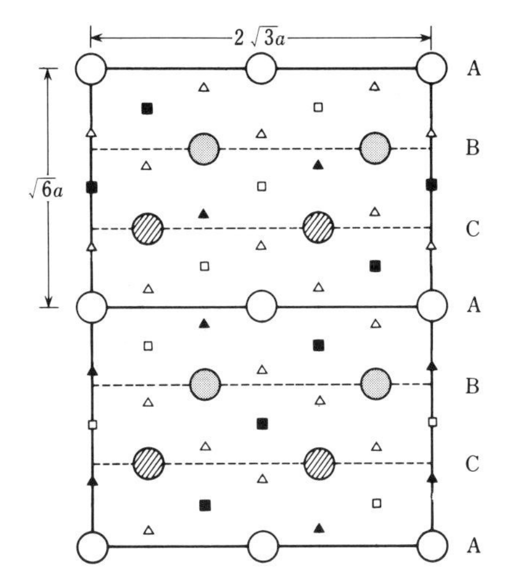

FIG. 2.82 Arrangement of oxygen (CCP), and occupied and unoccupied octahedral (■, □) and tetrahedral (▲, △) metal sites of the spinel structure projected on $(110)_h$.

Al occupy partially* the tetrahedral and octahedral sites. Figure 2.81 shows the cubic close packing of oxygen (. . . ABCABC . . .) projected on $(111)_{CCP}$ or $(001)_h$, where the subscript h stands for the hexagonal notation. Figure 2.82 shows the arrangement of oxygen, metal ion, and metal vacancy (octahedral and tetrahedral sites) of the spinel, projected on $(110)_h$. In the unit cell there are six layers of oxygen array (. . . ABCABC . . .).

* If N atoms form CCP, $2N$ tetrahedral and N octahedral sites are produced. Because there are eight molecules of $MgAl_2O_4$ in the unit cell, there are 64 tetrahedral and 32 octahedral sies. One-eighth of the tetrahedral and $\frac{1}{2}$ of the octahedral sites are regularly occupied by Mg (A site) and Al (B site).

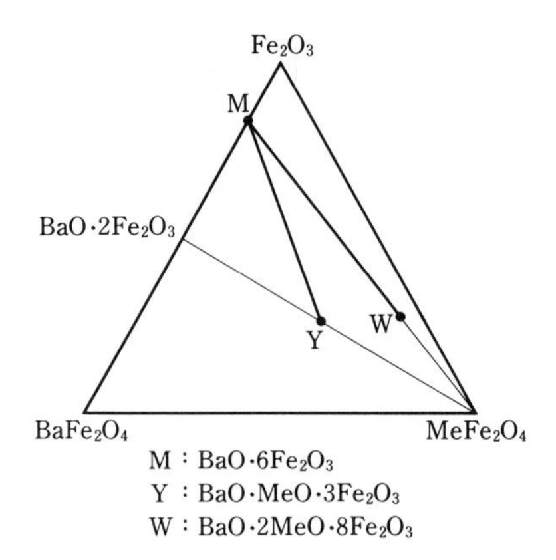

FIG. 2.83 Pseudo-ternary Fe_2O_3–$BaFe_2O_4$–$Me^{II}Fe_2O_4$ system. Compounds labelled M, Y, and W have the compositions $BaO \cdot 6Fe_2O_3$, $BaO \cdot MeO \cdot 3Fe_2O_3$, and $BaO \cdot 2MeO \cdot 8Fe_2O_3$, respectively.

Hexagonal ferrite appears in the pseudo-ternary Fe_2O_3–$BaFe_2O_4$–$Me^{II}Fe_2O_4$ system of the quaternary Ba–Fe–Me^{II}–O (Me^{II} = Zn, Mg, Fe, Co, Ni, Cu) system, as shown in Fig. 2.83. In this system three compounds labelled M, Y, W exist as typical hexagonal ferrites, whose compositions are $BaO \cdot 6Fe_2O_3$, $BaO \cdot MeO \cdot 3Fe_2O_3$, and $BaO \cdot 2MeO \cdot 8Fe_2O_3$, respectively. On the lines $\overline{MY}$ and $\overline{MW}$, there also appear many phases which usually have layered or intergrowth structures. The structure of phase M is shown in Fig. 2.84,there are ten layers of oxygen array in the unit cell. The closest packing layer of oxygen includes barium ions in an ordered arrangement as shown in Fig. 2.85 (O:Ba = 3:1). This may be because the ionic radius of a barium ion (1.35 Å) is nearly equal to that of an oxygen ion (1.40 Å). The oxygen layers, however, do not always contain barium ions.

The stacking of oxygen layers along the c-axis is as follows:

$$\ldots \underbrace{BC}_{S}\underbrace{Ac}_{R}\underbrace{AC}_{S^*}\underbrace{BA}_{R^*}bA \ldots$$

where the capital and small letters denote the oxygen-only layer (O_4) and the BaO_3 layer, respectively. The stacking can also be expressed by block notation as SRS*R* (see Fig. 2.84), where block S stands for the layer stacking of AB, BC, CA and block R stands for the layer stacking of ABA, ACA, BAB, BCB, CAC, CBC. Block S* is obtained by a 180° rotation of block S around the c-axis. The relation between R and R* is the same as that of S and S*. (The block stacking SRS*R* has to satisfy the layer stacking

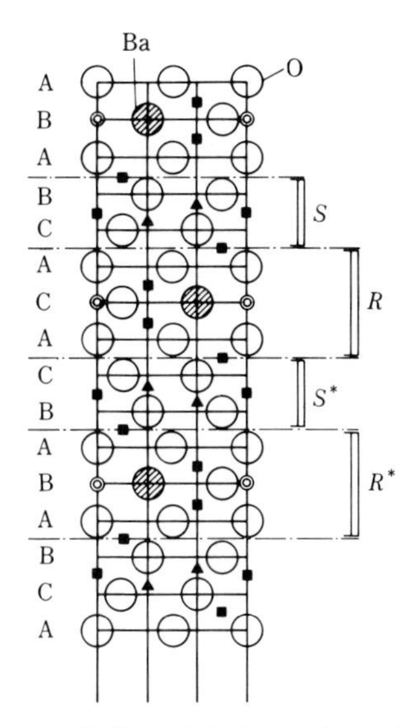

FIG. 2.84 Structure of phase M drawn in a similar manner to Fig. 2.82. Note that Ba atoms (hatched large circles) are incorporated in the oxygen CCP stacking, having the ordered structure shown in Fig. 2.85. This structure is composed of S, R, S*, and R* blocks (see text), and is denoted as SRS*R*$[(c^2h^3)^2]$.

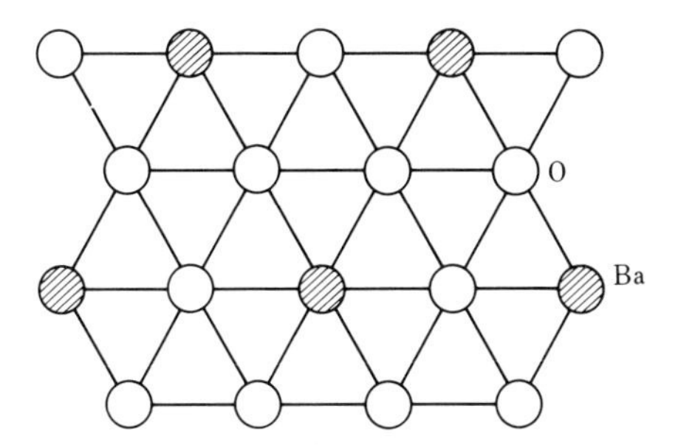

FIG. 2.85 Incorporation of Ba atoms in CCP oxygen layers with an ordered structure (O:Ba = 3:1).

of $[(c^2h^3)^2]$; this notation is referred to in Ref. 45.) Figure 2.86 shows the structure of the Y phase. The layer stacking is as follows:

. . . ABCbcBCABabABCAcaA . . .

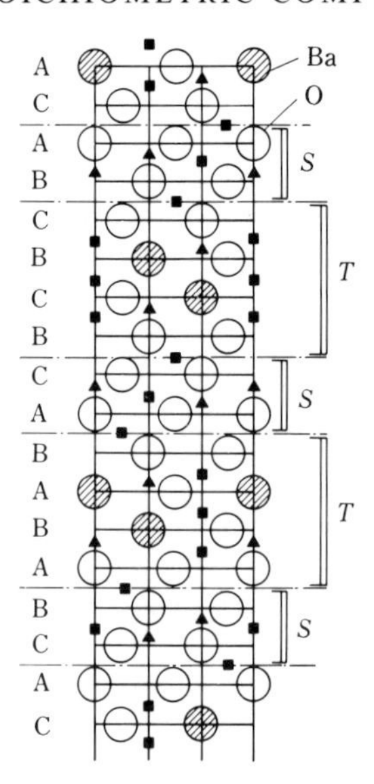

FIG. 2.86 Structure of phase Y drawn in a similar manner to Fig. 2.82. The structure is denoted as $(ST)_3[(c^2h^4)^4]$.

The block T stands for the layer stacking of ABAB, ACAC, BCBC, BABA, CACA, and CBCB. The structure is designated by the notation $(ST)_3[(c^2h^4)^4]$. Similarly the structure of the W phase is expressed as $S_2RS_2^*R^*[(c^4h^3)^2]$. The widths of the blocks S, R, T (length of the c-axis of each block) are about 5.0, 6.6, and 9.6 Å, respectively.

As mentioned above, many phases have been found on the line $\overline{MY}$ (Kohn and co-workers[46,47] have contributed greatly to these findings). These compounds, generally expressed as M_mY_n, have been revealed to have the intergrowth structure of M and Y. Table 2.5 shows some of the phases that have been confirmed. As seen from the table, many combinations of M and Y for a fixed (m, n) pair can exist as stable phases. For instance the compound M_2Y_{15}, whose composition is $Ba_{32}Me_{30}Fe_{228}O_{368}$, forms three types of structure $MYMY_{14}$, MY_4MY_{11}, and MY_6MY_9, which are three of the possible seven combinations of M and Y $[MY_pMY_{15-p}, 1 < p < 7]$. It is a surprising discovery that long period lattices such as M_8Y_{27} ($c = 1455$ Å) and M_4Y_{33} ($c = 1577$ Å) can exist in inorganic compounds.

Figure 2.87 shows the lattice images produced using electron microscopy for the compounds M_mY_n, from which the arrangement of M and Y, i.e. the structure, was inferred as tabulated in Table 2.6. To identify the arrangement of M and Y, the thin white lines (l_M) and thick white lines (l_Y) can be used as a guide for the eye, each of which corresponds to M and Y. For instance, the stacking of l_M and l_Y in (c) is ... $l_Ml_Ml_Yl_Ml_Ml_Y$ ..., from which the structure

Table 2.5

Compounds M_mY_n with block structure[47]

M: $BaO \cdot 6Fe_2O_3$, Y: $BaO \cdot MeO \cdot 3Fe_2O_3$
Me: Zn, Mg, Fe, Co, Ni, Cu

M:Y	Stacking of M and Y block	Length of c-axis (Å)	Ideal chemical composition
2:1	MMY	113.16	$Ba_4Me_2Fe_{36}O_{60}$
2:2	MYMY	52.25	$Ba_6Me_4Fe_{48}O_{82}$
2:3	$MYMY_2$	200.4	$Ba_8Me_6Fe_{60}O_{104}$
2:4	$MYMY_3$	244.0	$Ba_{10}Me_8Fe_{72}O_{126}$
2:4	MY_2MY_2	81.32	$Ba_{10}Me_8Fe_{72}O_{126}$
⋮	Twenty-four compounds		⋮
2:12	MY_3MY_9	197.5	$Ba_{26}Me_{24}Fe_{168}O_{302}$
2:12	MY_4MY_8	592.5	$Ba_{26}Me_{24}Fe_{168}O_{302}$
2:13	$MYMY_{12}$	636.1	$Ba_{28}Me_{26}Fe_{180}O_{324}$
2:13	MY_6MY_7	636.1	$Ba_{28}Me_{26}Fe_{180}O_{324}$
2:15	$MYMY_{14}$	723.2	$Ba_{32}Me_{30}Fe_{228}O_{368}$
2:15	MY_4MY_{11}	723.2	$Ba_{32}Me_{30}Fe_{228}O_{368}$
2:15	MY_6MY_9	241.1	$Ba_{32}Me_{30}Fe_{228}O_{368}$
2:18	$MYMY_{17}$	854.0	$Ba_{38}Me_{36}Fe_{252}O_{434}$
2:18	MY_8MY_{10}	854.0	$Ba_{38}Me_{36}Fe_{252}O_{434}$
2:21	MY_3MY_{18}	327.5	$Ba_{44}Me_{42}Fe_{276}O_{500}$
2:21	$MYMY_{20}$	982.5	$Ba_{44}Me_{42}Fe_{276}O_{500}$
4:3	MMYMYMY	269.9	$Ba_{10}Me_6Fe_{84}O_{142}$
4:5	$MYMYMYMY_2$	357.0	$Ba_{14}Me_{10}Fe_{108}O_{186}$
4:8	$MYMYMY_3MY_3$	162.6	$Ba_{20}Me_{16}Fe_{144}O_{252}$
4:8	$MYMYMY_2MY_4$	487.8	$Ba_{20}Me_{16}Fe_{144}O_{252}$
4:8	$MYMY_2MY_2MY_3$	487.8	$Ba_{20}Me_{16}Fe_{144}O_{252}$
4:9	$MYMYMYMY_6$	531.4	$Ba_{22}Me_{18}Fe_{156}O_{274}$
4:10	$MMYMY_2MY_7$	191.6	$Ba_{24}Me_{20}Fe_{168}O_{296}$
4:10	$MYMY_2MY_2MY_5$	574.9	$Ba_{24}Me_{20}Fe_{168}O_{296}$
4:13	$MYMY_5MY_2MY_5$	705.7	$Ba_{30}Me_{26}Fe_{204}O_{262}$
4:13	$MY_2MY_4MY_2MY_5$	705.7	$Ba_{30}Me_{26}Fe_{204}O_{362}$
4:15	$MYMY_2MY_3MY_9$	792.9	$Ba_{34}Me_{30}Fe_{228}O_{406}$
4:33	$MY_6MY_{10}MY_7MY_{10}$	1577	$Ba_{70}Me_{66}Fe_{444}O_{802}$
6:13	$MYMYMYMY_2MY_3MY_5$	775.2	$Ba_{32}Me_{26}Fe_{228}O_{400}$
6:14	$MYMYMY_2MY_3MY_2MY_5$	818.8	$Ba_{34}Me_{28}Fe_{240}O_{422}$
8:27	$MYMY_4MY_7MYMYMY_6MY_8$	1455	$Ba_{62}Me_{54}Fe_{420}O_{746}$

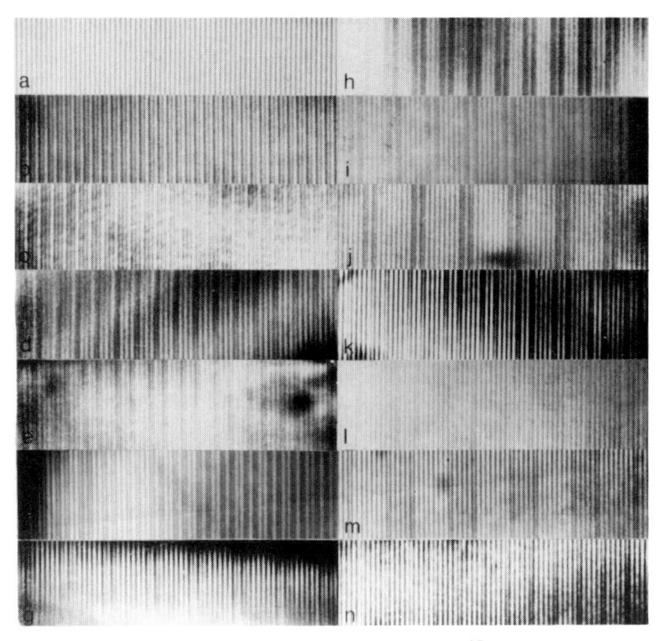

FIG. 2.87 Lattice images of M_mY_n cmpounds.[47] The results of the inferred arrangements of blocks from these lattice images are summarized in Table 2.6.

was estimated to be ... MMYMMY ... (It is noted that this structure M_2Y (=MMY) is a special one in the M_mY_n compounds, because M is usually isolated as seen from Tables 2.5 and 2.6 and Figs 2.88 and 2.89.) Next we show examples of longer period lattices. Figure 2.88(a) shows the complex stacking of M and Y: $\ldots l_M l_Y l_M 4l_Y l_M l_Y l_M 4l_Y l_M l_Y l_M 4l_Y l_M l_Y l_M 2l_Y \ldots$ (hereafter we use a simpler notation, ... 14141412 ..., for this structure; it is enough for the expression of the structure to write down the number of Y blocks sandwiched between M blocks, because M is always isolated except for the example of M_2Y shown in Fig. 2.87(c) as noted above). Figure 2.88(b) and (c) show the structures ... 533134 ... and ... 444333 ... with a stacking defect F, respectively. Another part of the crystal, whose lattice image is shown in Fig. 2.88(c), gives the disordered lattice images as shown in Fig. 2.89, which keeps, however, the rough arrangement of ... 444333 ... as a whole.

Thus, it has been confirmed that there are many phases on the line $\overline{MY}$ of Fig. 2.83, which originate from the ordered stacking of M and Y blocks along the *c*-axis. As is clear, infinite discrete compounds can exist between the M and Y phases in principle. This is a typical example of the intergrowth structure.

Table 2.6

Block (M, Y) *arrangement of* M_mY_n *compounds shown in Fig. 2.87*[47]

M:Y	Stacking of M and Y block	Corresponding alphabet in Fig. 2.87
1:∞	Y	a
2:1	MMY	c
2:2	MYMY	b
2:3	$MYMY_2$	f
2:4	$MYMY_3$	h
2:5	MY_2MY_3	e
2:6	$MYMY_5$	i
	MY_2MY_4	k
	MY_3MY_3	d
2:7	$MYMY_6$	g
2:8	$MYMY_7$	j
	MY_4MY_4	n
2:10	MY_3MY_7	l
2:11	MY_4MY_7	m

A second example of the intergrowth structure is tungsten bronze, M_xWO_3 (M = H, Li, Na, K, Rb, Cs, Ca, Sr, Ba, In, Tl, Ge, Sn, Pb, Cu, Ag etc.; nowadays many bronzes such as Ti, V, Mo are well known[48]), so named because its colour is similar to that of alloy bronze CuZn. It has been shown to have a kind of intergrowth structure in a limited composition (x) range. Generally, the stability of the structural types (hexagonal, tetragonal, and cubic) depends both on the ionic radius (r_i) and the composition (x) of M. For instance,

Li_xWO_3:	$0.31 < x < 0.57$	cubic
Na_xWO_3:	$0.28 < x < 0.38$	tetragonal
	$0.40 < x < 1.0$	cubic
K_xWO_3:	$0.13 < x < 0.31$	hexagonal
	$0.40 < x < 0.60$	tetragonal
	$0.81 < x < 0.92$	cubic

The structure of these phases is, in principle, based on ReO_3-type WO_3. WO_3 shows successive phase transitions on heating: triclinic → monoclinic → orthorhombic → tetragonal. At room temperature the monoclinic WO_3 changes to the cubic WO_3 (ReO_3-type) by incorporation of M.

Figure 2.90 shows the structure of the cubic M_xWO_3 (CTB, cubic tungsten bronze). The centre holes are partially and randomly occupied by M. This structure is the same as cubic $BaTiO_3$, apart from the partial occupation of M sites. The structure of the hexagonal M_xWO_3 (HTB, hexagonal tungsten

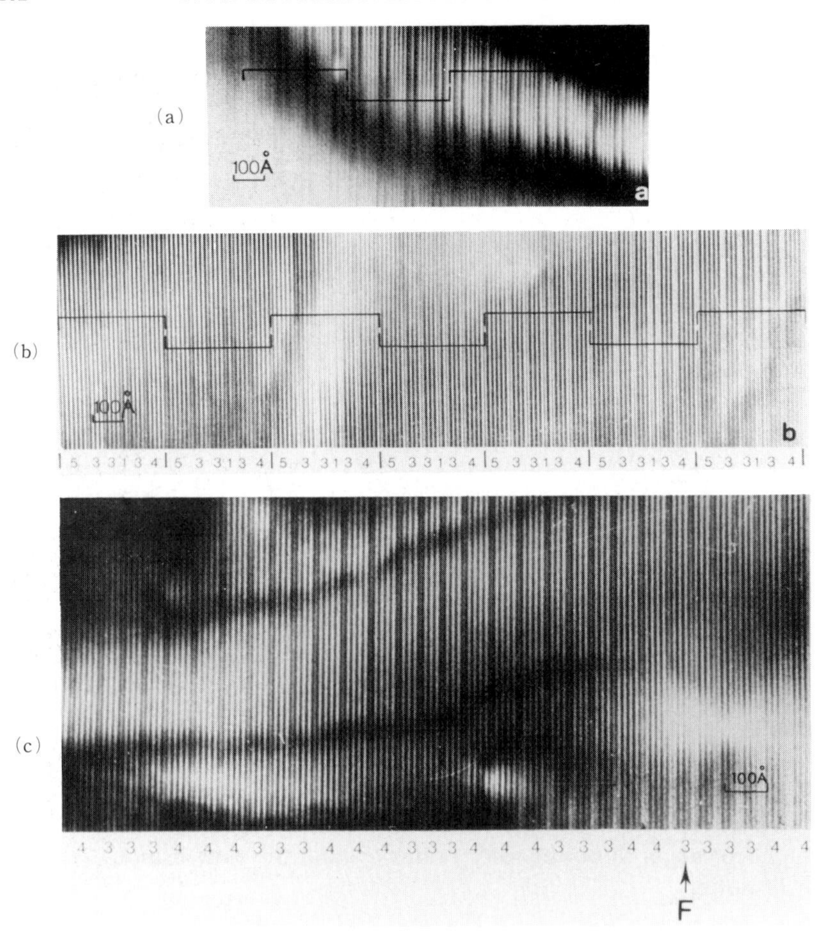

FIG. 2.88 Lattice images of the longer period lattices of M_mY_n compounds:[47] (a) ...14141412...; (b) ...533134...; (c) ...333444... (F denotes a stacking defect).

bronze) is shown in Fig. 2.91. Groups of six WO_6 form tunnels of hexagonal prisms by sharing corners, which are partially and randomly occupied by M. This structure is considered to be composed of the structural elements A (=B) and E, and can be derived from the ReO_3-type structure as follows. First a distorted renium tri-oxide type structure (DRO), (b) in Fig. 2.92, is obtained from a ReO_3-type structure (a) by rotating octahedra alternately by α degrees about $[001]_{ReO_3}$. The density of the DRO-type structure becomes higher than that of ReO_3, though the composition is unchanged. By putting $\alpha = 15°$ the arrangement of octahedra along the arrows A and

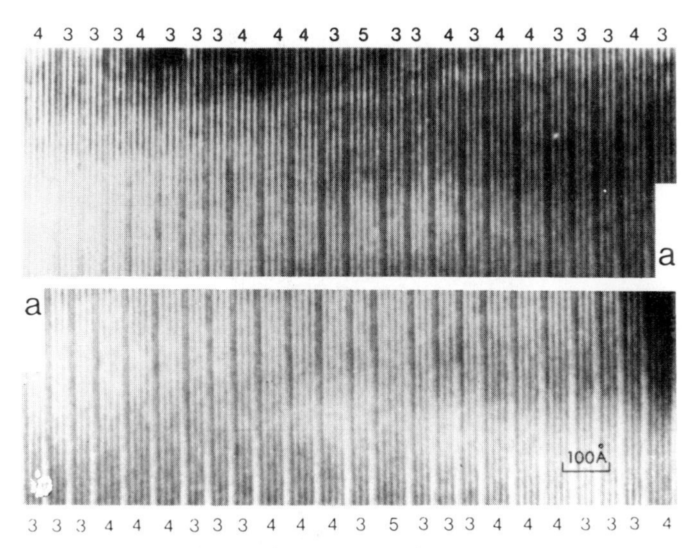

FIG. 2.89 Lattice images from a crystal fragment with a structure of . . . 333444 . . . containing stacking defects.[47] These pictures were obtained from another part of the same crystal as that of Fig. 2.88(c). The basic unit of structure MY_5 is seen.

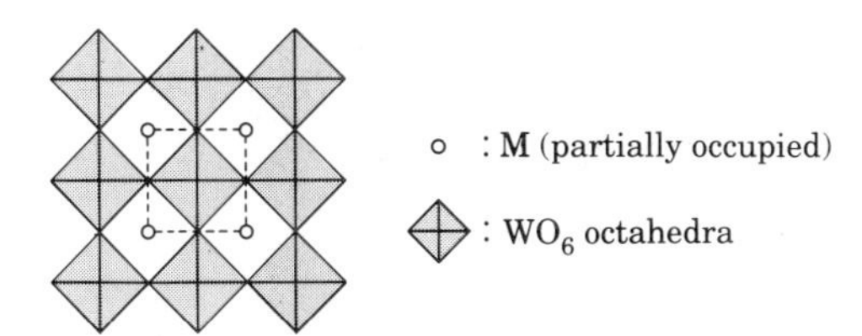

FIG. 2.90 Structure of cubic tungsten bronze (CTB) M_xWO_3, projected on (001). Metal sites (open circles) are partially and randomly occupied.

B in Fig. 2.92(b) becomes the same as that along A and B in Fig. 2.91. The HTB is derived from the DRO by inserting rows of extra octahedra (arrow E) between the A and B rows so as to form tunnels of hexagonal prisms.

With increasing ionic radius of M, the HTB phase is generally stabilized as the lowest x phase, as seen partly from the examples of the tungsten bronzes mentioned above. In fact the HTB phase is the only stable phase for M = Rb, Cs, In, and Tl in the composition range $0.13 < x < 0.33$. (It is to be noted that the stoichiometry of HTB is $M_{1/3}WO_3$. The homogeneity range extends down to $x = 0.13$, which means the partial occupation of hexagonal tunnel sites.)

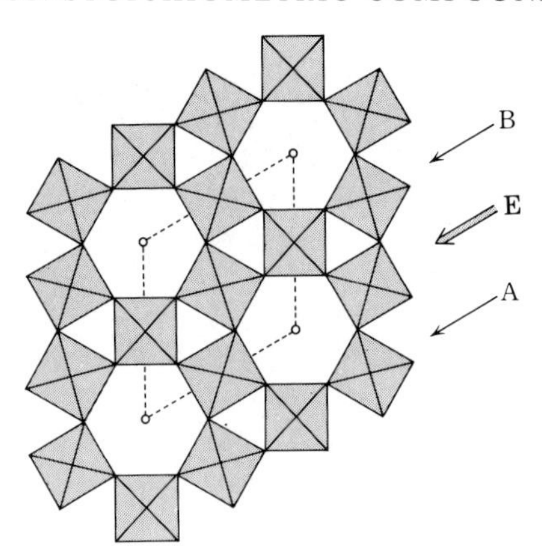

FIG. 2.91 Structure of hexagonal tungsten bronze (HTB) M_xWO_3. A group of six WO_6 forms tunnels of hexagonal prisms by sharing corners. This structure is composed of the structural elements A (=B) and E, as shown by the arrows. Metal sites (open circles) are randomly and partially occupied.

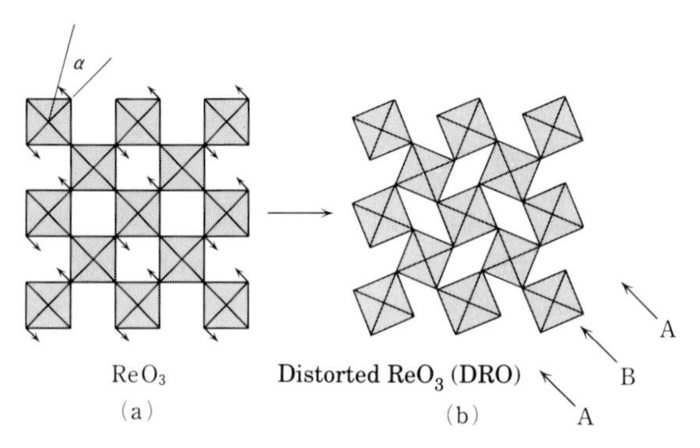

FIG. 2.92 Derivation of the distorted renium tri-oxide (DRO) structure from ReO_3. By rotating octahedra alternately about $[001]_{ReO_3}$ (a), the structure (b), DRO is derived. The rows A and B in (b) are the same as those in Fig. 2.91 ($\alpha = 15°$).

Kihlborg and co-workers studied the phase relation and the crystal structure for the pseudo-binary $M-WO_3$ systems having the HTB phase, focusing on the composition range $x < 0.13$.[49,50] As a result, a new homologous compound, named intergrowth tungsten bronze (ITB), was found in the substances prepared at high temperatures. For instance, Fig. 2.93 shows a diffraction pattern and structure image formed using an electron microscope for $K_{0.10}WO_3$, from which a model structure was inferred as

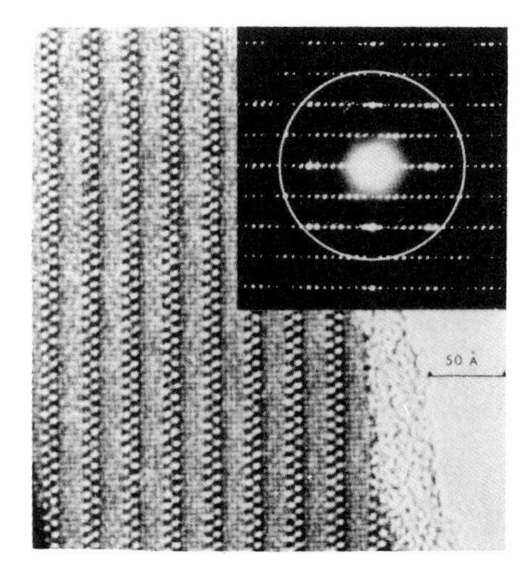

FIG. 2.93 Structure image of $K_{0.10}WO_3$.[49] The inset shows the electron diffraction pattern, the diffraction spots in the circle were used for structure imaging.

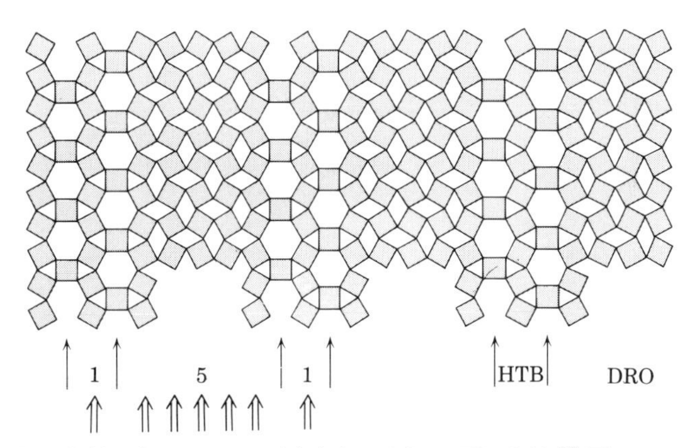

FIG. 2.94 Structure model deduced from Fig. 2.93.[49] The arrows ↑ show the rows of tunnels of hexagonal prisms. The structure can be regarded as the intergrowth of HTB and DRO, which was named intergrowth tungsten bronze (ITB) by Kihlborg and co-workers.

shown in Fig. 2.94.[49] This structure was also verified by X-ray analyses of a single crystal. The structure can be regarded as a regular intergrowth of HTB and DRO, and can also be derived by introducing rows of extra octahedra (shown by the arrows ↑ in Fig. 2.94) into a DRO structure so as to

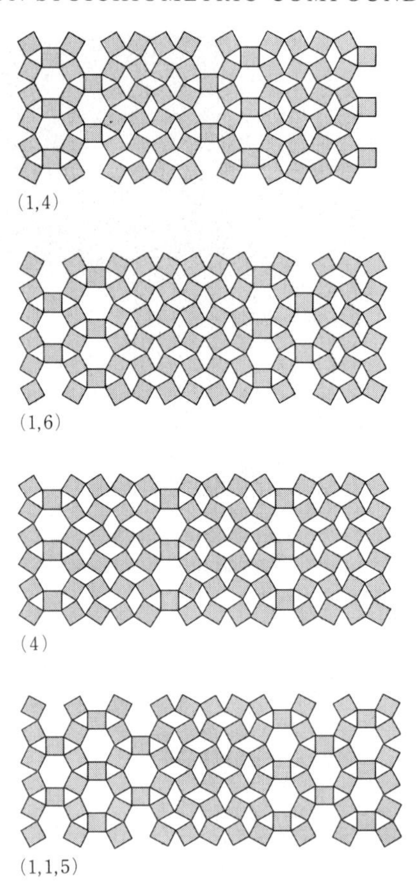

FIG. 2.95 Some examples of ITB.[49–52]

form the tunnels of hexagonal prisms, this being similar to the derivation of HTB. The width of the lamella of the two structural elements can vary, and various kinds of homologous series can therefore be formed. In other words the structure and composition of ITB depends on how the rows of extra octahedra are introduced into the DRO-type structure. The following notation for expressing the ITB structures was used by Kihlborg and co-workers.[49] The numbers of octahedral rows $\Uparrow$ with a DRO-type structure between the centres of the tunnels, indicated by the arrows ↑ in Fig. 2.94, are written down in parentheses as $(m, n, o, \ldots)$ in repeating sequences. For example the structure shown in Fig. 2.94 is designated as (1, 5). The phases like this with the double tunnel rows can be generally designated by $(1, n)$. The HTB structure is designated as (1).

Figure 2.95 shows some examples of ITB discovered by Hussain and

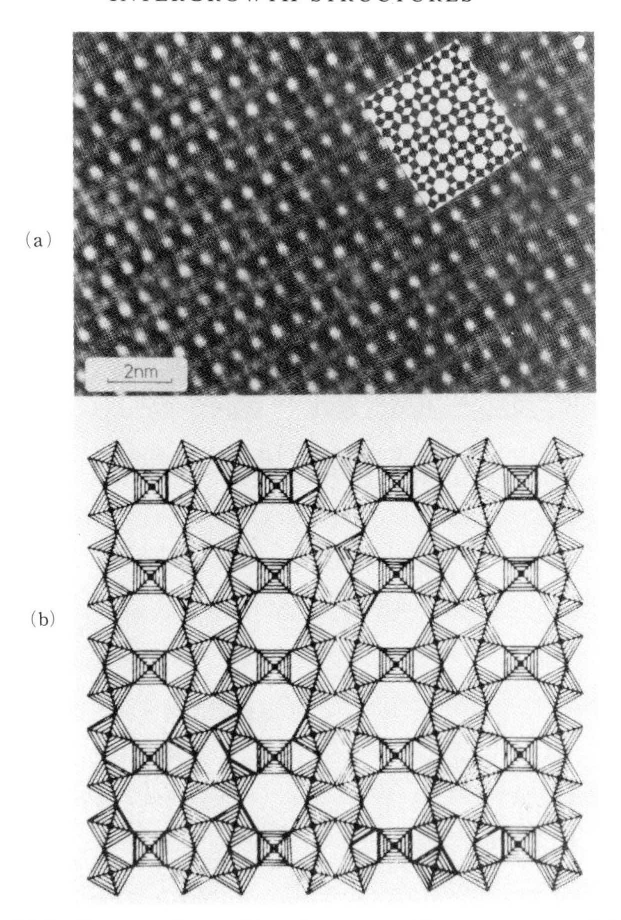

FIG. 2.96 Structure image (a) and corresponding structure model (b) for $Sb_{0.20}WO_3$.[53] This type of structure is designated (2).

Kihlborg,[49–52] who found the homologous compounds with n values ranging from 4 to 12 for the $(1, n)$ family. Since then various types of ITB have been revealed. Structure (2) observed in a crystal of $Sb_{0.20}WO_3$ is shown in Fig. 2.96,[53] as a structure image (a) and a model structure (b), deduced from the structure image. It has a single tunnel row and therefore is very close to the structure of HTB (1). This sample also showed other types of structure, i.e. (m) with $m = 3, 4, 6, 8$ as well as $m = 2$. An example of the double tunnel rows, designated as $(1, n)$, is shown in Fig. 2.93. ITB with more than three-tunnel rows was often observed in crystals of $M_xM'_xW_{1-x}O_3$, named bronzoids,[54,55] where $M' = V^{5+}$, Nb^{5+}, and Ta^{5+}. Figures 2.97(a) and (b) show the structure image of (1, 1, 4) and (1, 1, 1, 6) with the 3- and 4-tunnel rows observed in $Cs_xNb_xW_{1-x}O_3$.[56]

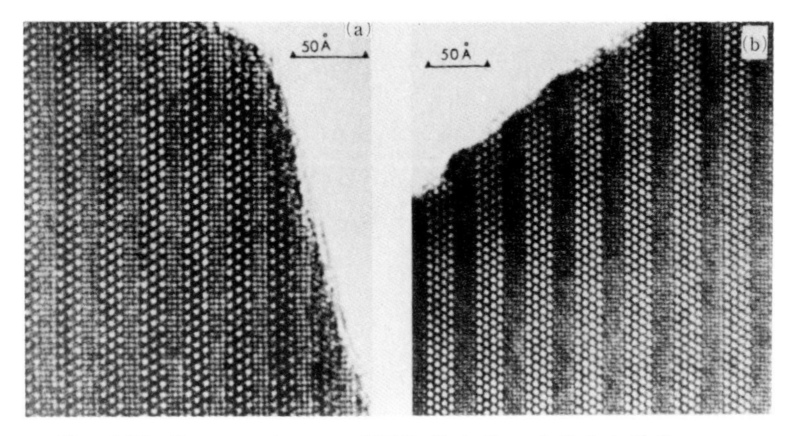

FIG. 2.97 Structure images of ITB (1, 1, 4) and (1, 1, 1, 6) from crystals of $Cs_xNb_xW_{1-x}O_3$.[56]

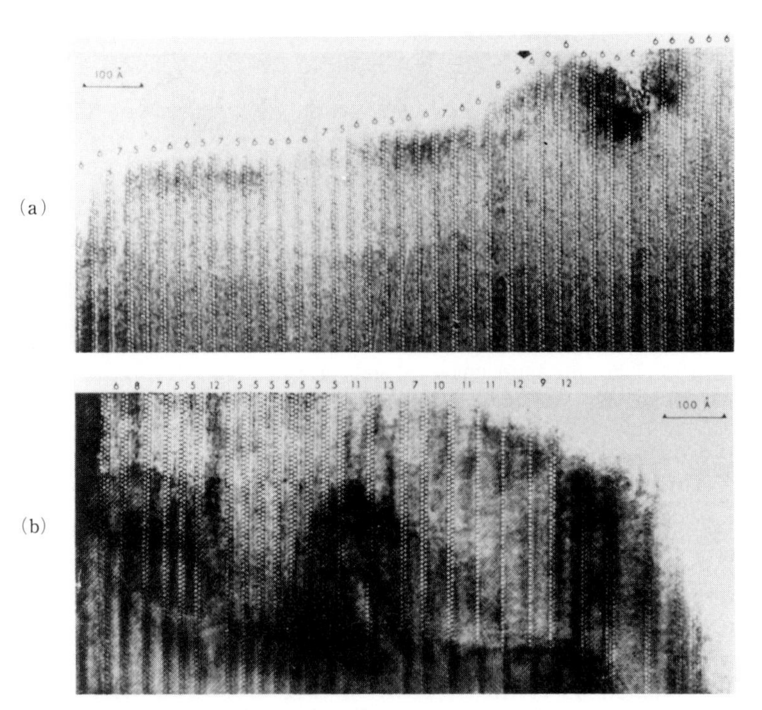

FIG. 2.98 Structure images of disordered ITB from crystals of $Rb_{0.03}WO_3$.[49] (a) Disorder of (1, n) structure. The width of DRO, n, ranges from 5 to 8. (b) Random intergrowth of (n), (1, n), and (1, 1, n), where n = 5–13.

Generally speaking, disorder is quite frequently observed in the intergrowth compounds, as seen in ITB. Figure 2.98(a) shows disorder of width of a DRO-type matrix for the (1, n) structure observed in $Rb_{0.03}WO_3$,[49] i.e. random distribution of n ranging from 5 to 8. Figure 2.98(b) also shows the random intergrowth of (n), (1, n), and (1, 1, n), where n ranges from 5 to 13.[49]

Thus we have studied two examples of the intergrowth structure. The concept of the intergrowth structure plays an important role in inorganic structural chemistry.

2.6 Adaptive structures

In the above sections, we have described the four types of non-stoichiometric compounds derived from extended defects based on the difference of structural characteristics. The concept of adaptive structure, which was proposed by the late Professor J. S. Anderson in 1973,[57] is a more general concept which explains some of the examples mentioned in the preceding sections. The compounds which have the adaptive structure are defined by Anderson as:

1. Within certain composition limits, every possible composition can attain a unique, fully ordered structure, without defects arising from solid solution effects and with no biphasic coexistence ranges between successive structures.

2. For most, but not all, of the infinitely adaptive structures there may be a multiplicity of discrete, related, fully ordered structures for any one composition and, for certain compositions, the number of possible structures is potentially infinite.

These compounds show two sets of reflections in their diffraction patterns: (a) strong reflection from a mother structure subcell, independent of composition; (b) weaker superstructure reflections, indicating irrational multiplicities (incommensurate), which change continuously with the composition. We cannot observe two sets of superlattice reflections.

Adaptive structures have a common structural principle. A large unit cell of the adaptive structure is built up from the ordered repetition of a set of basic sub-units, derived from the mother structure by systematic changes in site occupancy (see Section 1.4.11) or in stacking sequence, or by a shear operation. For example, the homologous compounds V_nO_{2n-1} ($n = 3$–9, see Section 2.2), which are derived from the mother compounds V_2O_4 (rutile) by the shear operation $(121)\frac{1}{2}[0\bar{1}\bar{1}]$, are able to be a set of basic sub-units.

If a superlattice (adaptive structure) consists of $a_1, a_2, a_3, \ldots, a_i, \ldots$ of these sub-units and these sub-units constitute a set with multiplicities

$m_1, m_2, m_3, \ldots, m_i, \ldots$ of the mother structure, which have compositions $x_1, x_2, x_3, \ldots, x_i, \ldots$, a superlattice multiplicity m^* and a composition x^* can be expressed by

$$m^* = \sum_i a_i m_i \tag{2.9}$$

$$x^* = \sum_i a_i x_i \tag{2.10}$$

To interpret the known systems, it is enough to consider only two consecutive coefficients a_i and a_{i+1}. Thus we get

$$m^* = a_i m_i + a_{i+1} m_{i+1} \tag{2.11}$$

$$x^* = a_i x_i + a_{i+1} x_{i+1} \tag{2.12}$$

As shown later, the infinite phases can exist, having the continuous composition between x_i and x_{i+1}.

Firstly we study the adaptive structure of the pseudo-binary Ta_2O_5–$11Ta_2O_5 \cdot 4WO_3$ system, related to the L-Ta_2O_5 structure, as a typical example. The phase relation of the Ta_2O_5–WO_3 system had not been derived up to 1970. Roth and his co-workers[58, 59] studied the phase diagram in detail and found out, by indexing the powder X-ray diffraction patterns, that there appears to be a continuous series of phases with various multiplicities of the basic orthorhombic subcell in the composition region Ta_2O_5–$11Ta_2O_5 \cdot 4WO_3$. The crystal structure determination was performed for some of the typical compositions.[60] It has also been confirmed that the U_3O_8-type structure is a mother structure of these compounds, which is composed of edge-shared pentagonal bipyramids and distorted octahedra. Consider a chain of edge-shared pentagonal bipyramids arranged so that adjacent pentagonal bipyramids point in opposite directions as shown in Fig. 2.99(a). An ideal U_3O_8-type structure is derived by fusion of the identical chains so as to share corners of pentagons in adjacent chains, as shown in Fig. 2.99(b). As is seen in this figure, distorted octahedra are created between fused chains. The unit cell of ideal U_3O_8-type is indicated in this figure.

A set of basic sub-units for our system is obtained by the operation of micro-twinning to the U_3O_8-type structure. The twinning plane is $(110)_{U_3O_8}$. (The twinning plane is called folding plane in the original paper by Roth *et al.*; hereafter we abbreviate folding plane (twinning plane) as FP.) The twinning operation reduces the ratio of anion to metal, due to the shared corners of octahedra. Figure 2.100 shows the twinning structure of $(110)_{U_3O_8}$. A set of basic sub-units is derived by regular twinning operation to the U_3O_8-type mother structure. The structure and composition of the sub-units depend on how the twinning plane is introduced into the mother structure,

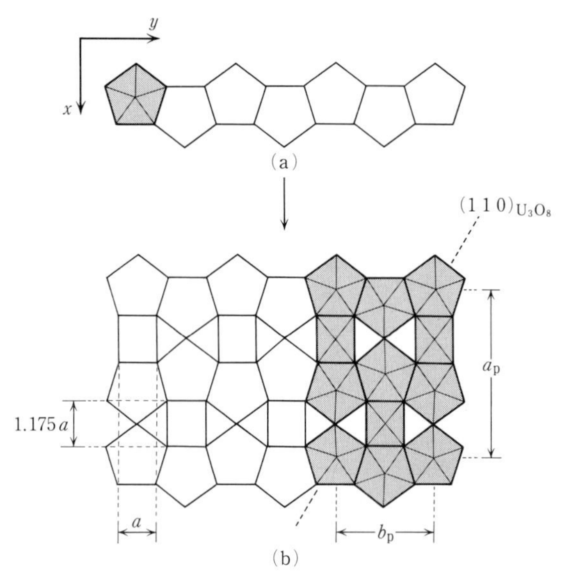

FIG. 2.99 Derivation of U_3O_8-type structure from chains of edge-sharing pentagonal bipyramids.[58] (a) A chain of edge-sharing pentagonal bipyramids arranged so that adjacent pentagons point in opposite directions. (b) A U_3O_8-type structure is derived by fusion of the chains, by which identical chains share the corners of pentagonal bipyramids in adjacent chains and distorted octahedra are created. The unit dimensions of the U_3O_8-type structure, a_p and b_p, are indicated.

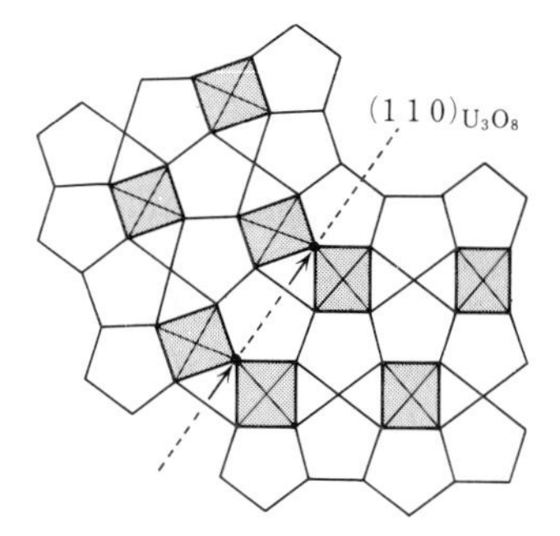

FIG. 2.100 Twinning operation to U_3O_8-type structure with (110) twin plane. It is noted that the pentagonal bipyramids on the twin plane are substantially distorted (see also Fig. 2.105).

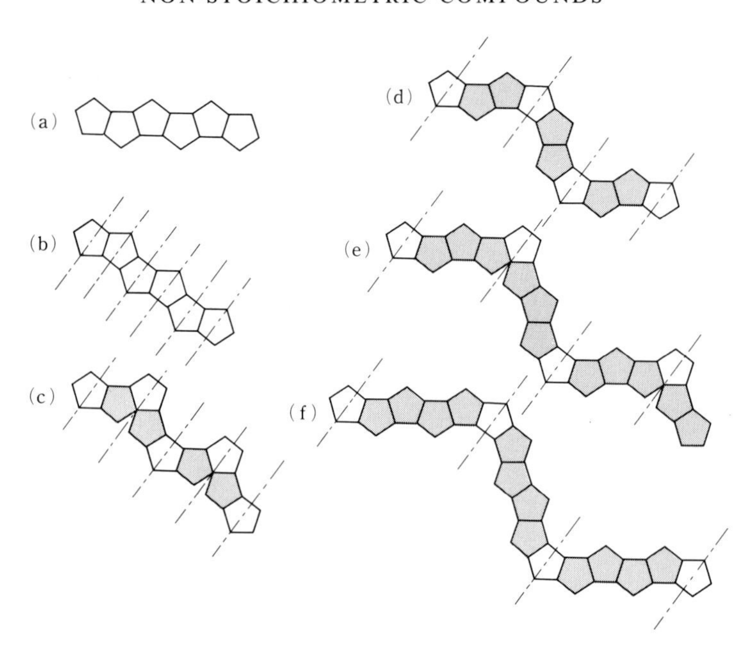

FIG. 2.101 Micro-twinning operation to U_3O_8-type structure. The structures are represented by a chain of pentagonal bipyramids for simplicity. The structures with a $(l, l)_{U_3O_8}$-type twinning operation are shown, where l is the number of pentagonal bipyramids between the twin planes (hatched ones). (a) U_3O_8-type structure (mother structure); (b) $l = 0$ (same as the mother structure); (c) $l = 1$; (d) $l = 2$; (e) $l = 3$; (f) $l = 4$. The chain lines indicate the folding planes.

as shown in Fig. 2.101, in which a type of $(l, l)_{U_3O_8}$ is depicted (for the notation, see Section 2.4). In this figure, the structures after the micro-twinning operation in every l pentagonal bipyramid (l is the number of pentagonal bipyramids between twin planes, hatched ones in this figure) are represented by a chain of pentagonal bipyramids for brevity. The structures with $l =$ odd differ from those with $l =$ even in the manner the pentagonal bipyramids are linked at the twin planes. Here we discuss only the structures with $l =$ even, because the structures of the Ta_2O_5–$11Ta_2O_5 \cdot 4WO_3$ system have been revealed to be derived from this type of structure. It was reported by Papiernik *et al.*[61] that it may be possible for the phases in the ZrO_2–ZrF_4 system to have structures with $l =$ odd. This paper also describes various kinds of structures derived from the U_3O_8-type structure by a micro-twinning operation.

A set of sub-units for the present system consists of the structures with

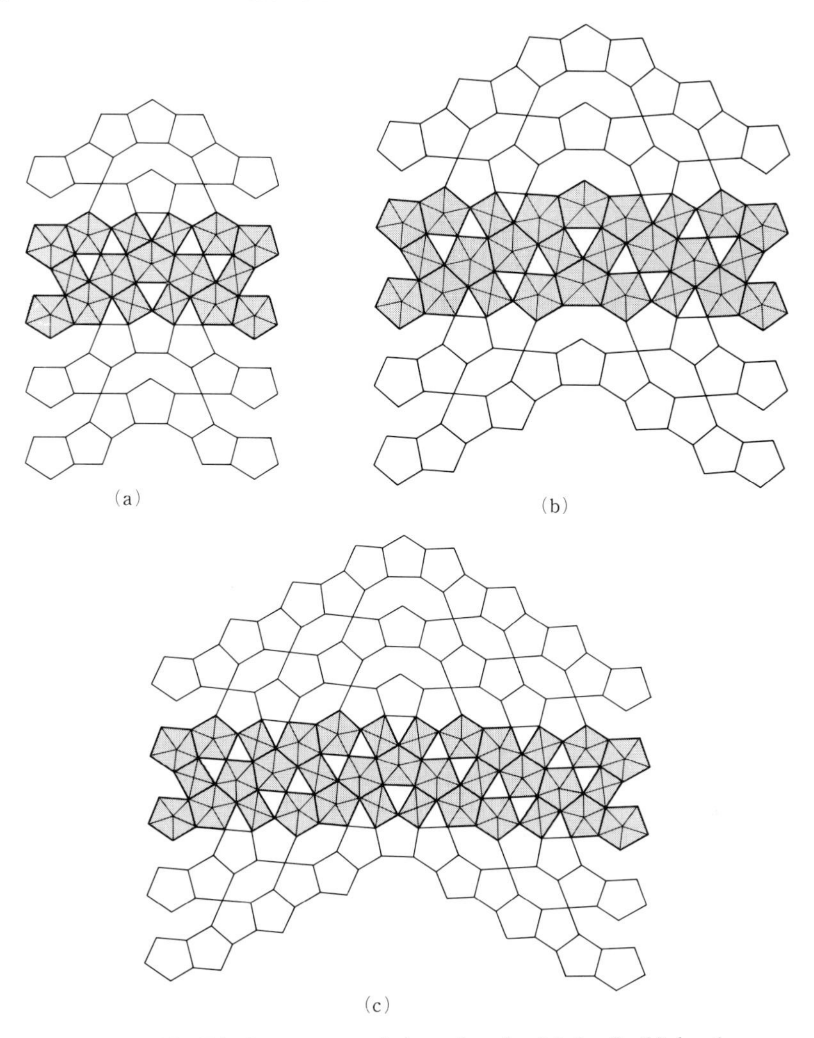

FIG. 2.102 Ideal structures of the sub-units (a) $l = 2$, (b) $l = 4$, (c) $l = 6$. Each structure has $5b'$, $8b'$, and $11b'$ lengths along the b-axis.

$l = 2i =$ (even). In Figs 2.102(a), (b), and (c), are shown the ideal structures with $l = 2$, 4, and 6. The length of sublattice unit b_{su} along the b-axis for these sub-units is expressed as

$$b_{su} = \tfrac{1}{2}b_p m_i = b' m_i$$
$$m_i = \tfrac{3}{2}l + 2 \tag{2.13}$$

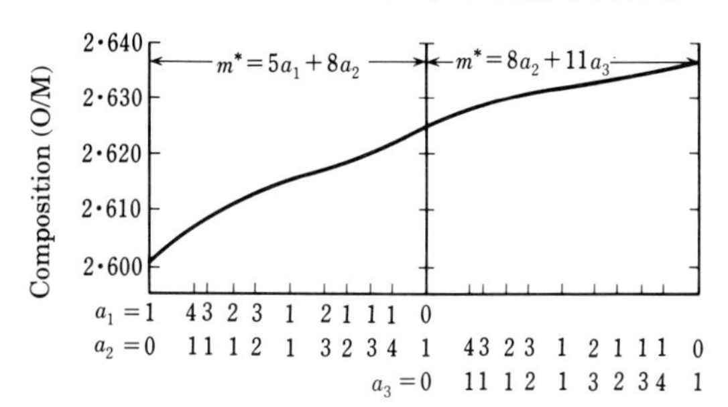

FIG. 2.103 Composition in the ratio of oxygen to metal (O/M) versus ratio of the number of sub-units (a_1, a_2, a_3). The left half is for $m^* = 5a_1 + 8a_2$ and the right half is for $m^* = 8a_2 + 11a_3$. Typical combinations of (a_1, a_2) or (a_2, a_3) are indicated on the abscissa.

where b_p stands for the unit cell length of the b-axis for the mother compound U_3O_8, and m_i stands for the multiplicity of the mother structure in a unit of $\frac{1}{2}b_p$ or b'.* Thus, the structures with $l = 2$, 4, and 6 have $5b'$, $8b'$, and $11b'$ of b-axis, i.e. $m_i = 3i + 2$ ($i = 1, 2, 3, \ldots$). Hereafter we use m_i as notation for structures instead of l. These m_is constitute a set of sub-units. The compositions of the structures with $m_1, m_2, m_3, \ldots, m_i, \ldots$ are $M_{10}O_{26}$, $M_{16}O_{42}$, $M_{22}O_{58}, \ldots, M_{2m_i}O_{(16m_i-2)/3}, \ldots$ The ordered intergrowth of these sub-units along the b-axis gives rise to infinite adaptive superlattices.

Only three members of the sub-unit, m_1, m_2, and m_3, are necessary for the interpretation of the structures of the present system Ta_2O_5–$11Ta_2O_5 \cdot 4WO_3$. In the composition range $2.600 < O/M < 2.625$, the superlattice multiplicity m^* is expressed as $m_1a_1 + m_2a_2$ $(=5a_1 + 8a_2)$ and in the composition range $2.625 < O/M < 2.636$, the m^* is expressed as $m_2a_2 + m_3a_3$ $(=8a_2 + 11a_3)$. The chemical composition of these two series is given as $M_{2m^*}O_{[16m^*-2(a_1+a_2)]/3}$ and $M_{2m^*}O_{[16m^*-2(a_2+a_3)]/3}$. In Fig. 2.103 is shown the composition dependence of these two series on a_1, a_2, and a_3.[57] For example, the ideal structure with $m^* = 19 = 8 \times 1 + 11 \times 1$ is shown in Fig. 2.104, the composition of which is $M_{38}O_{100}$.

As shown in Fig. 2.105 by a model structure,[58] the operation of microtwinning to fused linear chains of regular pentagonal bipyramids, i.e. U_3O_8-type structure, causes substantial distortion of regular pentagonal bipyramids near the twin plane (or folding plane). In real crystals, the distortion is relieved by reducing the coordination number of metals near

* Strictly speaking this is not correct. It is easily shown that b_p $(=2b')$ equals $(1 + 2\cos 36)a$, where a is the length of side of regular pentagonal bipyramids (see Fig. 2.99). The ideal length of the b-axis for the structures with l, b_{ideal}, is calculated to be $(l + 2)b' + la \cos 36$. The difference $(b_{ideal} - b_{su})$ becomes nearly $0.15la$.

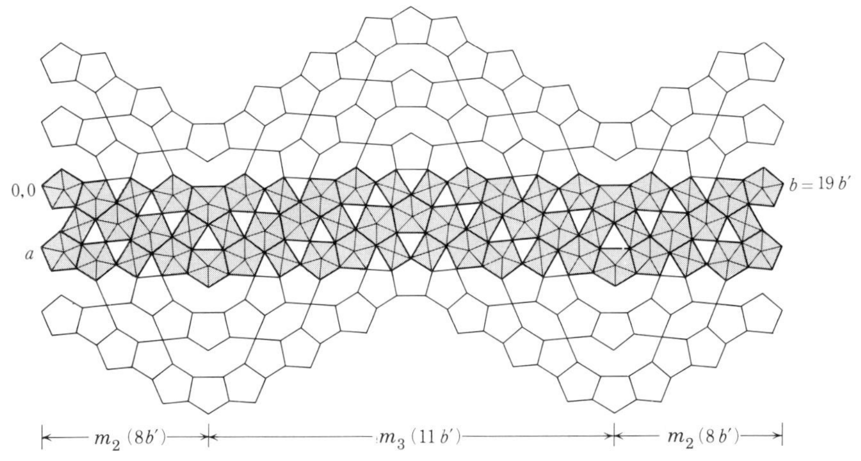

FIG. 2.104 Ideal structure with $m^* = 19 = 8 \times 1 + 11 \times 1$.

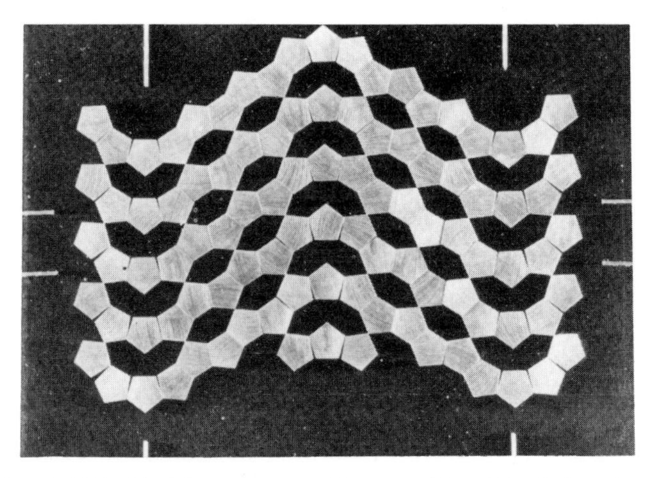

FIG. 2.105 Model structure with $m^* = 11 = 11 \times 1$ (see Fig. 2.102(c)).[58] Anion packing distortion, shown by gaps between adjacent pentagonal bipyramids, is evident near the folding plane.

the twin planes: elimination of an anion shared by two pentagonal bipyramids and by one octahedron, leading to three distorted octahedra. Metals coordinated by seven oxygen in pentagonal bipyramids are moved to octahedral positions (six coordination of oxygen). The relief of distortion is not always performed near the twin plane. Here we define the distortion plane or reduction plane as the plane (in the present case, parallel to (010)) on which eliminated anions (i.e. oxygens) were positioned. Figure 2.106

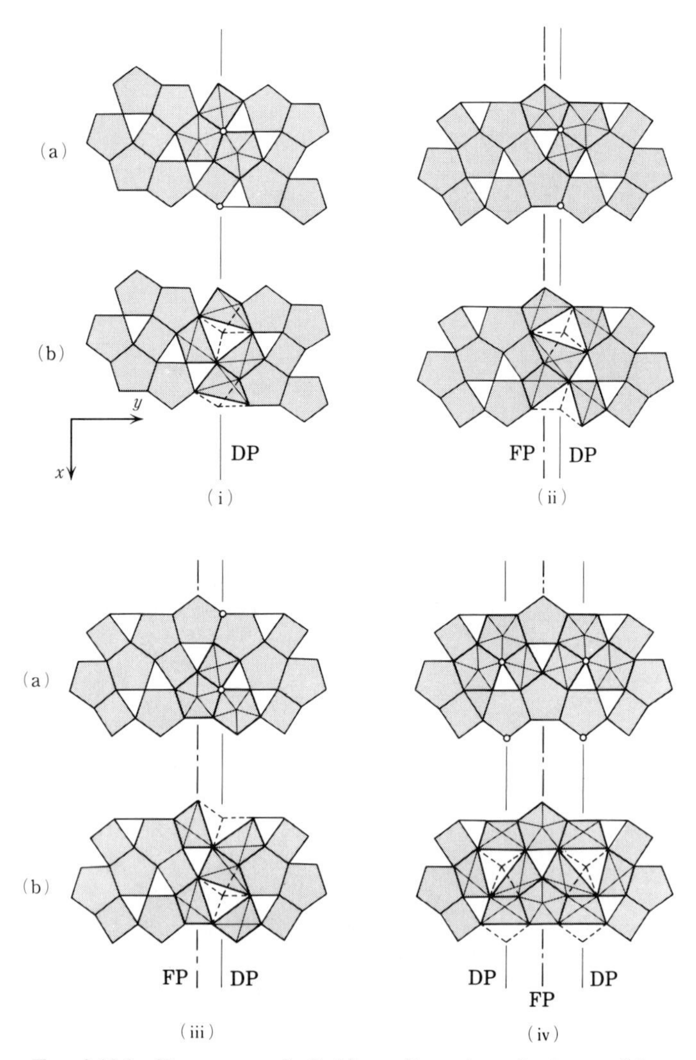

FIG. 2.106 Four types of relief from distortion of anion packings. In each (a) is shown the ideal structure. White circles are the oxygens on (010) to be eliminated. In (b) is the real structure after elimination of oxygens. Each of the structures (b) is denoted as DP(i), DP(ii), DP(iii), and DP(iv), respectively.

shows four examples of the structures near the distortion plane before (a) and after the above mentioned operation (b). In this figure the open circles in (a) indicate oxygen atoms to be eliminated. In Fig. 2.106 (i) is for the case when the distortion plane (DP) is far from the twin plane (FP); (ii) and (iii)

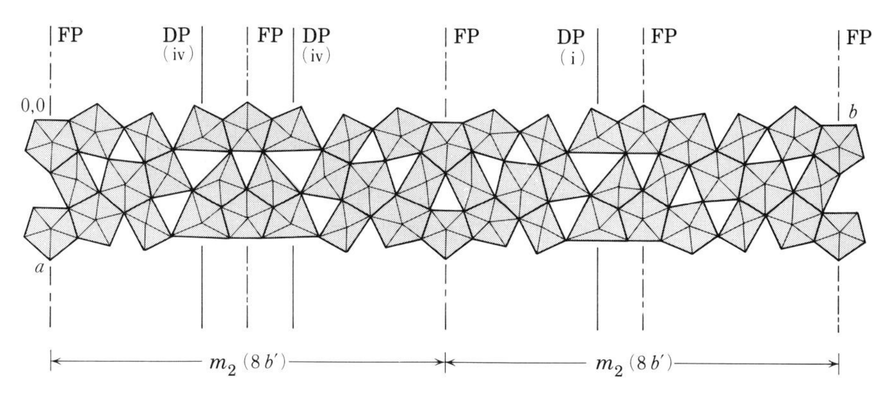

FIG. 2.107 Real structure of $W_2Ta_{30}O_{81}$ ($2WO_3 \cdot 15Ta_2O_5$) ($m^* = 16 = 8 \times 2$) with three distortion planes. This structure has two types of DP, i.e. (i) and (iv). The ideal structure is depicted in Fig. 2.102(b).

show the structure for the case of DP near FP; and (iv) shows the structure when there are two DPs near FP. As is obvious, the DP operation reduces the ratio of oxygen to metal. For example, the composition is expressed as $M_{2m^*}O_{\{[16m^* - 2(a_1 + a_2)]/3\} - d}$ for the case of $m^* = a_1m_1 + a_2m_2$, instead of $M_{2m^*}O_{[16m^* - 2(a_1 + a_2)]/3}$, where d is the number of DP in a unit cell. This indicates that the composition is variable even for fixed a_i and m_i values.

Figure 2.107 shows the real structure of $W_2Ta_{30}O_{81}$ ($2WO_3 \cdot 15Ta_2O_5$). The ideal structure for the compound is expressed as $m^* = m_2a_2 = 8 \times 2$, having the composition $M_{32}O_{84}$, which is shown in Fig. 2.102(b). The real structure has two types [(i) and (iv)] of DP, i.e. elimination of three oxygens per unit cell. Another example, $W_4Ta_{22}O_{67}$ ($4WO_3 \cdot 11Ta_2O_5$), is shown in Fig. 2.108(a), together with the ideal structure (b), having $m^* = a_1m_1 + a_2m_2 = 5 \times 1 + 8 \times 1$ ($M_{26}O_{68}$). The real structure has only one sheet of DP in a unit cell: the composition of the crystal is $M_{26}O_{67}$. The arrangement of polyhedra in the real crystal, however, seems to be very different from the ideal structure, aside from the arrangement near FP. This is due to the change from octahedra to pentagonal bipyramids (O → P) or pentagonal bipyramids to octahedra (P → O) as indicated by the dotted lines. The reason these drawings are reasonable comes from consideration of the result of structure determination. It is noted that these operations (O → P, P → O) do not change the composition.

If we express the average distance of DP, m^*/d in a unit of b', as $1/k$, composition (O/M) versus k shows a linear relation as shown in Fig. 2.109. (From the expression of $M_{2m^*}O_{\{[16m^* - 2(a_1 + a_2)]/3\} - d}$, we get the relation: composition (O/M) = $f(m^*, a_i) + (-\frac{1}{2})k$, where f is a function of m^* and a_i.) As mentioned above, the composition is variable even if m_i and a_i are

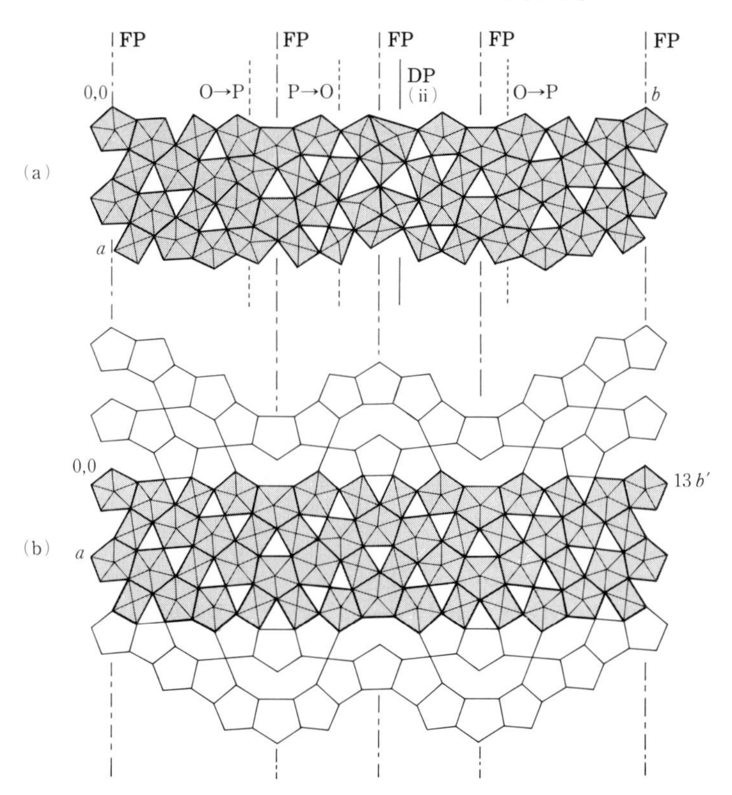

FIG. 2.108 (a) Real structure of $W_4Ta_{22}O_{67}$ ($4WO_3 \cdot 11Ta_2O_5$) ($m^* = 13 = 5 \times 1 + 8 \times 1$) with one distortion plane (see text for the notations O → P, P → O); (b) ideal structure with $m^* = 13 = 5 \times 1 + 8 \times 1$.

fixed. For the case $m^* = 13 = 5 \times 1 + 8 \times 1$ (the ideal structure is shown in Fig. 2.108(b)), for example, the composition change derived from the DP operation is shown in Table 2.7. Thus, a fixed m^* value provides a discrete set of compositions. In Fig. 2.109, the value of m^* is indicated on each line. Open circles on these lines corresponds to the composition derived from the DP operation to ideal structures, except for $k = 0$. Closed circles show the structures identified in the Ta_2O_5–WO_3 system. For observed structures the composition versus k relation seems to be on the heavy line. On this line L-Ta_2O_5 exists, which has a structure of $m^* = 11 = 11 \times 1$ with $d = 3$.

Figure 2.109 clearly shows that any change in composition (oxygen content) can be accommodated by passage to a nearby point in the continuum by a change in both m^* and k, and also that an infinite set of

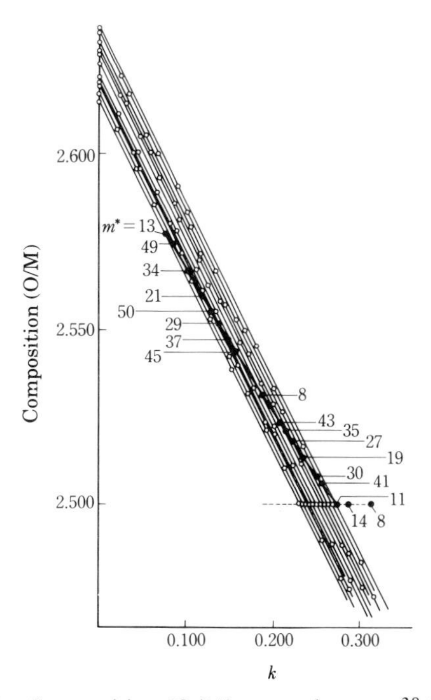

FIG. 2.109 Composition (O/M) versus k curves.[38] In this figure, k stands for the inverse of the mean distance of distortion planes. Each line corresponds to a fixed value of m^*. White circles on each line correspond to the compositions derived from DP operations. Closed circles are the compositions observed in this system. This figure clearly indicates the characteristics of the adaptive structure.

Table 2.7

Composition change by DP operation for fixed $m^* = 13$

d	Composition	O/M	k
0	$M_{26}O_{68}$	2.6153	0
1	$M_{26}O_{67}$	2.5769	0.0769
2	$M_{26}O_{66}$	2.5384	0.1538
3	$O_{26}O_{65}$	2.5000	0.2307
⋮	⋮	⋮	⋮

phases exists for a fixed composition. For discussion from a thermodynamic standpoint, the reader is referred to ref. 57.

Next we show another example of the adaptive structure, which is based on a one-dimensional shear structure. As discussed in Section 2.2.1, a homologous series of V_nO_{2n-1} (Magnéli phases, $n =$ positive integer) is derived from the mother structure rutile (TiO_2) by the shear operation $(121)\frac{1}{2}[0\bar{1}1]$. In this system, n can change in the range $3 < n < 9$, and the two-phase region can be observed in the intermediate composition of near neighbour phases. It was reported,[14] however, that a fragment with the composition of V_8O_{15} has the intergrowth structure of $2V_8O_{15}$ and V_7O_{13} (see Fig. 2.18). Hence there remains the possibility that the homologous compound V_nO_{2n-1} could be a set of sub-units for the adaptive structure of this system, depending on heat treatments.

On the other hand, two types of shear structure based on rutile appear in the Ti_2O_3–TiO_2,[62] Cr_2O_3–TiO_2,[63] and V_2O_3–TiO_2[64] systems. For the Ti_2O_3–TiO_2 (Ti_nO_{2n-1}) system, the shear operations of $(121)\frac{1}{2}[0\bar{1}1]$ and $(132)\frac{1}{2}[0\bar{1}1]$ are for $3 < n < 10$ and $16 < n < 36$, respectively. Between these compositions, i.e. between $n = 10$ and 16 ($TiO_{1.889}$ and $TiO_{1.937}$), the shear planes seem to pivot around in a continuous manner from (121) to (132), which is unambiguously indicated in the electron diffraction patterns. A similar phenomenon has been observed in V_2O_3–TiO_2,[64] as shown below.

Five samples with the nominal compositions shown in Table 2.8 were prepared by heating weighed mixtures of V_2O_3 and TiO_2 in evacuated silica tubes at 1200 °C for 1–10 days. Electron diffraction patterns were taken for these samples, and the patterns on the $[1\bar{1}1]$ zone could be classified as D_1, D_2, and D_3, as shown in Fig. 2.110. D_1 and D_3 patterns come from the structures derived by the shear operations of $(121)\frac{1}{2}[0\bar{1}1]$ and $(132)\frac{1}{2}[0\bar{1}1]$, respectively. D_2 patterns, on the other hand, show the arrays rotating between (121) and (132), which is very similar to the Ti_nO_{2n-1} system. Some of the examples with D_2 patterns on the $[1\bar{1}1]$ zone axis are shown in Fig. 2.111.

Table 2.8

Samples of the TiO_2–V_2O_3 system prepared in ref. 64

Samples	Nominal composition	Ratio of oxygen to metal
s-1	$Ti_8V_2O_{19}$	1.90
s-2	$Ti_{13}V_2O_{29}$	1.93
s-3	$Ti_{18}V_2O_{39}$	1.95
s-4	$Ti_{23}V_2O_{49}$	1.96
s-5	$Ti_{48}V_2O_{99}$	1.98

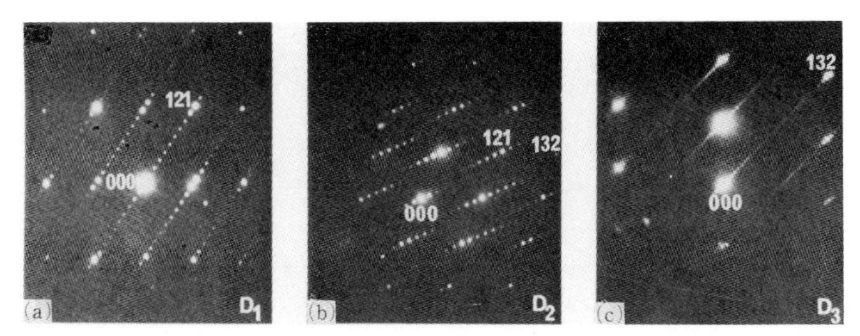

FIG. 2.110 Electron diffraction patterns (EDP) of the V_2O_3–TiO_2 system with $[1\bar{1}1]$ zone axis:[64] (a) D_1-type EDP: the structure for $(121)\frac{1}{2}[0\bar{1}1]$; (b) D_2-type EDP: see text; (c) D_3-type EDP: the structure for $(132)\frac{1}{2}[0\bar{1}1]$.

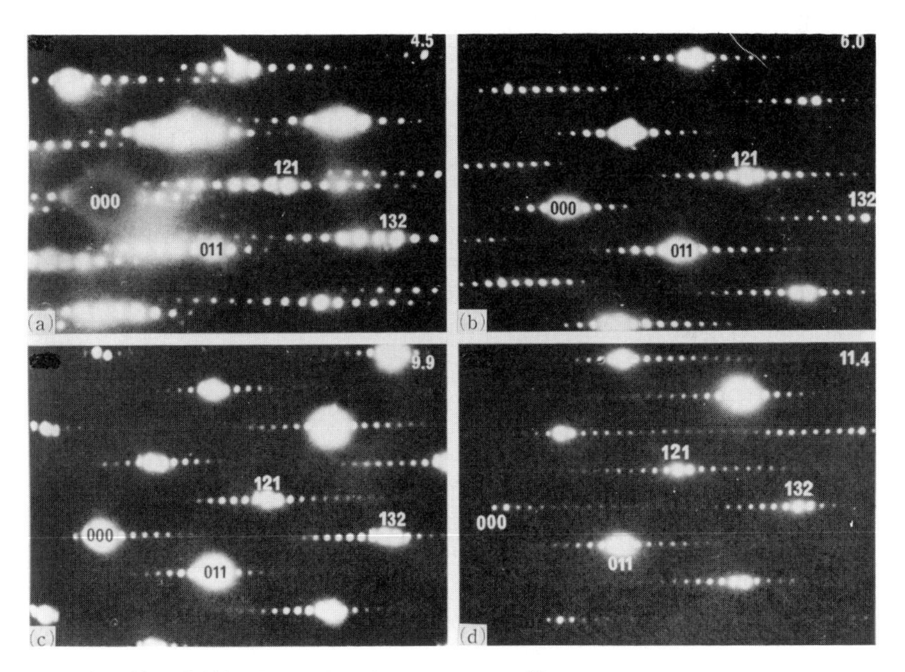

FIG. 2.111 Examples of D_2-type EDP.[64] Figures in the top right are the angles between **g**(121) and **g**(*hkl*).

A structure model was proposed for the compounds with D_2-type patterns by Bursill and Hyde.[65] Figure 2.112 shows a reciprocal lattice plane of rutile with a $[1\bar{1}1]$ zone axis. The arrays of superspots for D_1 and D_3 patterns are parallel to the vectors **g**(121) and **g**(132), respectively. Those for D_2 patterns

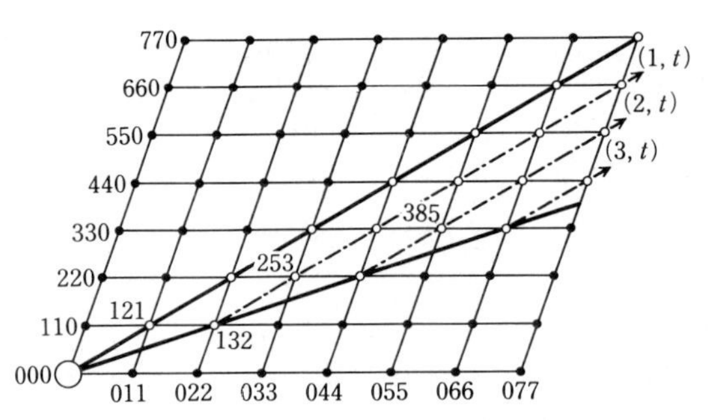

FIG. 2.112 Reciprocal lattice plane of rutile-type structure with $[1\bar{1}1]$ zone axis. The (s, t) series of shear planes are indicated by dashed lines (see text). White circles are the possible shear planes for this system.

pivot from **g**(121) to **g**(132), indicating that the shear planes have higher indices such as (2, 5, 3), (3, 8, 5), (4, 11, 7) ... $(p, 3p-1, 2p-1)$ (this is merely one series of examples; see below). Figure 2.113 shows some examples of the shear structures, $(hkl)\frac{1}{2}[0\bar{1}1]$, based on rutile, projected on $(100)_{\text{rutile}}$[64] (see also Figs 2.11 and 2.12). In this figure, the packing of anions is assumed to be HCP. Model structures with (121), (132), (253), and (385) shear planes are shown in Fig. 2.113(a_2–d). Figure 2.113(a_1) shows the model structure of (011)APB of rutile, having the composition M_6O_{12} (without the elimination of oxygen). In the plane the string of edge-sharing octahedra along $[001]_{\text{rutile}}$ show the steps, which we denote by oblique strokes, /. The structure of APB is expressed as

$$\ldots/////\ldots \tag{2.14}$$

The string of edge-sharing octahedra for the structure $(121)\frac{1}{2}[0\bar{1}1]$ (Fig. 2.113(a_2)), having the composition M_7O_{13} ($n = 7$ in M_nO_{2n-1}), shows a different type of step, which consists of an APB step and an extra cation. We denote this by Z, because the step is like a reverze Z shape. The structure is expressed as

$$\ldots \text{Z Z Z Z Z} \ldots \tag{2.15}$$

By use of the notation / and Z, the other structures shown in this figure can be denoted as

(132): ... / Z / Z / Z ...

(253): ... / Z Z / Z Z / Z Z ...

(385): ... / / Z Z Z / / Z Z Z / / Z Z Z ...

For example, structure (d) with the composition $M_{33}O_{63}$, $(385)\frac{1}{2}[0\bar{1}1]$, is

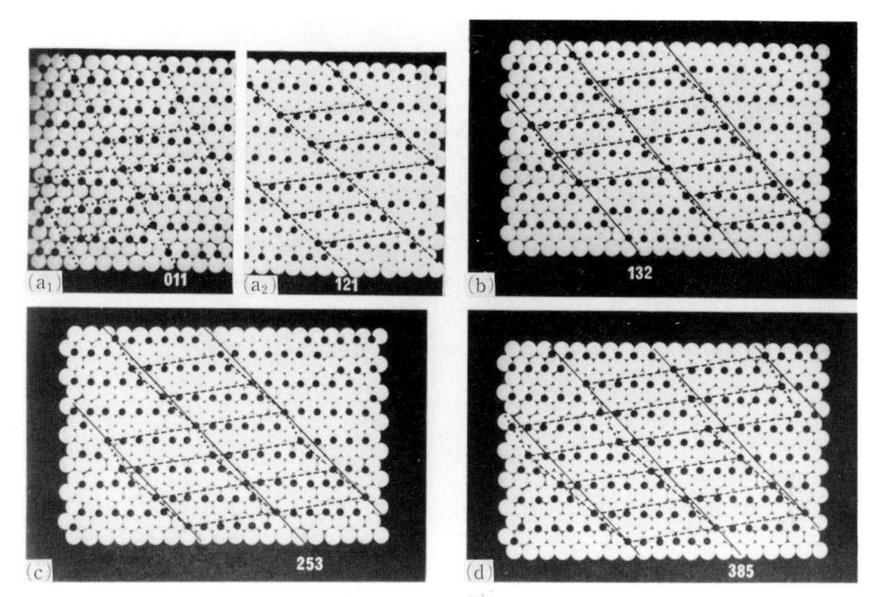

FIG. 2.113 Shear structure model based on rutile, by use of 'Go stone':[64] (a_1) (011)APB; (a_2) (121)SS; (b) (132)SS; (c) (253)SS; (d) (385)SS; (APB: anti-phase boundary, SS: shear structure, the shear vector $\frac{1}{2}[0\bar{1}1]$ is omitted in this notation). In (b), (c), and (d), the dotted lines show the decomposition of the structure into the sub-units, (a_1) and (a_2).

composed of $2(M_6O_{12})$ with the structure $(011)\frac{1}{2}[0\bar{1}1]$ (a_1) and $3(M_7O_{13})$ with the structure $(121)\frac{1}{2}[0\bar{1}1]$ (a_2). In other words, these have the regular intergrowth structure of $(011)\frac{1}{2}[0\bar{1}1]$ (a_1) and $(121)\frac{1}{2}[011]$ (a_2). In terms of the adaptive structure, the $(011)\frac{1}{2}[0\bar{1}1]$ (a_1) and $(121)\frac{1}{2}[0\bar{1}1]$ (a_2) structures are a set of sub-units.

In general, the shear structure $(hkl)\frac{1}{2}[0\bar{1}1]$ can be resolved into

$$(hkl)\tfrac{1}{2}[0\bar{1}1] = s(011)\tfrac{1}{2}[0\bar{1}1] + t(121)\tfrac{1}{2}[0\bar{1}1] \qquad (2.16)$$

where $h = t$, $k = s + 2t$, $l = s + t$. The stoichiometric composition of this structure is expressed as

$$M_nO_{2n-t} = sM_{n_1}O_{2n_1} + tM_{n_2}O_{2n_2-1} \qquad (2.17)$$

where $n = sn_1 + tn_2$. The shear plane spacing $D_{sp}(hkl)$ is calculated as

$$D_{sp}(hkl) = d_{hkl}(n - t/2) \qquad (2.18)$$

For example, the structure of $(385)\frac{1}{2}[0\bar{1}1]$ has to be $t = 3$ and $s = 2$ from eqn (2.16). Then, $D_{sp}(385) = d_{385}(n - \frac{3}{2})$, from which we can get the value of

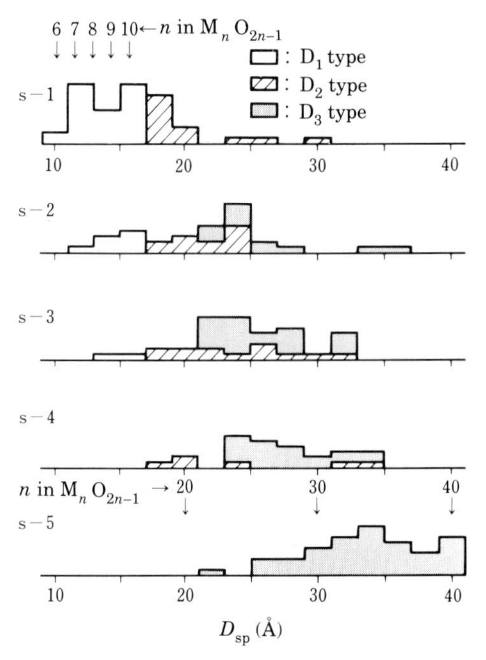

FIG. 2.114 Histogram of D_{sp} for the samples s-1, s-2, s-3, s-4, and s-5 (see Table 2.8).[64]

n, because the value of $D_{sp}(385)$ is experimentally obtained (d_{385} = known value).

As shown in Fig. 2.112, the shear plane is able to change from (121) to (132) continuously, depending on (*s*, *t*) values. These structures derived from the above-mentioned principle are regarded as the intergrowth structures of $(011)\frac{1}{2}[0\bar{1}1]$ and $(121)\frac{1}{2}[0\bar{1}1]$.

Figure 2.114 shows the histogram of $D_{sp}(hkl)$ for the samples of s-1, s-2, . . . , s-5, in addition to the type of diffraction pattern, D_1, D_2, and D_3. Generally the diffraction patterns show the successive change $D_1 \rightarrow D_2 \rightarrow D_3$ with increasing ratio of oxygen to metal. The D_2-type diffraction patterns were observed in s-1, s-2, s-3, and s-4 with the shear plane spacing range from 17 to 35 Å. In Fig. 2.115 is shown the relation between the shear plane spacing $D_{sp}(hkl)$ and the angle $\theta_{121}^{(hkl)}$ between **g**(121) and **g**(*hkl*). Thus, the V_2O_3–TiO_2 system as well as the Cr_2O_3–TiO_2 system is fully understood based on the concept of the adaptive structure. This concept can be applied to all non-stoichiometric compounds derived from extended defects; however, problems remain from the thermodynamical point of view.

The structures described in this chapter, especially on micro-twin structure, are referred to in *Inorganic crystal structures* by B. G. Hyde and S. Andersson.[66]

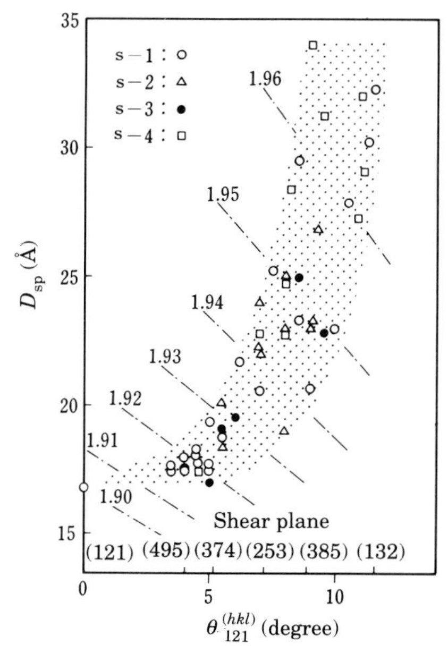

FIG. 2.115 Relation between D_{sp} and $\theta_{121}^{(hkl)}$ for the samples with D_2-type EDP.[64] Chain lines show the ratio of oxygen to metal.

References

1. G. Andersson, *Acta Chem. Scand.*, 1954, **8**, 1599.
2. S. Andersson and A. Magnéli, *Naturwiss.*, 1956, **43**, 495.
3. S. Andersson, B. Collen, U. Kuylenstierna, and A. Magnéli, *Acta Chem. Scand.*, 1957, **11**, 1641.
4. S. Andersson, A. Sundholm, and A. Magnéli, *Acta Chem. Scand.*, 1959, **13**, 989.
5. S. Andersson, *Acta Chem. Scand.*, 1960, **14**, 1161.
6. H. Horiuchi, M. Tokonami, N. Morimoto, K. Nagasawa, Y. Bando, and T. Takada, *Mat. Res. Bull.*, 1971, **6**, 833.
7. J. S. Anderson and B. G. Hyde, *J. Phys. Chem. Solids*, 1967, **28**, 1393.
8. S. Kachi and K. Kosuge, in *Nonstoichiometric-metallic compounds*, Japan Institute of Metals, Maruzen, 1975, Chapter 6 (in Japanese).
9. R. J. D. Tilley, *Chem. Scr.*, 1978–79, **14**, 147.
10. R. Pickering and R. J. D. Tilley, *J. Solid State Chem.*, 1976, **16**, 247.
11. S. Iijima, *J. Solid State Chem.*, 1974, **14**, 52.
12. K. Kosuge, H. Okinaka, S. Kachi, K. Nagasawa, Y. Bando, and T. Takada, *Jap. J. Appl. Phys.*, 1970, **9**, 1004.
13. J. Van Landuyt and S. Amelincks, *Mat. Res. Bull.*, 1970, **5**, 267.

14. Y. Hirotsu, Y. Tsunashima, S. Nagakura, H. Kuwamoto, and H. Sato, *J. Solid State Chem.*, 1982, **43**, 33.
15. R. S. Roth and A. D. Wadsley, *Acta Crystallogr.*, 1965, **19**, 42.
16. S. Iijima and J. G. Allpress, *J. Solid State Chem.*, 1973, **7**, 94.
17. S. Kimura, *J. Solid State Chem.*, 1973, **6**, 438.
18. S. Iijima, S. Kimura, and M. Goto, *Acta Crystallogr.*, 1973, **A29**, 632.
19. S. Horiuchi, *Chem. Scr.*, 1978–79, **14**, 75.
20. M. O'Keeffe, P. R. Buseck, and S. Iijima, *Nature*, 1987, **274**, 322.
21. A. D. Wadsley and S. Andersson, in *Perspectives in structural chemistry*, ed. J. D. Dunitz and J. A. Ibers, J. Wiley, 1970, Vol. 3, Chapter 1.
22. L. Eyring and L. Tai, in *Treatise on solid state chemistry*, ed. N. B. Hanny, Plenum Press, New York, London, 1976, Vol. 3, Chapter 3.
23. L. Eyring, in *Nonstoichiometric oxides*, ed. O. T. Sørensen, Academic Press, New York, 1981, Chapter 7.
24. J. S. Andersson, *Chem. Scr.*, 1978–79, **14**, 129.
25. J. S. Andersson, in *Intercalation chemistry*, ed. M. S. Whittingham and A. J. Jacobson, Academic Press, New York, 1982, Chapter 15.
26. D. J. M. Bevan and A. W. Mann, *Acta Crystallogr.*, 1975, **B31**, 1406.
27. B. G. Hyde, A. N. Bagshaw, S. Andersson, and M. O'Keeffe, *Ann. Rev. Sci.*, 1974, **4**, 43.
28. J. Galy and R. S. Roth, *J. Solid State Chem.*, 1973, **7**, 277.
29. W. Jung and R. Juza, *Z. Anorg. Allg. Chem.*, 1973, **399**, 129.
30. E. Makovicky and B. G. Hyde, *Structure and Bonding*, 1981, **46**, 101.
31. I. E. Grey, *Acta Crystallogr.*, 1975, **B31**, 45.
32. J. T. Hoggins and H. Steinfink, *Acta Crystallogr.*, 1977, **B33**, 673.
33. N. Nakayama, K. Kosuge, and S. Kachi, *J. Solid State Chem.*, 1981, **36**, 9.
34. N. Nakayama, K. Kosuge, and S. Kachi, *Chem. Scr.*, 1982, **20**, 174.
35. R. Ridder, G. Tendeloo, and S. Amelincks, *Phys. Stat. Sol.*, 1976, **A33**, 383.
36. A. C. Hollady and L. Eyring, *J. Solid State Chem.*, 1986, **64**, 113.
37. N. Nakayama, K. Kosuge, and S. Kachi, unpublished work.
38. J. S. Anderson, *J. Chem. Soc., Dalton Trans.*, 1973, 1107.
39. S. Andersson and B. G. Hyde, *J. Solid State Chem.*, 1974, **9**, 92.
40. B. G. Hyde, S. Andersson, M. Bakker, C. M. Plug, and M. O'Keefe, *Prog. Solid State Chem.*, 1979, **12**, 273.
41. H. H. Otto and H. Strunz, *N. Jb. Miner. Abh.*, 1968, **108**, 1.
42. T. J. White and B. G. Hyde, *Phys. Chem. Minerals*, 1982, **8**, 55.
43. T. J. White and B. G. Hyde, *Phys. Chem. Minerals*, 1982, **8**, 167.
44. T. J. White and B. G. Hyde, *Am. Mineral.*, 1983, **68**, 1009.
45. A. F. Wells, *Structural inorganic chemistry*, Oxford University Press, Oxford, 1984.
46. J. A. Kohn, D. W. Eckart, and C. F. Cook, Jr, *Science*, 1971, **172**, 519.
47. J. van Landuyt, S. Amelinckx, J. A. Kohn, and D. W. Eckart, *J. Solid State Chem.*, 1974, **9**, 103.
48. S. Kachi and K. Kosuge, in *Nonstoichiometric-metallic compounds*, Japan Institute of Metals, Maruzen, 1975, Chapter 2 (in Japanese).
49. A. Hussain and L. Kihlborg, *Acta Crystallogr.*, 1976, **A32**, 551.
50. A. Hussain, *Acta Chem. Scand.*, 1978, **A32**, 479.
51. A. Hussain, *Chem. Commun. Univ. Stockholm*, 1978, 1.

52. L. Kihlborg, *Chem. Scr.*, 1979, **14**, 187.
53. T. Ekstrom, M. Parmentier, and R. J. D. Tilley, *J. Solid State Chem.*, 1980, **34**, 397.
54. A. Deschanvres, M. Frey, B. Raveau, and J. Thomazeau, *Bull. Soc. Chim. Fr.*, 1968, **13**, 3519.
55. B. Darriet and J. Galy, *C. R. Acad. Sci. Paris*, 1971, **273**, 1173.
56. L. Kihlborg and R. Sharma, *J. Microsc. Spectrosc. Electron.*, 1982, **7**, 387.
57. J. S. Anderson, *J. Chem. Soc., Dalton Trans.*, 1973, 1107.
58. R. S. Roth and N. C. Stephenson, in *The chemistry of extended defects in non-metallic solids*, ed. L. Eyring and M. O'Keeffe, North-Holland, Amsterdam, London, 1970, p. 167.
59. R. S. Roth, J. L. Waring, and H. S. Parker, *J. Solid State Chem.*, 1970, **2**, 445.
60. N. C. Stephenson and R. S. Roth, *Acta Crystallogr.*, 1971, **B27**, 1010, 1018, 1031, 1037.
61. R. Papiernik, B. Gaudreau, and B. Frit, *J. Solid State Chem.*, 1978, **25**, 143.
62. L. A. Bursill, B. G. Hyde, S. Terasaki, and D. Watanabe, *Philos. Mag.*, 1969, **20**, 347.
63. L. A. Bursill, B. G. Hyde, and D. K. Phillip, *Philos. Mag.*, 1971, **23**, 1501.
64. K. Kosuge and S. Kachi, *Chem. Scr.*, 1975, **8**, 70.
65. L. A. Bursill and B. G. Hyde, *Prog. Solid State Chem.*, 1972, **7**, 177.
66. B. G. Hyde and S. Andersson, *Inorganic crystal structures*, J. Wiley and Sons, New York, 1989.

3

EXAMPLES OF THE PRACTICAL USE OF NON-STOICHIOMETRIC COMPOUNDS

3.1 Introduction

In this chapter, we describe four kinds of non-stoichiometric compound, which are or will be in practical use, from the viewpoint of preparation methods or utility. As a first example, the solid electrolyte $(ZrO_2)_{0.85}(CaO)_{0.15}$ is described, which are discussed in Sections 1.4.6–1.4.8 from the viewpoint of basic characteristics. The second example is the magnetic material Mn–Zn ferrite, for which the control of non-stoichiometry and the manufacturing process will be described. Then the metal hydrides or hydrogen absorbing alloys, which are one of the most promising materials for storing and transporting hydrogen in the solid state, are described, mainly focusing on the phase relation. Finally, we describe the relation between the control of composition and the growth of a single crystal of the semiconductive compound GaAs, which is expected to give electronic materials for IC and LSI etc.

3.2 Ionic conducting materials—use of $(ZrO_2)_{0.85}(CaO)_{0.15}$

Solid electrolytes, which show ionic conductivity in the solid state, are considered to be potential materials for practical use, some are already used as mentioned below. Solid electrolytes have characteristic functions, such as electromotive force, ion selective transmission, and ion omnipresence. Here we describe the practical use of calcia stabilized zirconia (CSZ), $(ZrO_2)_{0.85}(CaO)_{0.15}$, the structure and basic properties of which are discussed in detail in Sections 1.4.5–1.4.8.

The most simple practical application of CSZ is for the gauge of oxygen partial pressure, as mentioned in Sections 1.4.7 and 1.4.8. The oxygen partial pressure $P^2_{O_2}$ in the closed system as shown in Fig. 3.1 can be measured, taking the air as the standard oxygen pressure $P^1_{O_2}$. The electromotive force (EMF) of this concentration cell is expressed as

$$E = (RT/4F)\ln(P^1_{O_2}/P^2_{O_2}) \tag{3.1}$$

This principle is applied in the measurement of oxygen partial pressure in

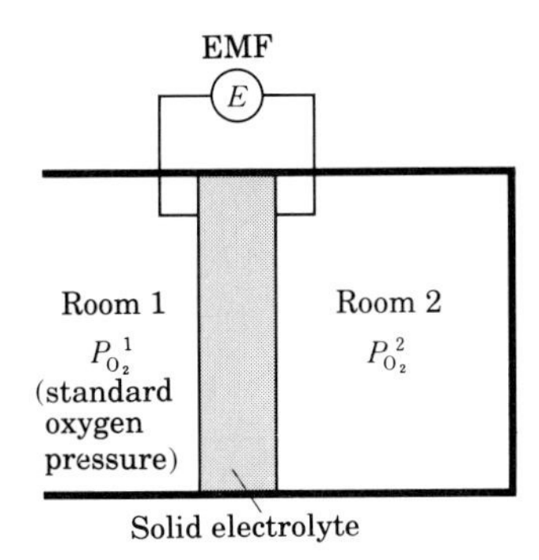

FIG. 3.1 Cell for measurement of oxygen partial pressure, $P^2_{O_2}$, in a closed system. $P^1_{O_2}$ is a standard oxygen pressure (air).

laboratory experiments and of the oxygen activity of slag in refineries. Based on the principle of coulometric titration (see Section 1.4.8), the oxygen partial pressure of a closed system can be kept constant by feedback of the EMF, in the oxygen pressure range 1 to 10^{-7} atm. By use of this closed system, investigations on redox reactions of metals and also enzyme reactions have been carried out. The CSZ can also be used as a separator of oxygen from mixed gases and as a purifier of oxygen.

The most popular use of CSZ is as a gas sensor (λ sensor) for automotive exhaust, the structure of which is shown in Fig. 3.2.[1] The sensor with the standard electrode (air) is inserted in exhaust pipes. The relation between the ratio of air to fuel (RAF) and the EMF is depicted in Fig. 3.3. The EMF shows a sharp drop at RAF = 15 (equivalent composition), left of which is called the rich burn region and right of which is called the lean burn region. Because the 'bad' gases, such as CO and NO_x, in exhausted gas are minimized at the point of equivalent composition, the value of RAF is controlled by the feedback of EMF.

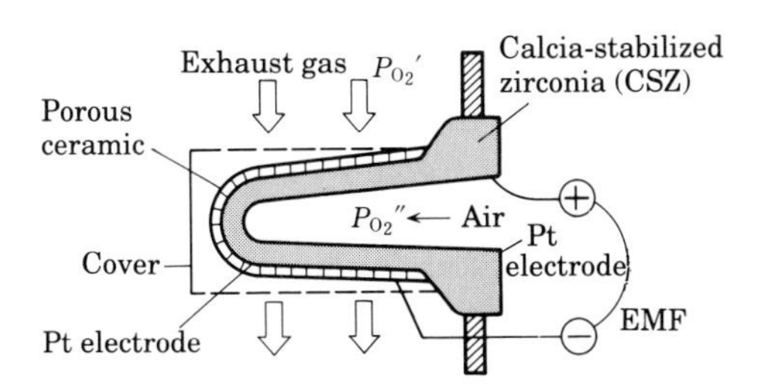

FIG. 3.2 Structure of λ sensor for automotive exhaust.[1]

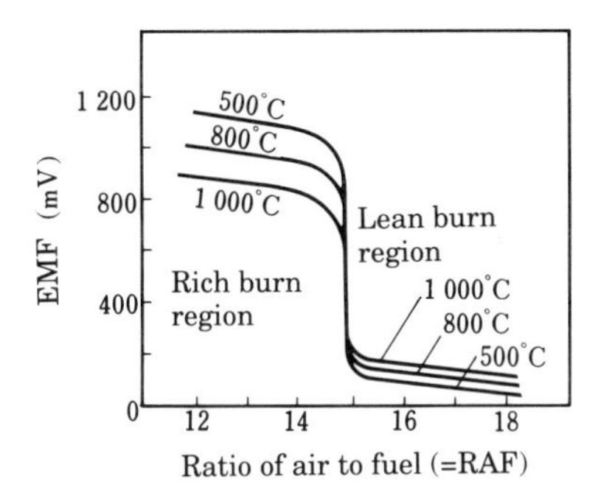

FIG. 3.3 EMF versus the ratio of air to fuel (RAF) curves at various temperatures.[1] RAF = 15 is called the equivalent composition.

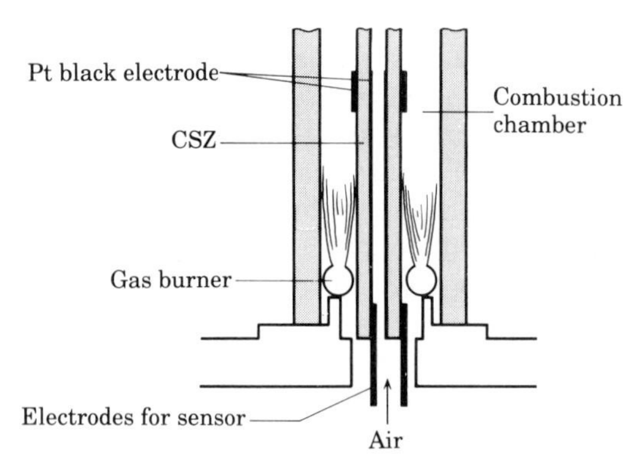

FIG. 3.4 Warning sensor for imperfect combustion in gas stoves.[2]

A similar application is the warning sensor for imperfect combustion in gas stoves.[2] As shown in Fig. 3.4, a CSZ tube goes through the centre part of the stove. Fresh air always flows through the inner part of the tube, which is the standard oxygen pressure, and there is a burning chamber around the tube. Imperfect combustion due to oxygen deficiency, which generates gases such as CO and NO_x, can be detected by measurement of the EMF.

Another practical use of a CSZ, which is of interest at the present time, is as a fuel cell, which is schematically illustrated in Fig. 3.5. Gases such as H_2, CO, and hydrocarbons are used as fuel. The theoretical value of the EMF for the following cell reaction is about 1.1 V at 1000 °C,

$$H_2 + \tfrac{1}{2}O_2 = H_2O$$

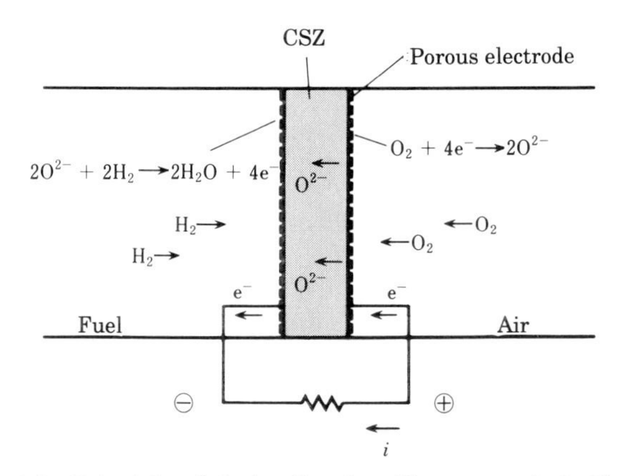

FIG. 3.5 Principle of fuel cell using H_2 gas as fuel. The cell reaction is: $H_2 + \frac{1}{2}O_2 \rightarrow H_2O$.

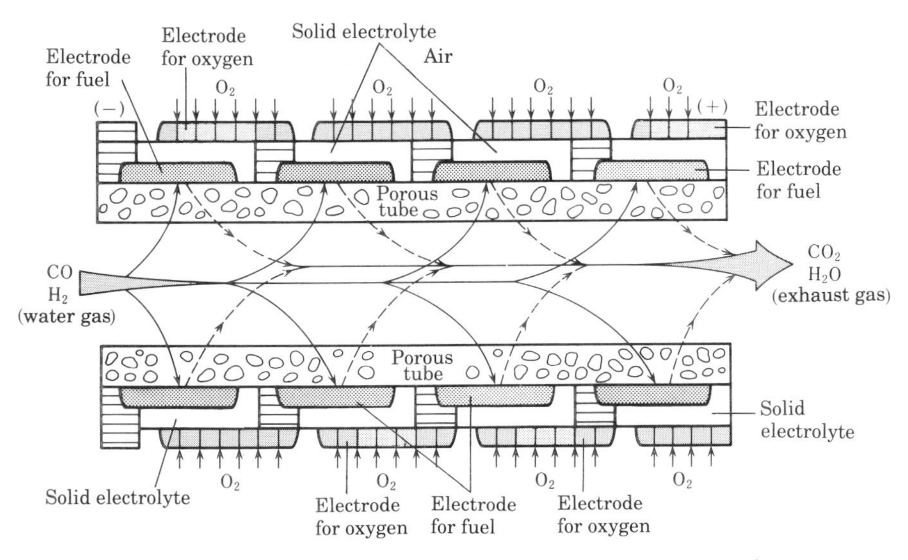

FIG. 3.6 Fuel cell for water gas ($CO + H_2$) utilizing CSZ.[2]

The characteristics of this type of fuel cell are as follows:

1. It is unnecessary to use expensive materials such as Pt as electrode catalysts.
2. This technology is corrosion-free and also pollution-free.

Figure 3.6 shows an example of a fuel cell using water gas ($H_2 + CO$),

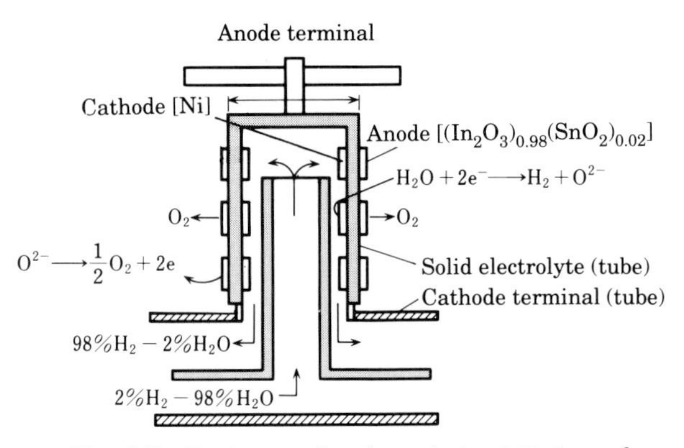

FIG. 3.7 Equipment for electrolysis of H_2O gas.[2]

which was devised by Westing House Co., America. The fuel electrode (porous anode) is a porous ceramic tube on which Ni and ZrO_2 are coated and the oxygen electrode (porous cathode) is made from $Sr_xLa_{1-x}MnO_3$ or Sn-doped In_2O_3 (electron conducting oxides). A thin film of CSZ is sandwiched between these electrodes. The following chemical reaction takes place in the cell,

$$H_2 + CO + O_2 \rightarrow CO_2 + H_2O$$

As is mentioned in Section 3.4, the use of hydrogen gas as a source of clean energy is gaining attention. Based on the ion selectivity of CSZ, hydrogen gas can be produced by electrolysis of H_2O gas. In principle, the reverse reaction of the fuel cell shown in Fig. 3.5, can be used for this purpose, that is, the electrochemical extraction of hydrogen from H_2O gas by an external current. In Fig. 3.7 the equipment for the electrolysis of H_2O gas, devised by General Electric Co., is shown.

The fuel cell and the electrolysis of H_2O vapour using CSZ, mentioned here, are very promising from the viewpoint of economy and overcoming pollution problems.

3.3 Magnetic materials—control of oxygen composition and physical properties of Mn–Zn ferrites

Magnetic materials can be classified into two groups: metal and alloy materials, and metal oxide materials represented by ferrites. We here describe the Mn–Zn ferrites, which are used, for example, as the magnetic heads in magnetic recording, focusing on the preparative processes and physical properties, which are closely related to the oxygen non-stoichiometry.

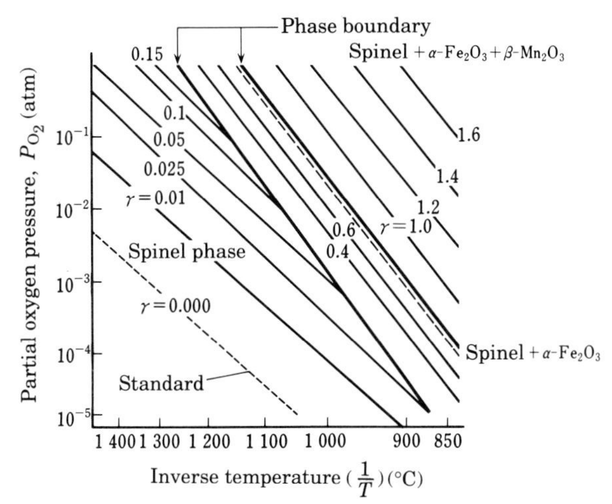

FIG. 3.8 Relation between temperature, oxygen partial pressure, and oxygen non-stoichiometry, γ, for the starting composition $(ZnO)_{18.3}(MnO)_{26.8}(Fe_2O_3)_{54.9}$.[4] γ denotes the weight gain from the standard composition.

First, we discuss the phase relation from the viewpoint of the phase rule. Because Mn–Zn ferrite is composed of four elements (Mn, Zn, Fe, O), the freedom F in the phase rule is expressed as $F = 6 - p$.* Considering one solid phase with a spinel structure (see Section 2.5 and Ref. 3), equilibrated with the oxygen (partial) pressure P_{O_2}, we get the relation $F = 4$. The starting compounds for the preparation of Mn–Zn ferrite are ZnO, MnO, and Fe_2O_3, in the molar ratio $\alpha:\beta:3 - (\alpha + \beta)$. This mixture is heated at fixed temperature and partial oxygen pressure, and the product obtained has the composition $Zn_\alpha Mn_\beta Fe_{3-(\alpha+\beta)}O_{4+\delta}$. Under conditions of fixed temperature, fixed oxygen partial pressure and fixed mole ratio of the starting compounds, we have $F = 0$, that is, the oxygen non-stoichiometry is automatically fixed. Thus we can regard this phase relation in the spinel phase as a pseudo-binary system (alloy $Zn_\alpha Mn_\beta Fe_{3-(\alpha+\beta)}$–O system). For the Mn–Zn ferrite system, the phase diagram has been investigated in limited composition regions, i.e. $10 < ZnO < 30$, $10 < MnO < 50$, $40 < Fe_2O_3 < 70$ (mole per cent).

As an example, Fig. 3.8 shows the relation between oxygen partial pressure,

* If the composition of oxygen in Mn–Zn ferrite is controlled by flowing mixed gases such as $(Ar + O_2)$ and $(CO_2 + H_2)$, i.e. open system, the phase rule $F = c + 2 - p$ has to be changed as $F = c + 3 - p$. Therefore, we get $F = 7 - p$ for the Mn–Zn ferrite system. In the one solid phase (spinel) plus oxygen gas region, we have $F = 5$. Then temperature T, total pressure P, oxygen partial pressure P_{O_2} and the composition α, β in the expression of $Zn_\alpha Mn_\beta Fe_{3-(\alpha+\beta)}O_{4+\delta}$ are fixed, we get $F = 0$. In the text, however, the description is for the closed system, similar to that mentioned in Section 1.2.2.

temperature, and oxygen non-stoichiometry for the starting composition $(ZnO)_{18.3}(MnO)_{26.8}(Fe_2O_3)_{54.9}$.[4] In this figure, γ denotes the weight gain ($\Delta W/W$) from the standard composition ($\gamma = 0$). The value of γ is proportional to δ in the expression $Zn_\alpha Mn_\beta Fe_{3-(\alpha+\beta)}O_{4+\delta}$. It is to be noted that in this figure only the spinel region can be treated as a pseudo-binary system. In this region, we can see the following relation between $\log P_{O_2}$ and $1/T$ for fixed γ,

$$\log P_{O_2} = -(A/T) + B \tag{3.2}$$

where constant A is independent of oxygen non-stoichiometry, whereas constant B depends on δ. This enables us to control the oxygen non-stoichiometry in the spinel phase.

In order to use Mn–Zn ferrites as magnetic head materials, the materials have to show the following characteristics:

(1) high value of magnetic permeability;

(2) high performance of frequency and temperature dependence of the magnetic permeability;

(3) high quality of strength, workability, and anti-wear.

Many papers on the preparation of this material, aiming at improvement of magnetic and mechanical properties, have been published.[4–9] The following results were obtained mainly by Tanaka.[10]

Figure 3.9 shows the lattice parameter versus oxygen partial pressure

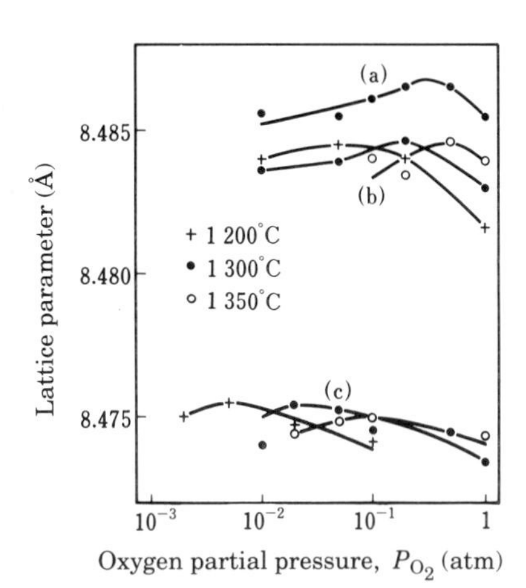

FIG. 3.9 Lattice parameter versus P_{O_2} curves for the samples (a), (b), and (c) annealed at 1200, 1300, and 1350 °C.[10] (a) $(MnO)_{0.298}(ZnO)_{0.189}(Fe_2O_3)_{0.522}$; (b) $(MnO)_{0.278}(ZnO)_{0.199}(Fe_2O_3)_{0.523}$; (c) $(MnO)_{0.258}(ZnO)_{0.173}(Fe_2O_3)_{0.569}$.

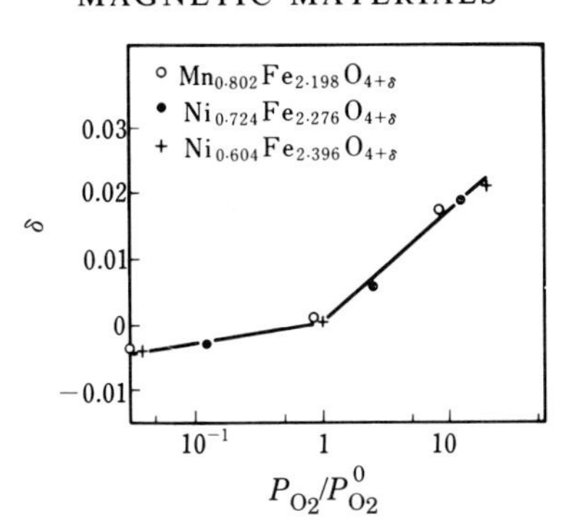

FIG. 3.10 Non-stoichiometry, δ, versus $P_{O_2}/P^0_{O_2}$ for Mn- and Ni-ferrites. $P^0_{O_2}$ is the oxygen partial pressure where the lattice parameter versus P_{O_2} exhibits a maximum.[10]

curves for samples having the starting compositions (a) $(MnO)_{0.298}$ $(ZnO)_{0.189}$ $(Fe_2O_3)_{0.522}$, (b) $(MnO)_{0.278}(ZnO)_{0.199}(Fe_2O_3)_{0.523}$, (c) $(MnO)_{0.258}(ZnO)_{0.173}$ $(Fe_2O_3)_{0.569}$, heated at 1200, 1300, and 1350 °C in the oxygen partial pressure range of 10^{-3} to 1 atm.[10] Under the condition of fixed temperature, the lattice parameter shows a maximum value at an oxygen partial pressure, dependent on the starting composition and also the annealing temperature. Similar results were found for other ferrites such as Mn- and Ni-ferrite. It is believed that in these ferrites the cation vacancies are dominent for $\delta > 0$ and the anion vacancies are dominant for $\delta < 0$, and that the lattice parameter exhibits a maximum value at $\delta = 0$. Figure 3.10 shows the non-stoichiometry δ versus oxygen partial pressure for these ferrites, where $P^0_{O_2}$ denotes the value of P_{O_2} at $\delta = 0$.[10] This figure suggests that all the experimental values of δ are scaled by the value of $(P_{O_2}/P^0_{O_2})$, that is, $\delta = A \log(P_{O_2}/P^0_{O_2})$, where $A = 0.003$ for $(P_{O_2}/P^0_{O_2}) < 1$ and 0.017 for $(P_{O_2}/P^0_{O_2}) > 1$. It was postulated that this relation is valid for the case of Mn–Zn ferrite.[10] The estimated value of δ, based on this assumption, for the samples $(MnO)_{0.30-x}(ZnO)_{0.20}(Fe_2O_3)_{0.50+x}$ ($x = 0.01, 0.02, \ldots, 0.05$) ranges from -0.005 to 0.02. The temperature dependence of $P^0_{O_2}$ for each sample is shown in Fig. 3.11.[10]

For the practical use of ferrite it is necessary to get high-density material. In the firing process, the starting compound mixture has to be 'ferritized' and 'densified' by solid–solid and solid–gas reactions. In this process, the value of δ and the grain size also have to be controlled. For this purpose, a heat-treatment, which was called 'atmosphere controlled two step firing' by Tanaka, was devised, as shown schematically in Fig. 3.12 and in detail in Table 3.1. For reference, the usual heat-treatment for preparation of ferrites is shown in Fig. 3.12(b). The new heat treatment is composed of the following six processes. In process (1), the starting mixed compounds are heated to

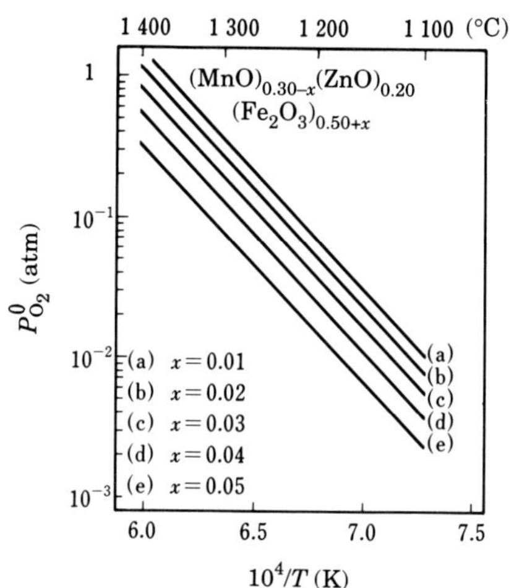

FIG. 3.11 $P^0_{O_2}$ versus $1/T$ curves for $(MnO)_{0.30-x}(ZnO)_{0.20}(Fe_2O_3)_{0.50+x}$.[10]

Table 3.1
Process of atmosphere controlled two step firing[10]

Process	Heating or cooling rate	Atmosphere	Remarks
1	300 °C h^{-1} $\xrightarrow{\text{heating}}$ 1000 °C	N_2	Ferritization (ρ = 70 per cent)
2	50 °C h^{-1} $\xrightarrow{\text{heating}}$ 1100 °C annealing (1 h)	Controlled P_{O_2}	Control of oxygen composition Densification (ρ = 99 per cent)
3	300 °C h^{-1} $\xrightarrow{\text{heating}}$ 1250 °C	Controlled P_{O_2}	Densification Control of oxygen composition
4	1300–1400 °C annealing (2 h)	Controlled P_{O_2}	Grain growth (ρ = 99.8 per cent)
5	300 °C h^{-1} $\xrightarrow{\text{cooling}}$ 1250 °C	Controlled P_{O_2}	Densification
6	300 °C h^{-1} $\xrightarrow{\text{cooling}}$ R.T.	N_2	

1000 °C at a heating rate of 300 °C h^{-1} in flowing nitrogen gas ferritization. In process (2), the product is heated to 1100 °C at a rate of 50 °C h^{-1}, in which the oxygen partial pressure P_{O_2} is controlled in order to obtain the required oxygen content. It is to be noted that the P_{O_2} has to be varied with temperature to get the correct composition of oxygen (see Fig. 3.8).

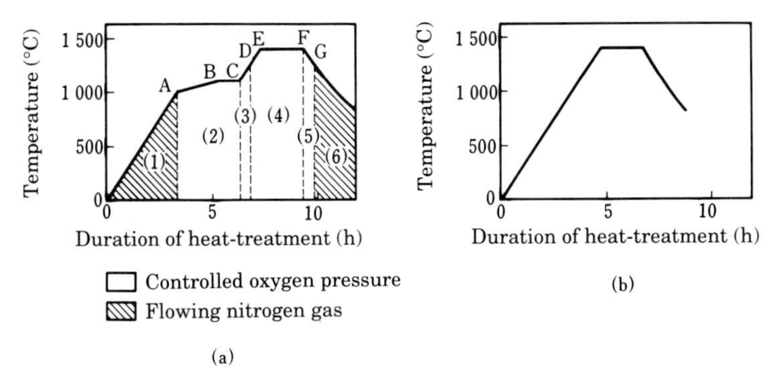

FIG. 3.12 Manufacturing processes for the ferrites:[10] (a) atmosphere controlled two step firing; (b) usual heat-treatment (in N_2 gas). In the hatched regions (1) and (6) in (a), the samples are heated or cooled under flowing N_2 gas. In the other regions, P_{O_2} has to be controlled to obtain the required oxygen content.

At 1100 °C, the sample is annealed for about 1 h for homogenization. In process (3), the product is further heated to 1250 °C at a rate of 300 °C h^{-1} under controlled P_{O_2} for densification. In process (4), the product is heated to 1400 °C and annealed for 2 h under controlled P_{O_2} for promoting grain growth. In this step, the density of the product usually reaches up to 99.8 per cent of the theoretical value. Finally, the sample is cooled under controlled P_{O_2} (process (5)) and then in flowing nitrogen gas (process (6)) to room temperature.

The thus obtained high-density Mn–Zn ferrite was investigated in detail from the view of physical and mechanical properties, that is, the relationships between the composition of metals (α, β) and δ; the magnetic properties such as temperature and frequency dependence of initial permeability, magnetic hysteresis loss and disaccommodation; and the mechanical properties such as modulus of elasticity, hardness, strength, and workability. Figures 3.13(a) and (b) show the optical micrographs of the samples prepared by the processes depicted in Fig. 3.12(a) and (b), respectively. The density of the sample shown in Fig. 3.13(a) reached up to 99.8 per cent of the theoretical value, whereas the sample shown in Fig. 3.13(b) which was prepared without a densification process, has many voids.

Figure 3.14 shows the frequency dependence of initial permeability for various compositions of $(MnO)_{0.26}(ZnO)_{0.20}(Fe_2O_3)_{0.54}$ ((a) $\delta = -0.002$, (b) $\delta = 0$, (c) $\delta = 0.01$) at room temperature.[10] The initial permeability for each sample is constant below 500 kHz and that for sample $\delta = 0$ is highest.

Figure 3.15 shows the relation between δ and Vickers hardness for the

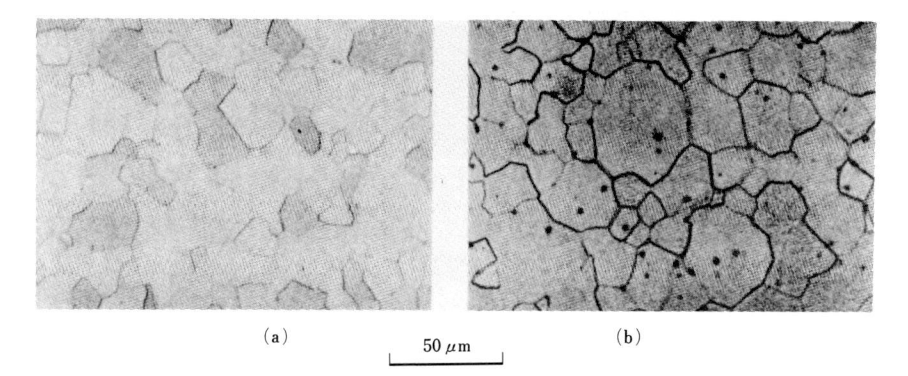

FIG. 3.13 Typical optical micrographs of Mn–Zn ferrite:[10] (a) micrograph of sample prepared by process (a) of Fig. 3.12; (b) micrograph of sample prepared by process (b) of Fig. 3.12.

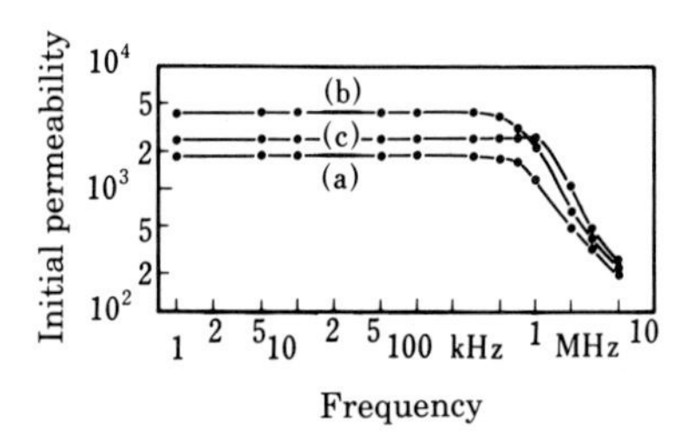

FIG. 3.14 Frequency dependence of initial permeability of $(MnO)_{0.26}(ZnO)_{0.20}(Fe_2O_3)_{0.54}$ with $\delta = -0.002$ (a), 0 (b), and 0.01 (c).[10]

sample $(MnO)_{0.26}(ZnO)_{0.20}(Fe_2O_3)_{0.54}$;[10] with increasing δ, the hardness increases gradually.

Previously, polycrystalline Mn–Zn ferrite had been prepared by solid state reaction *in vacuo* or by hot-press methods. Using these methods densification and the size of grains could be controlled, but control of the δ value was a problem. Tanaka succeeded in preparing high performance Mn–Zn ferrites by controlling P_{O_2}.

The results mentioned above are just some of the results obtained by Tanaka, for a complete account the reader is referred to Refs 10–16.

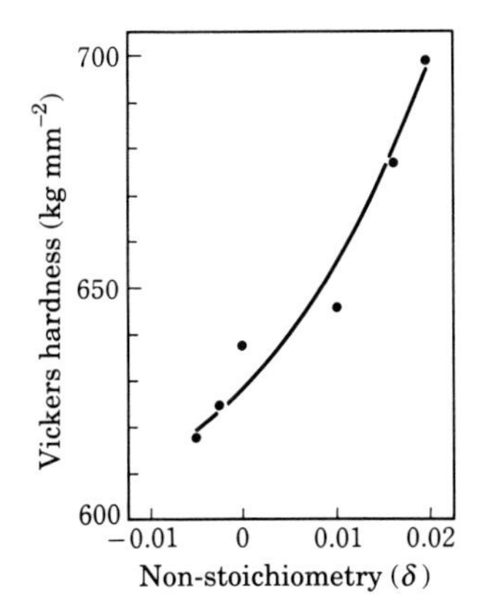

FIG. 3.15 Relation between Vickers hardness and non-stoichiometry, δ, for the starting composition $(MnO)_{0.26}(ZnO)_{0.20}(Fe_2O_3)_{0.54}$.[10]

3.4 Hydrogen absorbing alloys—the phase relation of the $LaNi_5$–H_2 system

It is well known that most metals and alloys absorb and desorb hydrogen gas reversibly. This phenomenon is typically shown by phase diagrams, for example, Fig. 3.16 shows the phase diagram of the Pd–H system determined by Bruning and Sieverts in 1933.[17]* In the Pd–H system, there exist two phases α and β in the composition range $0 < H/Pd < 0.5$. In the one-phase solid regions, i.e. α and β, the pressure of the equilibrium hydrogen gas P_{H_2} increases with increasing x in PdH_x, and in the two-phase region, i.e. $(\alpha + \beta)$, P_{H_2} is constant at fixed temperature.

Hydrogen gas H_2 is promising as a source of clean energy, although at present there are many problems in its production. Use of the usual bottle or cylinder for storage and transport of hydrogen gas is not satisfactory, and so research, since the beginning of the 1970s, has been carried out not only of the use of hydrogen as a source of clean energy, but also on the use of metals and alloys as storage and transport media for hydrogen.

The metal–H or alloy–H system can be regarded as a binary or pseudo-binary system, i.e. $c = 2$ in the phase rule. Figure 3.17 shows schematically the relation between P_{H_2} and solid phase composition H/M for various values of T, assuming two solid phases α and β. These phases correspond to the

* Lacher[18] and Anderson[19] discussed this phase diagram from the viewpoint of statistical thermodynamics based on the Fowler–Guggenheim treatment. On the other hand, Libowitz[20] and Simon and Flanagan[21] also calculated this phase diagram based on Libowitz's method mentioned in Section 1.3.7. The result obtained by the latter is the same as eqn (1.124).

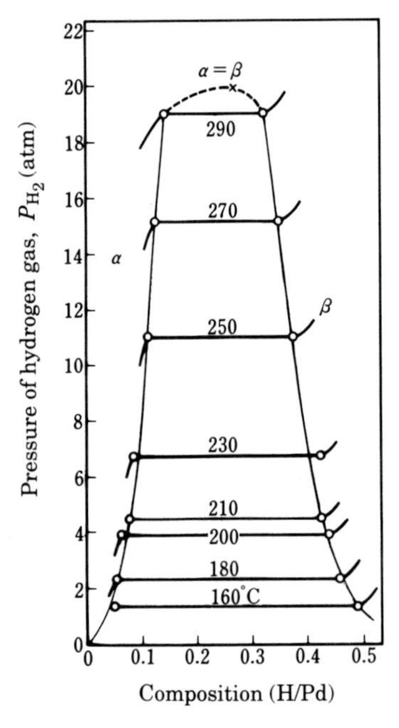

FIG. 3.16 Phase diagram of the Pd–H system as a function of hydrogen pressure P_{H_2} and composition (H/Pd) at various temperatures.[17]

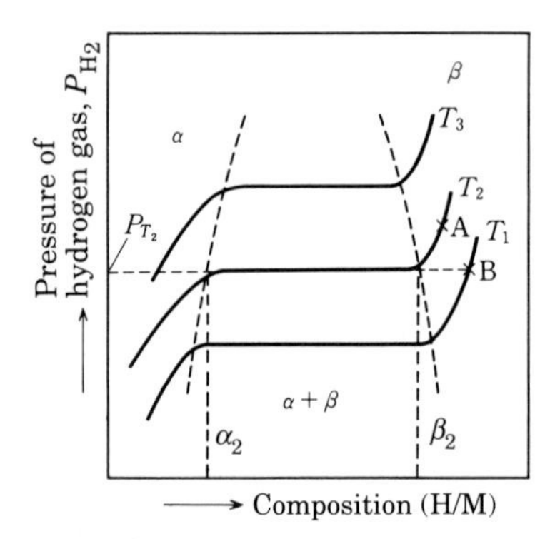

FIG. 3.17 Schematic drawing of the relation between P_{H_2} and solid phase composition H/M for various values of T.

non-stoichiometric compounds. At a temperature of T_1, P_{H_2} increases with increasing H/M in the α phase region and then P_{H_2} becomes constant, that is, it is impossible to enhance P_{H_2}, in the $(\alpha + \beta)$ region, and P_{H_2} again increases with increasing H/M in the β phase region. At temperatures of T_2 and T_3, similar curves are obtained as shown in Fig. 3.17. In the equilibrium

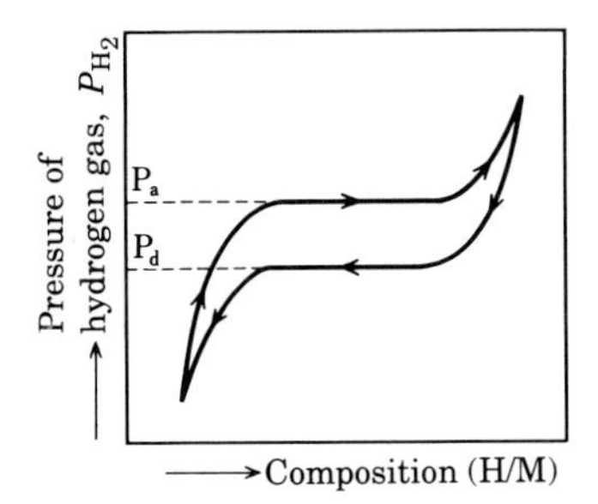

FIG. 3.18 Schematic drawing of the hysteresis of the absorption–desorption cycle at constant temperature.

state, there must be no hysteresis in the absorption–desorption curves. For example, let us start from point A at T_2. With decrease of P_{H_2} down to P_{T_2} (the plateau pressure of hydrogen gas at temperature T_2), the desorption of hydrogen gas with constant pressure P_{T_2} should continue from the composition β_2 to α_2 of H/M. In practice, however, hysteresis is usually observed, which may originate from the difference of reaction rate between absorption and desorption.

Figure 3.18 shows a schematic representation of the hysteresis of the absorption–desorption curve at constant temperature. In the plateau region, absorption occurs if $P_{H_2} > P_a$ and desorption occurs if $P_{H_2} < P_d$. For practical use, we have two types of operation: pressure of hydrogen gas at constant temperature (mentioned above) and at varying temperature. The latter type of operation, in which absorption–desorption is controlled by temperature, is of great advantage in accelerating the desorption reaction and also in controlling the desorption pressure of hydrogen gas.

For practical use hydrogen absorbing materials must have the following characteristics:

1. The quantity of absorbed hydrogen per unit weight of material is as high as possible.
2. The absorption–desorption cycle works below 100 °C.
3. The absorption–desorption cycle works below a hydrogen pressure of 10 atm.
4. The materials are inexpensive.
5. The heat of reaction for absorption–desorption is reasonable.
6. The reaction rate of absorption–desorption is reasonable.
7. The materials are stable under working conditions.

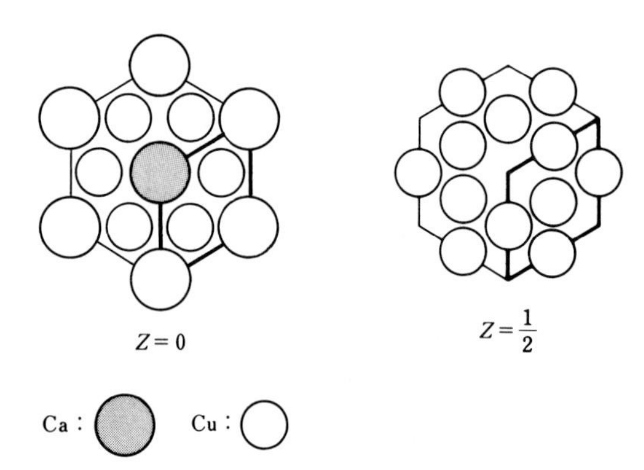

FIG. 3.19 Crystal structure of $CaCu_5$. Planes $Z = 0$ and $Z = \frac{1}{2}$ are alternately stacked along the c-axis.

Typical examples of hydrogen absorbing materials for practical use are the transition metal alloys such as the Fe–Ti, V–Nb, and Mg–Ni–Cu systems with a $CaCu_5$-type structure. Here we shall concentrate on the compounds $CaNi_5$ and $LaNi_5$ with a $CaCu_5$-type structure. The crystal structure of $CaCu_5$ is shown in Fig. 3.19. This structure has hexagonal symmetry with space group $P6/mmm$, and consists of alternate (Ca + Cu) planes ($z = 0$) and Cu-only planes ($z = \frac{1}{2}$) stacked along the c-axis. As mentioned below, there are two sets of sites for hydrogen. For the case of $LaNi_5D_x$ ($x \leq 6$), the phase transition from $P6/mmm$ to $P31m$ takes place at a fixed composition of hydrogen.[22–24] This relation is tabulated in Table 3.2.

Figure 3.20 shows two sets of interstitial atomic sites for H (D) in a $CaCu_5$-type structure: one is marked by ■ (I_1, octahedral site, $z = 0$) and the other is marked by + (I_2, tetrahedral site, $z = \frac{1}{2}$). The numbers of each

Table 3.2

Phase transition of $LaNi_5H_x(D_x)$ from $P6/mmm$ to $P31m$

		$P6/mmm$	⟶	$P31m$
La	1(a)	$0\,0\,0$	1(a)	$0\,0\,z$
Ni_5	2(c)	$\frac{1}{3}\,\frac{2}{3}\,0;\ \frac{2}{3}\,\frac{1}{3}\,0$	2(b)	$\frac{1}{3}\,\frac{2}{3}\,z;\ \frac{2}{3}\,\frac{1}{3}\,z$
	3(g)	$\frac{1}{2}\,0\,\frac{1}{2};\ 0\,\frac{1}{2}\,\frac{1}{2};\ \frac{1}{2}\,\frac{1}{2}\,\frac{1}{2}$	3(c)	$x\,0\,z;\ 0\,x\,z;\ \bar{x}\,\bar{x}\,z$
D_1	3(f)	$\frac{1}{2}\,0\,0;\ 0\,\frac{1}{2}\,0;\ \frac{1}{2}\,\frac{1}{2}\,0$	3(c)	$x\,0\,z;\ 0\,x\,z;\ \bar{x}\,\bar{x}\,z$
D_2	6(m)	$x\,2x\,\frac{1}{2};\ \overline{2x}\,\bar{x}\,\frac{1}{2};\ x\,\bar{x}\,\frac{1}{2}$ $\bar{x}\,\overline{2x}\,\frac{1}{2};\ 2x\,x\,\frac{1}{2};\ \bar{x}\,x\,\frac{1}{2}$	6(d)	$x\,y\,z;\ y\,x-y\,z;\ y-x\,\bar{x}\,z$ $x\,y\,z;\ \bar{x}\,y-x\,z;\ x-y\,\bar{y}\,z$

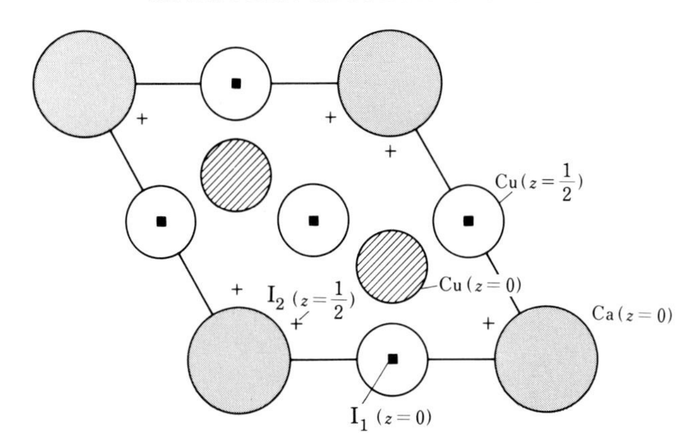

FIG. 3.20 Sites for H or D in the $CaCu_5$-type structure: I_1 (■) at $Z = 0$ and I_2 (+) at $Z = \frac{1}{2}$.

site in the unit cell are 3 and 6, respectively. As discussed later in detail, $LaNi_5$ can absorb hydrogen (deuterium*) up to $LaNi_5D_9$. Up to $LaNi_5D_{0.2}$, the structure has a symmetry of *P6/mmm*, and deuterium is sited randomly at octahedral positions. Around the composition $LaNi_5D_6$, the structure has a symmetry of *P31m*, and octahedral sites are fully occupied, also three out of six sites of I_2 are randomly occupied. The atomic positions of deuterium for the compound $LaNi_5D_6$ are given in Table 3.3.

Figure 3.21 shows a typical example of the absorption–desorption curves observed in the $CaNi_5$–H_2 system at 25 °C.[24] From this figure, the following

Table 3.3
Atomic parameters for $LaNi_5D_6$ (P31m)[a]

		x	*y*	*z*
La	1(a)	0	0	0
Ni	2(b)	$\frac{1}{3}$	$\frac{2}{3}$	0.94 (1)
Ni	3(c)	0.480 (0.5)	0	0.482 (0.50)
D_1	3(c)	0.470 (0.5)	0	0.077 (0)
D_2	6(d)	0.180	0.182	0.560

[a] Figures in parentheses correspond to those of *P6/mmm*.

* For structure determination by neutron diffraction, deuterium is used instead of hydrogen.[23] It is noted that the absorption–desorption curve for deuterium is slightly different from that of hydrogen (isotope effect; see Fig. 3.23).

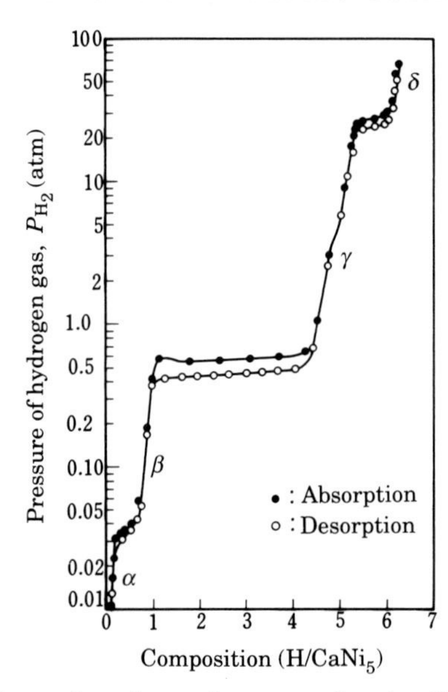

FIG. 3.21 Absorption–desorption curve for the $CaNi_5$–H_2 system at 25 °C. There are four phases, α, β, γ, and δ.[22]

results are obtained:

1. In this system, there appear the following four phases, which are non-stoichiometric.

α phase	$x < 0.2$
β phase	$0.75 < x < 1.2$
γ phase	$4.5 < x < 5.5$
δ phase	$6.0 < x < 6.2$

2. In the plateau region, which indicates the coexistence of two solid phases, hysteresis of the curve can be observed.

The temperature dependence of the plateau hydrogen pressure in the $(\alpha + \beta)$ region of the $CaNi_5$–H_2 system is shown in Fig. 3.22.[22] With increasing temperature, the plateau pressure increases.

As mentioned above, there exist two types of interstitial site for H: three equivalent sites with eight coordination (I_1) and six equivalent sites with four coordination (I_2) in a unit cell. Accordingly, it is likely that there are two non-stoichiometric compounds around the compositions $CaNi_5H_3$ and $CaNi_5H_9$. However, this is not observed for the $CaNi_5$–H_2 system. In this system, three phases appear at c. $x = 1, 5, 6$ (or 7). This result suggests that

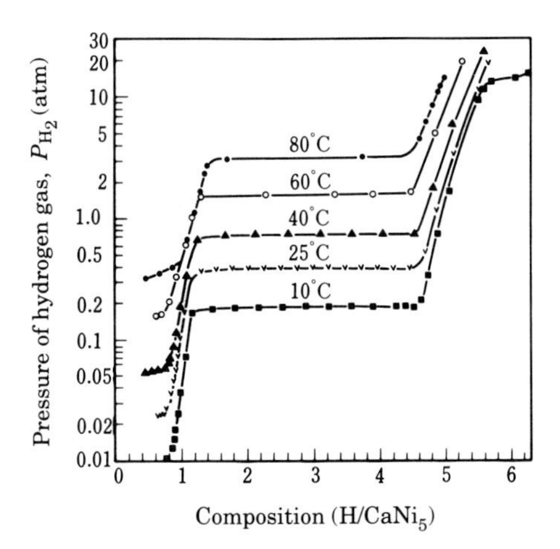

FIG. 3.22 Temperature dependence of the plateau pressure, P_T, for the $CaNi_5$–H_2 system.[22]

the real structure of this system is not as simple as that shown in Fig. 3.20. This may be due to the interaction among H–H, Ca–H, and Ni–H. In fact the composition at which a new phase appears is strongly dependent on the mother compound AB_5, where A = La, Ce, Pr, and Er, and B = Co and Ni.[24]

Figure 3.23 shows the isotope effect on hydrogen absorption of $CaNi_5$.[25] In the plateau regions, we can see a remarkable difference of gas pressure in the absorption curves between H and D, except for the $(\beta + \gamma)$ region. In the single phase regions, on the other hand, we cannot see an isotope effect, though the homogeneity region of each phase seems to depend on the isotopes. These facts have been explained by the entropy effect.[26]

The material $LaNi_5$, which is being applied to practical use in Europe, was found to be useful as a hydrogen absorbing material in the course of an investigation of hydrogen embrittlement on rare earth alloy magnets.[27,28] Figure 3.24 shows the absorption–desorption curves of the $LaNi_5$–H_2 system as a function of temperature.[29] In this system, there appear two phases α and β, where β is a non-stoichiometric compound having a stoichiometric composition $LaNi_5H_7$. These curves show hysteresis on absorbing and desorbing. From these curves, the enthalpy changes, ΔH, of absorption and desorption were found to be -6.76 and $6.56\ \text{kcal mol}^{-1}$. The smaller the value of ΔH, the more suited the material for practical use. On the other hand, the rate of absorption and desorption is another important factor in practical use. Figure 3.25 shows the relations between the quantity of absorbed hydrogen, hydrogen pressure, and time at various temperatures for the $LaNi_5$–H_2 system.[29] The absorption rate increases with increasing hydrogen pressure. Figure 3.26 also shows the relation between the quantity

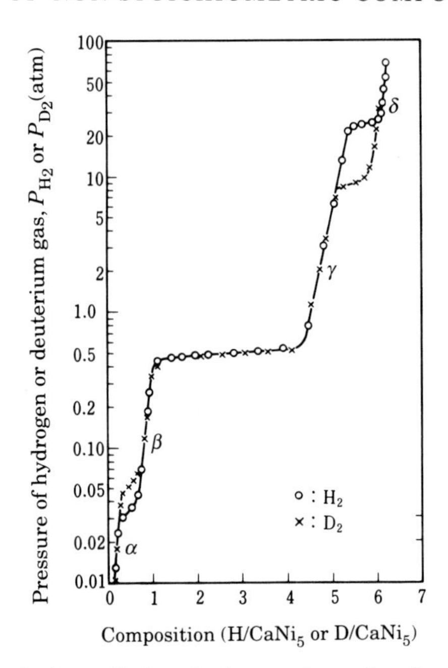

FIG. 3.23 Isotope effect on hydrogen absorption for the $CaNi_5$–H_2 system at 25 °C.[25]

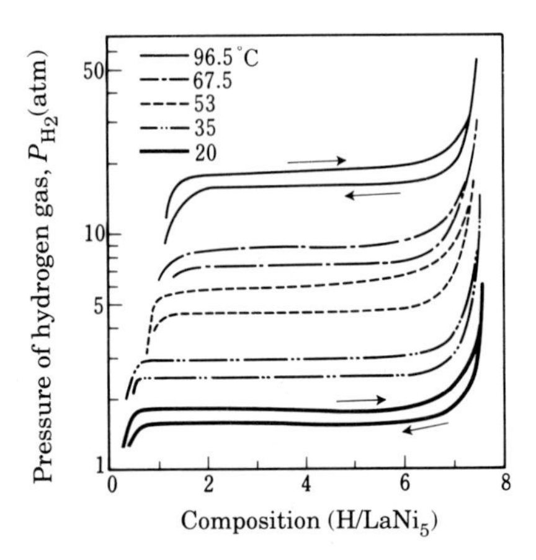

FIG. 3.24 Absorption–desorption curves for the $LaNi_5$–H_2 system as a function of temperature.[29]

of desorbed hydrogen, temperature, and time for this system[29] (the initial quantity of absorbed hydrogen is 200 cc g^{-1}).

Lakner *et al.*[24] studied the $LaNi_5$–$H_2(D_2)$ system for pressures as high as 1500 atm, as shown in Fig. 3.27 (see also Fig. 3.24). The highest hydrogen

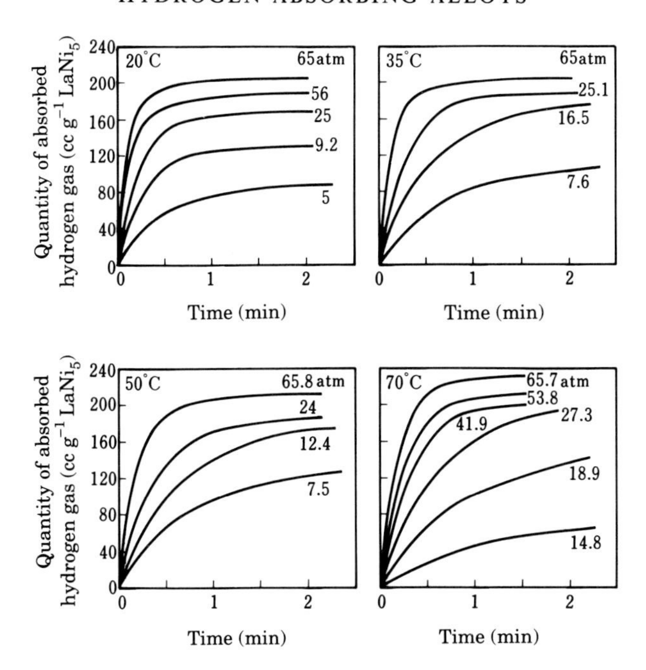

FIG. 3.25 Relation between the quantity of absorbed hydrogen, pressure, and time at various temperatures for the $LaNi_5$–H_2 system.[29]

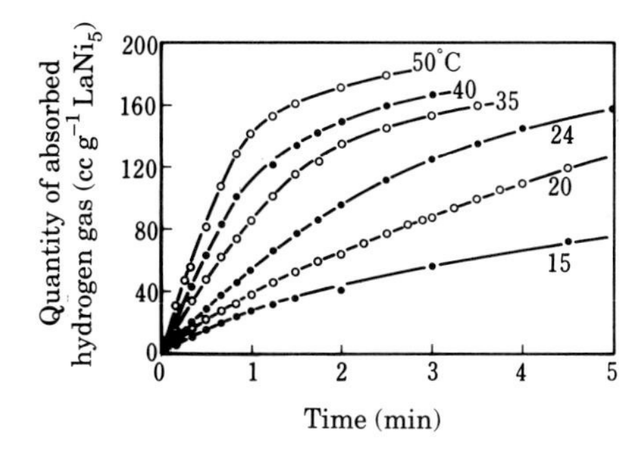

FIG. 3.26 Relation between the quantity of desorbed hydrogen, temperature, and time.[29] At the start, the quantity of absorbed hydrogen is 200 cc g^{-1} $LaNi_5$.

value gives up to $LaNi_5H_{8.35}$, which is very near the ideal composition of $LaNi_5H_9$. In the composition region $LaNi_5H_6$ to $LaNi_5H_8$, many phases appear to exist.

Practical use of hydrogen-absorbing metals can be classified into the

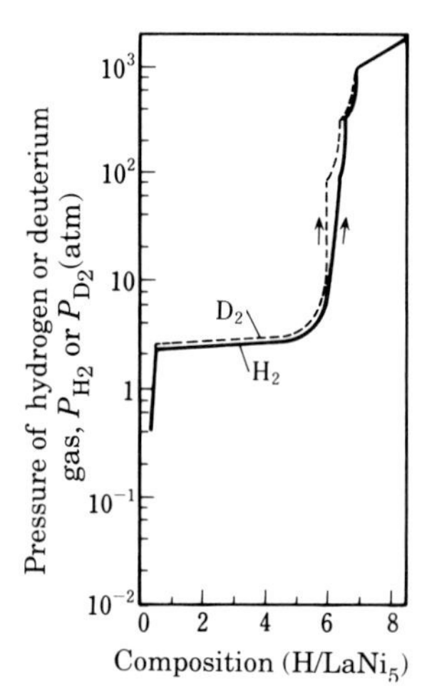

FIG. 3.27 Absorption curve for the $LaNi_5$–H_2 (D_2) system[24] (see also Fig. 3.24).

following types:

(1) hydrogen storage;

(2) enthalpy change, ΔH, of absorption and desorption;

(3) purification of H_2 gas;

(4) separation of H, D, and T.

Type (1) is the most popular use, the principles of which are studied above. The densities of hydrogen, ρ_H, in a hydrogen bottle and in various kinds of substances are listed in Table 3.4. This indicates that the alloys TiFe and $LaNi_5$ are useful from the point of view of hydrogen storage (the alloy VH_2

Table 3.4

Density of hydrogen in various forms

	Density of whole body	Density of hydrogen
H_2 gas (100 atm)	0.008	0.008
Liquid hydrogen	0.070	0.070
Water	1.000	0.111
VH_2	4.5	0.17
$TiFeH_{1.93}$	5.47	0.10
$LaNi_5H_6$	6.818	0.093

is not suitable for hydrogen storage, because the absorption–desorption characteristics are not appropriate). The value of ρ_H for $LaNi_5H_6$ is higher than that of liquid hydrogen, which has to be kept below 20 K. The value of ρ_H for $LaNi_5H_6$ is also more than 10 times that of a high pressure hydrogen bottle. Consequently the total weight of $LaNi_5H_6$ carrying 7 m^3 H_2 is $\frac{1}{5}$–$\frac{1}{3}$ that of a high pressure hydrogen bottle having a volume of 7 m^3, and the total pressure of gas is less than $\frac{1}{10}$ of that in the bottle.

There are problems in the practical use of hydrogen-absorbing metals such as $LaNi_5$. The first is that the absorption curve is strongly affected by impurity gases contained in hydrogen gas.[30] For example, if the hydrogen contains several per cent of CO_2 or H_2O this results in a decrease of absorbed quantity of 10–20 per cent, and hydrogen gas containing 1 per cent CO cannot be absorbed. This problem is of importance in the practical use of hydrogen-absorbing metals for hydrogen storage, because hydrogen gas used as a source of clean energy has to be of high purity. The second problem is that the hydrogen-absorbing metal is degraded due to hydrogen embrittlement, by cycling of absorption and desorption, and the desorbed hydrogen gas contains a fine powder of the metal, which is inconvenient for practical use. As mentioned above, the substance $LaNi_5$ was found to be a good hydrogen-absorbing metal in the course of an investigation into the effect of hydrogen gas on rare earth magnets.

The use of type (2) is to push the enthalpy change ΔH of the absorption or desorption cycle to practical use. This type of heat can be utilized as a heat pipe system, as the heating and cooling system used in solar heating, and as a heat exchange system. As mentioned above, the phenomena of absorption and desorption also accompany the decrease or increase of hydrogen pressure. This pressure change ΔP can be converted to mechanical energy, which can be used for freezing machines, compressors, and pumps.

As seen in Fig. 3.23, the absorption–desorption curves for H are different from those for D. This phenomena is used in types (3) and (4). By use of this phenomena, the separation of H and D, and enrichment of H and D from mixed gas are possible. The absorption–desorption curve for T (tritium) also differs from those for H and D;[31] thus we can separate and enrich H or D or T from the mixed gases by use of the absorption–desorption curves. D and T, which are used in nuclear reactors and nuclear fusion reactors, can be very efficiently separated and enriched by this principle.

Figure 3.28 shows the role of hydrogen-absorbing metals in the hydrogen energy system, proposed by Kitada.[32]

In the near future, hydrogen gas will undoubtedly play an important role in the energy media for effective use of primary energies. The production of high purity hydrogen gas, and the storage and transport of hydrogen gas are very important for this purpose.

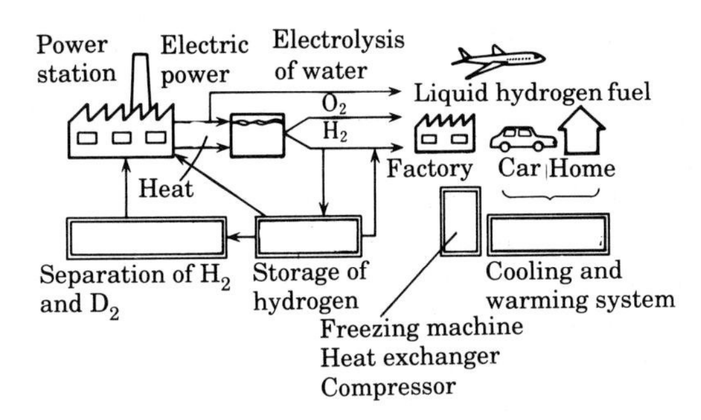

FIG. 3.28 Role of hydrogen-absorbing metals in hydrogen energy system.[32]

3.5 Electronic materials—phase diagram and crystal growth of GaAs

Semiconductive elements Si and Ge (Group IVB or **13** in the periodic table) have become very important electronic materials since development of a purification method. The electronic properties of semiconductive elements of high purity can be controlled by the species and concentration of defects and impurity elements. On the other hand, in the case of semiconductive compounds, that is, III–V and II–VI compounds, we have to consider not only control of the purity of constituent elements but also the non-stoichiometry, both of which have much influence on the electronic properties. In this sense, control of the electrical properties of semiconductive compounds is more difficult than that of semiconductive elements.

We shall now discuss the method of crystal growth and the electronic properties of GaAs, a typical example of a III–V compound which is expected to become more useful than Si and Ge in the near future, concentrating on the relation between non-stoichiometry and physical properties. GaAs has a zinc blende type structure, which can be regarded as an interpenetration of two structures with face centred cubic lattices, as shown in Fig. 3.29. Disregarding the atomic species, the structure is the same as a diamond-type

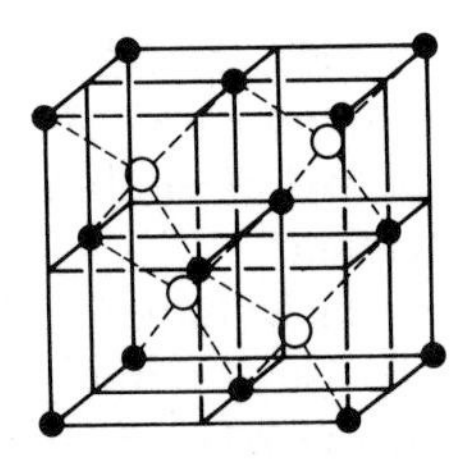

FIG. 3.29 Crystal structure of zinc blende (ZnS).

structure. Generally crystals with a diamond structure have covalent bonding, but crystals with the zinc blende structure exhibit ionic bonding due to the difference of electronegativity of each element. Therefore, GaAs is represented as $Ga^{-}As^{+}$.

GaAs is expected to be used for ICs (Integrated Circuit), FETs (Field Effect Transistor), LEDs (Light Emitting Diode), semiconductor lasers, and as the base material for IC. To put GaAs to practical use, the following points have to be examined in detail:

1. The purity of the starting elements Ga and As is important for practical use. The higher the purity, the better the practical use. At present, the highest purity of Ga and As are 7–9 and 6–9, respectively (the purity of 7–9 and 6–9 denotes 99.99999 and 99.9999 per cent, respectively).
2. In the growing process of single crystals of GaAs, the kind of impurities that dissolve easily into the crystal, and the impurity levels in the band structure for dissolved elements have to be made clear.
3. How to control the non-stoichiometry of GaAs, which is highly correlated to atomic and electronic defects.

A scheme showing the impurity levels of various elements in GaAs crystals is presented in Fig. 3.30.[33] The impurity levels below the centre of the forbidden band (Fermi level of the semiconductor) act as acceptors and those above the Fermi level act as donors, except O and Se. Ge and Si function as amphoimpurities.

The phase diagram of the Ga–As system is shown in Fig. 3.31,[34] which indicates the existence of the compound GaAs. In this figure, the GaAs phase is represented by a line, suggesting a line phase or stoichiometric compound. Straumanis and Kim[35] demonstrated that GaAs is a non-stoichiometric

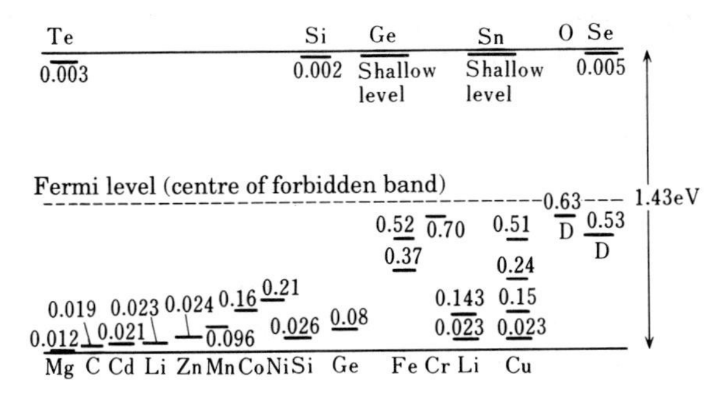

FIG. 3.30 Scheme of impurity levels of various kinds of element for GaAs crystals at room temperature.[33] Impurity levels above and below the centre of the forbidden band act as donors and acceptors, respectively.

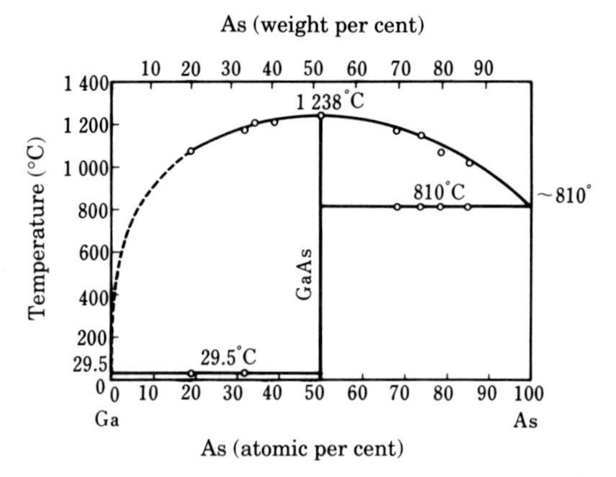

FIG. 3.31 Phase diagram of the Ga–As system.[34]

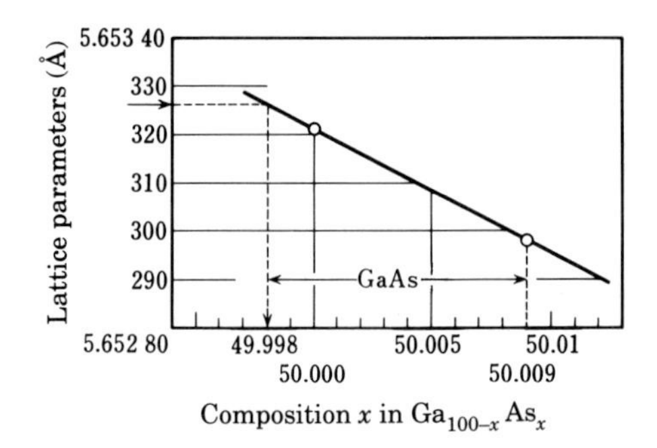

FIG. 3.32 Lattice parameter versus composition x in $Ga_{100-x}As_x$.[35]

compound, by accurate measurements of lattice parameter and density against nominal composition $Ga_{100-x}As_x$. The results are shown in Fig. 3.32, which suggests the one-phase region to be between $Ga_{50.002}As_{49.998}$ and $Ga_{49.991}As_{50.009}$, i.e. $-8.0 \times 10^{-5} \leq \delta \leq 3.6 \times 10^{-4}$ in the expression $GaAs_{1+\delta}$. The value of non-stoichiometry seems to be very small, but this non-stoichiometry substantially influences the electrical properties. For example, a value of $\delta = 3.6 \times 10^{-4}$ in the As excess region corresponds to a value of 8×10^{18} cm^{-3} for the defect concentration, assuming the excess As is cited at interstitials (As_i). This defect concentration is equivalent to that of low quality semiconductors. For the practical use of GaAs as a semiconductor, it is necessary to control the value of δ.

In Fig. 3.33 is shown the detailed phase diagram around the GaAs composition, which was constructed by Harris *et al.*[36] (this figures was estimated, not determined experimentally, from the analysis of electronic properties of GaAs crystal with various sources). It is important that

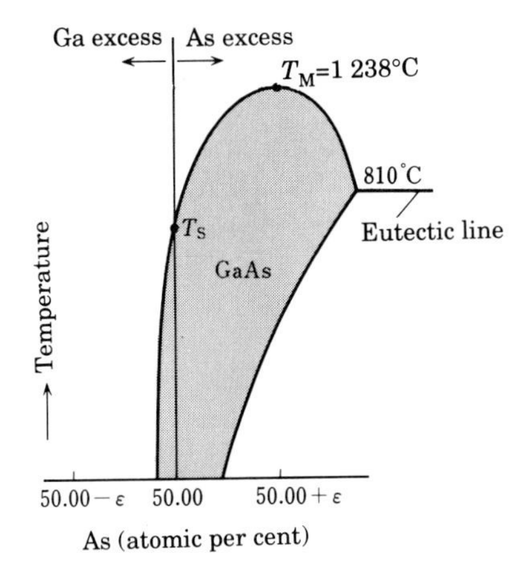

FIG. 3.33 Schematic drawing of phase diagram around GaAs composition.[36] T_S denotes the incongruent melting temperature for stoichiometric GaAs and T_M denotes the congruent melting temperature. The congruent melting composition is assumed to be in the As excess region.

stoichiometric GaAs shows an incongruent melting point at T_S, which is lower than T_M (congruent melting temperature). The congruent melting composition is in the As excess region. This figure shows that crystals grown at a higher solidification temperature than T_S have the excess As composition and those grown at a lower solidification temperature than T_S have the excess Ga composition.

Information on the relation of temperature–composition–pressure of As gas* is necessary in order to control the non-stoichiometry of GaAs. Figure 3.34 shows a schematic drawing of the phase relation of the Ga–As system, as a function of pressure of As gas (P_{As}). For example, P_{As} for stoichiometric GaAs depends on temperature. The relation between temperature, composition, and P_{As} has not yet been determined, this may be due to the narrowness of the homogeneity range of the GaAs phase. Boomgard and Schol[38] investigated the relation between temperature, phase, and P_{As} by the following method, which is a general method for controlling P_{As}. The device for this purpose is depicted in Fig. 3.35. A reaction vessel made of transparent quartz is placed in a two-zone furnace. The reaction container has a large block of carbon at one end, on which metal Ga is placed, and a reservoir for As at the other end. The temperature is controlled by the furnaces (I)

* The gas phase equilibrated with the solid phase of GaAs may contain various kinds of gas molecules of Ga and As. It was confirmed that As_2 and As_4 are the predominant gas molecules in this system.[37]

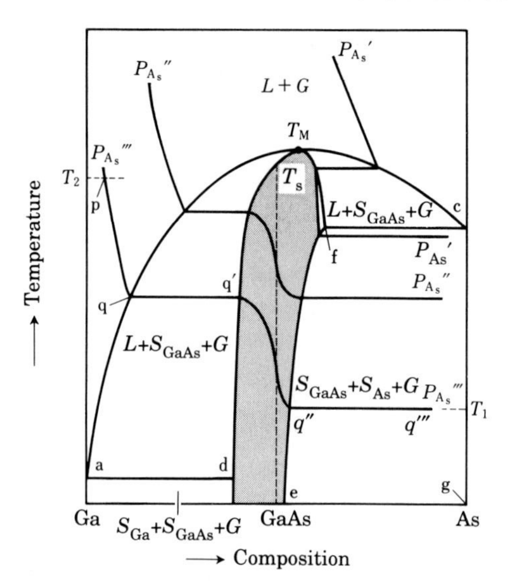

FIG. 3.34 Relation between composition (phase), temperature, and pressure of As gas P_{As} for the Ga–As system (schematic). In this figure, the isobars, P'_{As}, P''_{As}, . . . , are plotted.

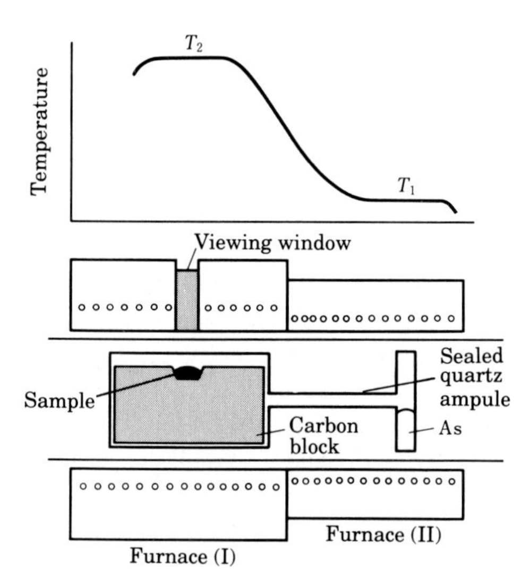

FIG. 3.35 Device for growing single crystal GaAs under controlled pressure of As gas.[38] The graph indicates the temperature distribution in the furnace.

and (II), so as to maintain T_1 and T_2 in each zone as shown in the figure ($T_2 > T_1$). The values of P_{As} equilibrated with solid As or liquid As were previously determined precisely as a function of temperature. Accordingly the P_{As} of this system can be controlled by the cool zone temperature T_1.

The experimental procedure adopted by Boomgard is as follows: First T_1 is controlled, which determines P_{As} in the whole closed system. For convenience of explanation, P_{As} at T_1 is assumed to be P'''_{As} in Fig. 3.34. Then Ga metal is heated to T_2 and the reaction vessel is placed under the controlled temperature gradient shown at the top of Fig. 3.35. At the equilibrium state, As dissolves in liquid Ga (point 'p' in Fig. 3.34). The hot zone (Ga metal zone) is then gradually cooled. The solubility of As into liquid Ga increases with decreasing temperature along curve p–q in Fig. 3.34. The point q (liquidus temperature) can be determined precisely by monitoring the precipitation of solid phase, which has the composition q′, from the viewing window. Thus the liquidus temperature was determined as a function of P_{As}. Then, by decreasing the temperature of the hot zone, a curve q′–q″–q‴ is obtained. A similar experiment has to be carried out to obtain a phase diagram.

Figure 3.36 shows the phase relation against inverse temperature ($1/T$) and P_{As}. The curve a–T_M–c, determined by this experiment, corresponds to that in Fig. 3.34. The curve c–g indicates the temperature dependence of

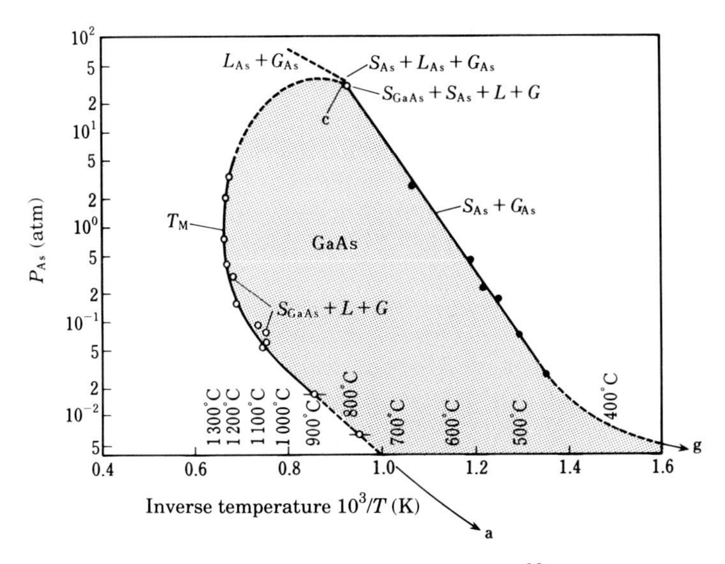

FIG. 3.36 P_{As} versus $1/T$ for the Ga–As system.[38] The shadowed area indicates the GaAs region; a ~ g correspond to those in Fig. 3.34.

P_{As} equilibrated with solid As. On the curve a–T_M, the three phases, i.e. liquid-Ga, solid-GaAs, and gas, coexist, i.e. the region a–T_M–d in Fig. 3.34 is represented by the curve a–T_M in this figure. From the phase rule there remains one freedom in the three-phase region for the binary system; therefore, either temperature or P_{As} is an independent variable for this case. Similarly, the region T_M–c–f in Fig. 3.34 is represented by the curve T_M–c in this figure and the region c–g–e–f in Fig. 3.34 is represented by the curve c–g. Diagrams such as that in Fig. 3.36 are called P–T diagrams, and have no information about composition. The existing region of solid GaAs is represented by the shadowed region a–T_M–c–g in this figure. For example, GaAs can exist at 800 °C in the pressure range from 8×10^{-3} to 3×10 atm of As gas. The composition of GaAs continuously changes with P_{As} and at some P_{As} the composition may be the stoichiometric one.

Using this knowledge on the phase relation of the Ga–As system, many types of methods for growing single crystals of GaAs have been proposed. The methods are classified into two types. In the first the composition or non-stoichiometry is controlled by reheating single crystals, which were grown without controlling As atmosphere, under an atmosphere of controlled As gas pressure. In the second single crystals are grown under controlled P_{As} in order to obtain a predetermined composition. The latter method is a more desirable practical producing process.

Before introducing experimental results for crystals grown by these methods, we shall consider the possible crystal defects of GaAs for a better understanding of experimental results. It is expected that at higher P_{As}, vacancies of Ga lattice sites or interstitial As may occur and at lower P_{As} vacancies of As or interstitial Ga may occur. Because GaAs is considered to be an ionic compound $Ga^{-}As^{+}$, these defects at higher P_{As} act as donors (n-type) and those at lower P_{As} act as acceptors (p-type). As shown below, the experimental results are not so simple.

In the first place, let us consider the experimental results for crystals grown by the former method. In Fig. 3.37 is shown the relation of carrier

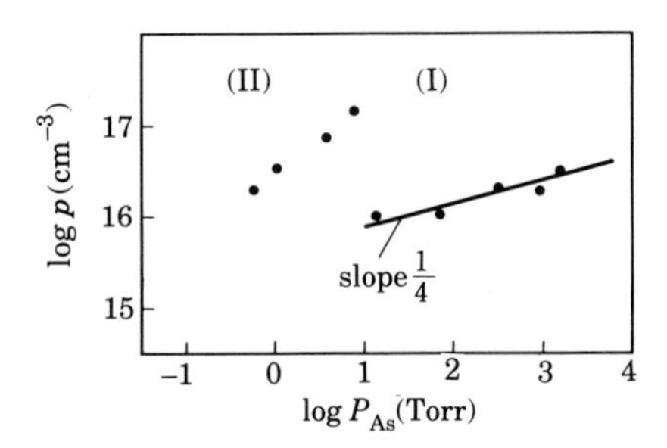

FIG. 3.37 Carrier concentration p versus pressure of As gas, P_{As}, for the crystals prepared by reheating undoped n-GaAs at 900 °C.[39]

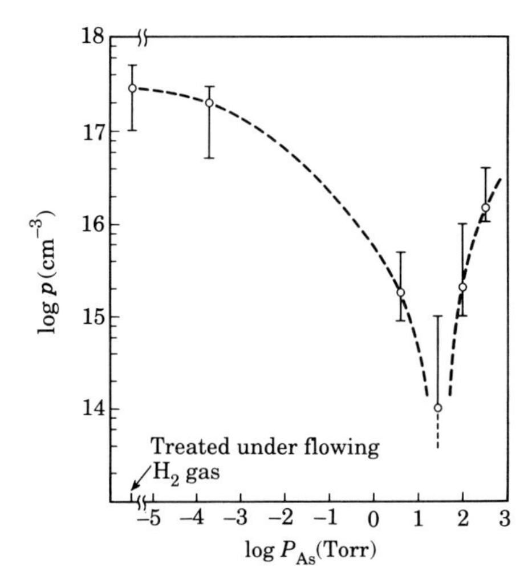

FIG. 3.38 Carrier concentration p versus pressure of As gas P_{As} for the crystals obtained by reheating undoped n-GaAs, which was prepared by the method of liquid phase epitaxy (LPE) at 850 °C.[40]

concentration, p, and P_{As} for crystals obtained by reheating undoped n-GaAs* crystals at various As gas pressures.[39] The method of controlling P_{As} is similar to that shown in Fig. 3.35. The carrier concentration n and carrier mobility μ_n of the starting GaAs, i.e. undoped n-GaAs, were 3.6–6.3 × 10^{16} cm^{-3} and 4700–5200 cm^2 V s^{-1}, respectively. The sign of the carrier is changed from n to p-type by prolonged heating under various pressures of As gas. In the pressure range $P_{As} > 1.43 \times 10$ torr (1 torr $= \frac{1}{760}$ atm) (region I), the hole concentration p is proportional to $P_{As}^{1/4}$, and in the pressure range $P_{As} < 10$ torr (region II), p shows a different dependence on P_{As}. In region I, the hole mobility, μ_p, equals c. 35–45 cm^2 V^{-1} s^{-1}, and in region II, it is 200–300 cm^2 V^{-1} s^{-1}. This indicates that in these two regions, there is an essential change in electrical properties, which may be due to the difference in crystal defects.

A similar experiment was performed on the crystals grown by the method of liquid phase epitaxy (LPE) as the starting crystal.[40] The starting crystals were reheated under various pressures of As gas at 850 °C for 30 min. Figure 3.38 shows the hole concentration, p, dependence on P_{As} at the surface of the heat-treated crystals. (The starting crystals showed n-type conduction, and after heat-treatment under this condition, only 7 μm of the surface of

* The electrical properties of semiconductive elements are usually controlled by impurity doping. For the case of a semiconductive compound, however, as-grown crystals usually show n-type or p-type characters without impurity doping. 'Undoped n-GaAs' denotes the as-grown crystal of GaAs showing n-type character without doping.

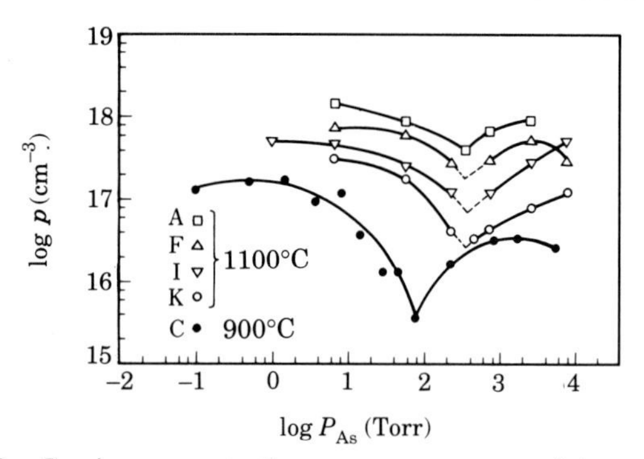

FIG. 3.39 Carrier concentration p versus pressure of As gas, P_{As}, for the crystals prepared by reheating Te-doped n-GaAs at 900 and 1100 °C.[42]

the crystal showed p-type conduction.) The result is similar to that shown in Fig. 3.37. In comparison with Fig. 1.67, which shows the relation between carrier concentration and equilibrium gas pressure for undoped and doped PbS, the pressure of $P_{As} \doteqdot 10$ torr seems to correspond to the stoichiometric GaAs composition.

Figure 3.39 shows the carrier concentration versus P_{As} curves for Te-doped n-GaAs treated at 900 and 1100 °C,[41,42] which is quite similar to that shown in Fig. 3.38. All the samples treated under various P_{As} values show p-type conduction. The value of P_{opt}, defined as the P_{As} showing a minimum value of carrier concentration, depends on the temperature of heat treatment. At P_{opt}, the crystal may be considered to have the stoichiometric composition, and if this were correct, the sign of carrier has to change from p to n at P_{opt}. However, this is not the case, this material shows p-type conduction in all the measured pressure range of As gas. Below and above P_{opt}, the main impurity level is considered to be an acceptor level, judging from the sign of the carrier. To explain this difference, the temperature dependence of the Hall coefficient was measured for the samples in both regions. The activation energy calculated from the measurement was found to be about 0.18 eV in both regions. Thus it was impossible to identify the origin of carrier concentration dependence on P_{As} for this system by this method.

Figure 3.40 shows the relation between the donor concentration, N_D, of the starting crystals and the acceptor concentration, N_A, of the re-treated crystals for various values of P_{As} for Te-doped n-GaAs.[42] This figure shows a linear relation between N_D and N_A, correctly $N_A \propto N_D^{3/4}$, which indicates that the defect acceptors formed by the heat-treatment are complex defects with the Te-donors.

In the pressure range $P_{As} < P_{opt}$, the defect may be considered to be an As vacancy, V_{As}. This defect combines with the Te-donors, leading to complex

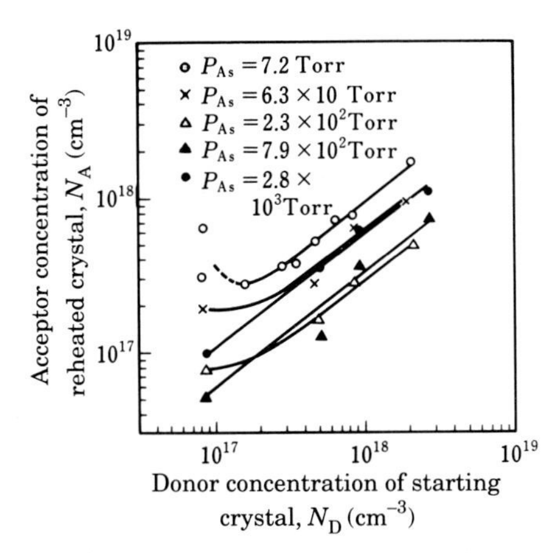

FIG. 3.40 Relation between donor concentration, N_D, of virgin Te-doped n-GaAs and acceptor concentration, N_A, of reheated Te-doped n-GaAs.[42]

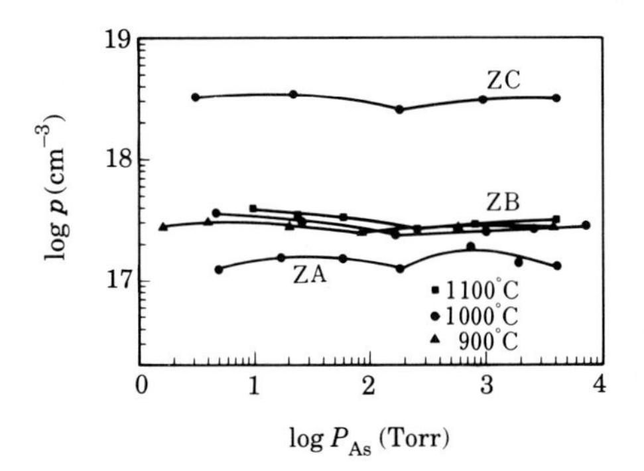

FIG. 3.41 Carrier concentration p versus pressure of As gas, P_{As}, for the crystals (ZA, ZB, ZC) prepared by reheating Zn-doped p-GaAs at 900, 1000, and 1100 °C.[42]

acceptor defects, Te-donor·V_{As}. In the pressure range $P_{As} > P_{opt}$, on the other hand, the defect may be considered to be a Ga vacancy, V_{Ga}, or interstitial As, As_i. These defects also combine with the Te-donors, leading to complex acceptor defects, Te-donor·V_{Ga} and Te-donor·As_i. These two acceptor levels may be very near in the forbidden band, judging from the measurement of Hall coefficient.

In Fig. 3.41 the carrier concentration p versus P_{As} for Zn-doped p-GaAs is shown, where ZA, ZB, and ZC denote the different starting crystals.[42] (For the case of Zn-doped GaAs, the starting crystals show p-type conduction (see Fig. 3.30), having $p \doteqdot 10^{17}$–10^{18} cm^{-3}.) This figure also suggests the existence of P_{opt}, although the P_{As} dependence of carrier concentration is very

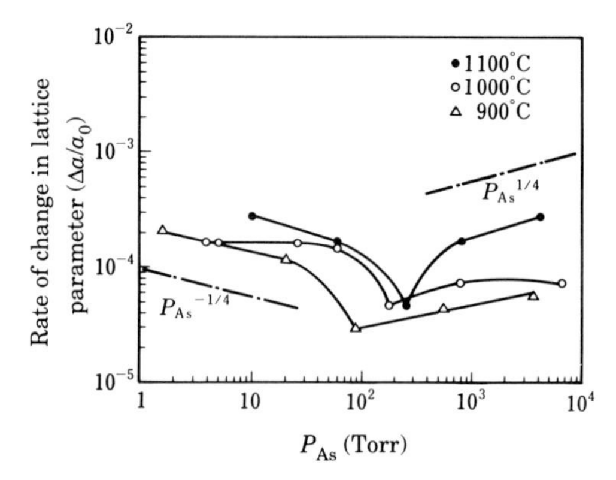

FIG. 3.42 Lattice parameter change ($\Delta a/a_0$) versus pressure of As gas, P_{As}, for the crystals prepared by reheating Zn-doped p-GaAs at 900, 1000, and 1100 °C.[42]

moderate, which may originate from the nature of the starting crystals (Zn-doped p-GaAs).

The rate of change of lattice parameter ($\Delta a/a_0$) with P_{As} is shown for the Zn-doped p-GaAs system in Fig. 3.42.[42] It clearly shows a similar anomaly at P_{opt} to that shown in Fig. 3.41. In the pressure region $P_{As} > P_{opt}$, the value of ($\Delta a/a_0$) decreases and in the pressure region $P_{As} < P_{opt}$ the value increases, with increasing P_{As}. This behaviour was also seen in the Te-doped n-GaAs system. The most important fact shown in Fig. 3.42 is that the lattice parameter takes a minimum value at P_{opt}, where the concentration of crystal defects seems to have a minimum value, i.e. the lattice parameter increases with increasing defect concentration in both sides of P_{opt}. The defects formed in Zn-doped p-GaAs are considered to be V_{As}, V_{Ga}, and As_i. If vacancies such as V_{As} and V_{Ga} were predominant in the whole P_{As} region, the density of the crystal ρ should take a maximum value at P_{opt}. Figure 3.43 shows the

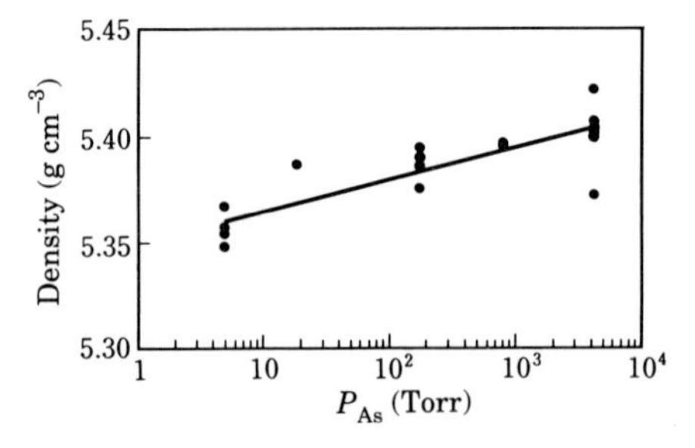

FIG. 3.43 Density versus pressure of As gas, P_{As}, for the crystals prepared by reheating Zn-doped p-GaAs at 1000 °C.[42]

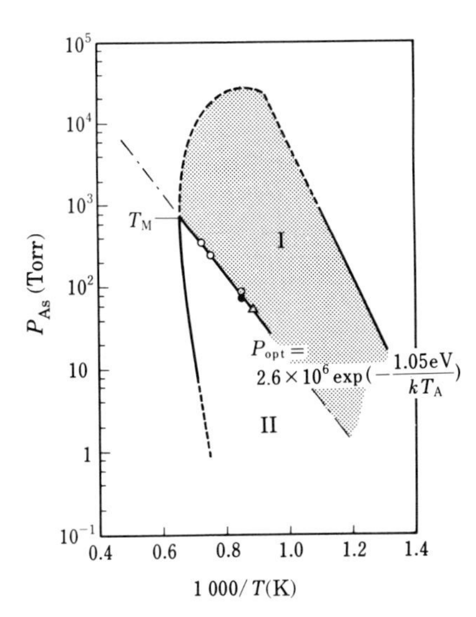

FIG. 3.44 Equilibrium As gas pressure, P_{As}, versus inverse temperature $1/T$ for the GaAs region (see Fig. 3.36). Equation (3.3) is also plotted. T_M is the congruent melting temperature.

ρ versus P_{As} curve for this system reheated at 1000 °C. This indicates a linear increase of ρ with increasing P_{As}. It seems likely that the vacancy model is adequate only in the pressure region $P_{As} < P_{opt}$. In the pressure region $P_{As} > P_{opt}$, interstitial As is considered to be predominant, and these As_i may cluster in a crystal, which was corroborated by the measurement of X-ray anomalous transmission.[41,42]

It has, therefore, been confirmed that the non-stoichiometry and electrical properties of GaAs can be controlled by heat-treatment under various pressures of As gas. This method is simple in principle, but is not applicable to practical processes. By these experiments[39,41,42] the temperature dependence of P_{opt} was found to be

$$P_{opt} = 2.6 \times 10^6 \exp -[1.05(\mathrm{eV})]/[kT_A] \quad (\mathrm{torr}) \tag{3.3}$$

where T_A denotes the annealing temperature. This relation is common in undoped and doped GaAs. Equation (3.3) is plotted in Fig. 3.44.[42] This line divides the GaAs region into two regions, I and II. In region I, the composition of As is in excess and in region II, the composition of Ga is in excess. It is to be noted that the congruent melting temperature T_M lies just on this line. This is in conflict with the phase diagram (Fig. 3.33) proposed by Harris *et al.*[36] If the stoichiometric composition was coincident with the congruent melting composition, the growth of a single crystal having

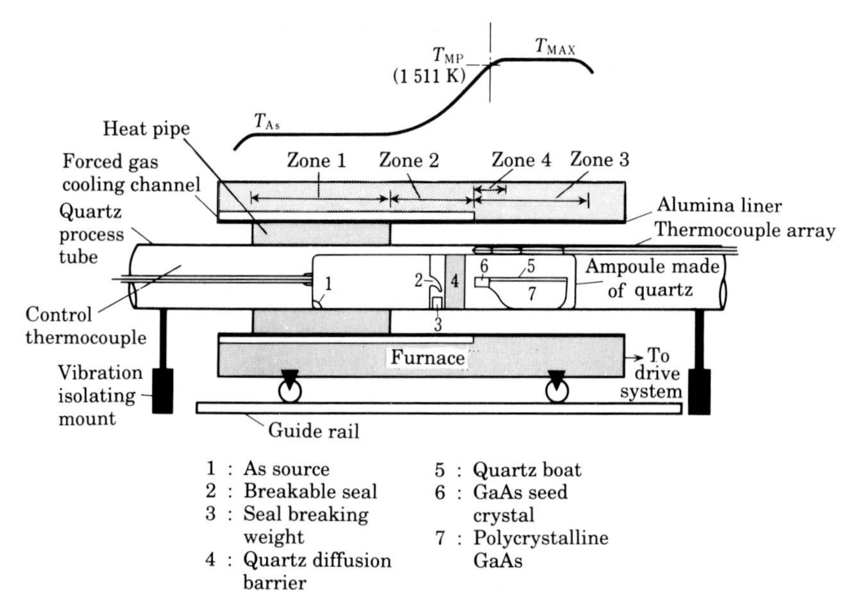

FIG. 3.45 Horizontal Bridgman apparatus for growing GaAs single crystals.[43]

stoichiometric composition should become rather easy. At present, it is not clear which is correct.*

We now describe the method of crystal growth combined with *in-situ* control of P_{As} for obtaining a predetermined composition. The methods of crystal growth for GaAs are not special in principle, they are classified into two types, Bridgman and Czochralski methods. In the Bridgman method, two types, horizontal and vertical, are usually adopted. We here consider an example of the former type. In the growth of GaAs single crystals from the melt, it is necessary to control a number of parameters such as As gas pressure, thermal gradients in the grown crystal and the melt, melt composition (the ratio of Ga to As), and growth rate. To meet the requirements of independent control of these parameter, a unique Bridgman apparatus was designed and constructed as shown in Fig. 3.45 by Parsey *et al.*[43] This apparatus has the following salient features:

1. To attain control over the thermal profiles, four independent resistance heaters are utilized.

* As is obvious, if the stoichiometric GaAs shows a congruent melting, the composition of the melt, from which a solid crystal is grown, has to be coincident with that of the grown crystal, i.e. stoichiometric. This seems not to be the case (see Figs 3.50 and 3.51).

2. To control the gas pressure of As precisely, a sodium heat pipe and an internal cooling gas distribution system are installed in the cool zone, which serve to establish a flat thermal profile and enhanced temperature stability.

3. In a typical Bridgman method, a viewing window is used for seeding the crystal and visual monitoring of the growth process, as shown in Fig. 3.35. The presence of a window, however, creates an asymmetric thermal field in the solid–liquid interface region, leading to convection in the melt and to generation of stress in the growing crystal. To facilitate seeding without a viewing window, thermal monitoring and control of the pertinent thermal characteristics are utilized to position the solid–liquid interface.

Thermal uniformity in the cold zone was found to be from 0.01 to 0.02 °C, and that in the hot zone was found to be better than ± 0.5 °C vertically and ± 0.1 °C horizontally. Thermal gradients near the solid–liquid interface were achieved in excess of 30 °C cm^{-1} in the crystal region and up to 20 °C cm^{-1} in the melt. The growth of crystals was performed in a sealed transparent silica ampoule, which has two rooms for As source and GaAs polycrystalline, respectively, separated by a quartz diffusion barrier. For details of the growth process the reader is referred to Ref. 43. In this experiment the As source temperature T_{As} was systematically reduced by 2 °C at 3 h intervals from 620 °C to 614 °C.

The effect of P_{As} on the dislocation density is shown in Fig. 3.46.[43] It can be concluded that there is an optimum P_{As} or T_{As} (T_{opt}) ($\doteqdot 617$ °C) for growing low dislocation density (less than 500 cm^{-2}) crystals (at the time this work was carried out, the dislocation density of commercial GaAs with the highest quality was more than 2×10^3 cm^{-2}). Above and below this

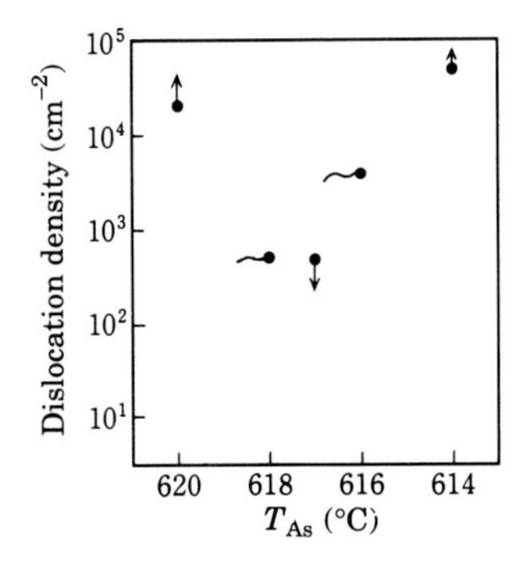

FIG. 3.46 Dislocation density versus As source temperature, T_{As}, for GaAs crystals grown by the Bridgman method.[43] (Marks ↑, ↓ and ~ denote more than, less than, and nearly equal, respectively).

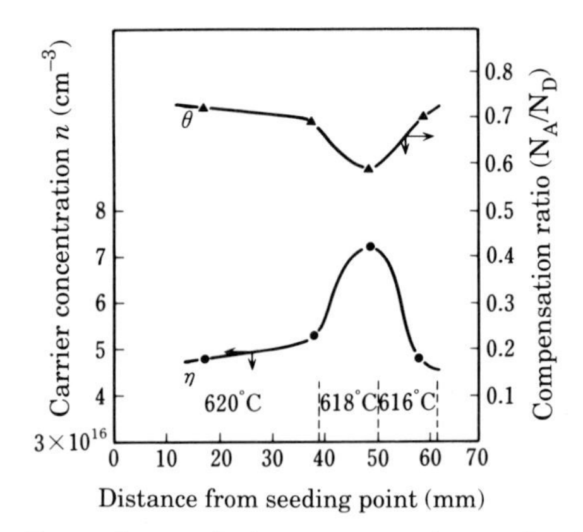

FIG. 3.47 Dependence of the compensation ratio θ and free carrier concentration n ($=N_D - N_A$) on As source temperature, T_{As}, for GaAs crystals grown by the Bridgman method.[43]

temperature, the dislocation density increases drastically. A crystal grown at $T_{As} = 617$ °C can be regarded as a stoichiometric crystal, because the dislocations formed in the compounds mainly originate from non-stoichiometry or crystal defects. (In Fig. 3.36, there must be an incongruent melting temperature T_S for the stoichiometric composition on the curve T_M–a, and the value of P_{As} at $T_{As} = 617$ °C must equal that at T_S, i.e. a line $T_{As}T_5$ is parallel to the abscissa.)

The carrier concentration n ($=N_D - N_A$) and the compensation ratio θ (N_A/N_D), which were obtained by measurements of the Hall coefficient and carrier mobility, respectively, were found to be functions of T_{As} or P_{As}. The compensation ratio θ exhibits a minimum and the carrier concentration n a maximum, at the optimum temperature 617 °C, shown in Fig. 3.47.[43] From the compensation ratio and the carrier concentration, the concentrations of the ionized donors, N_D, are calculated as a function of T_{As}, as shown in Fig. 3.48.[43] At the optimum temperature 617 °C, N_D shows a maximum and N_A a minimum. The total concentration of ionized impurities, $N_{imp} = N_A + N_D$, remains essentially constant in the measured P_{As} range. This result suggests that vacancy-related mechanisms are associated with the formation of dislocations during growth and the compensation process. Thus, high quality crystal GaAs with a low density of dislocations has been grown by precise control and high stability of P_{As} or T_{As}.

The following two points are to be noted. The first is that, as shown in Fig. 3.45, the T_{MP} equals the congruent melting temperature T_M, T_{MP} being the temperature where the melt is coexistent with the solid phase. The composition of the melt, C_{melt}, which decides the composition of grown crystals, changes continuously with changing T_{As}. Only at C_{melt} = composition

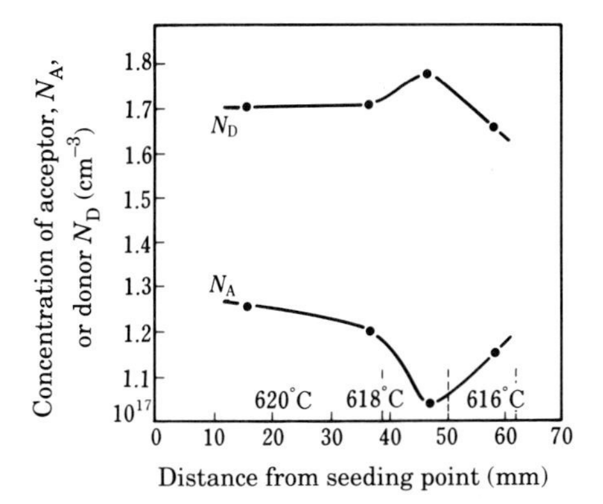

FIG. 3.48 Donor, N_D, and acceptor, N_A, concentration versus As source temperature T_{As} for GaAs crystals grown by the Bridgman method.[43]

of T_M, the composition of grown crystals has the same composition as C_{melt}. In the other composition of C_{melt}, the composition of grown crystals differs from C_{melt}, depending on liquidus curves in Fig. 3.34. In any case, the composition of crystals grown in the condition of source temperature $T > T_{opt}$ has to be excess in As and that grown in the condition of source temperature $T < T_{opt}$ has to be excess in Ga. Accordingly the sign of change carrier must change from n to p or p to n at T_{opt}. However, experimentally the sign of the charge carrier never changes through T_{As}.

The second point is that the composition of the crystal is controlled mainly by the composition of the liquid phase equilibrated with the solid phase. A crystal grown at the interface of the solid and liquid is cooled under the same P_{As}, but the composition of the grown crystal does not change during this process except at the surface of the crystal. This may be due to the difference of reaction rate between the gas–liquid and gas–solid phases, i.e. the reaction rate between gas and liquid is faster than that between gas and solid.

We next describe the Czochralski method of single crystal growth for GaAs. A single crystal is grown by pulling a seed crystal from the melt, as schematically shown in Fig. 3.49. This method is essentially similar to the vertical Bridgman method. (Usually a seed crystal is not used in the Bridgman method, but in the growth of GaAs as mentioned above, a seed crystal was used in order to direct the growth axis of the crystal.) The growing crystal and crucible are rotated around the growth axis to homogenize the composition of the melt and the distribution of temperature. For the case of a high pressure gas phase such as GaAs, the following apparatus was devised.

In order to avoid the vaporization of the melt of GaAs and also to control

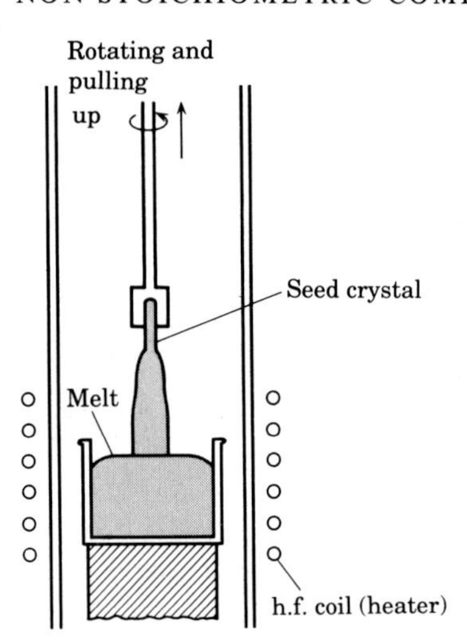

FIG. 3.49 Schematic drawing of the Czochralski method.

the P_{As} the melt is covered by a melt of B_2O_3 (melting point: 450 °C, boiling point: 1500 °C).* A rotating bar with a seed crystal goes through the melt of B_2O_3. In order to avoid the vaporization of melted B_2O_3 and As from the crystal, the vessel is filled with an inert gas such as Ar up to 20–60 atm. This method is called the liquid encapsulated Czochralski (LEC) method, and has been put on the market as LEC apparatus (in this the crucible is made of pyrolytic boron nitride, and Si from the crucible, or the ampoule, made of transparent silica, dissolves into the GaAs crystals giving the impurity donor level). By the LEC method the composition (ratio of As to Ga) of the crystal is controlled by choosing the composition of the melt of GaAs. It is clearly demonstrated in Fig. 3.34 that the liquidus temperature is a function of the melt composition. Hence the melt composition controls the composition of the crystal and also the solidification temperature.

Holmers *et al.*[44] investigated the electrical properties of GaAs crystals grown by the LEC method. They grew single crystals as a function of the melt composition, i.e. As/Ga = 0.43–0.54. The value of elctrical resistivity ρ (ohm cm) at room temperature is shown against the melt composition in Fig. 3.50.[44] In the Ga excess region, crystals show a p-type conduction and have a constant value of $\rho = 1$–10 ohm cm, at As/Ga = 0.475 (called the critical composition of As) ρ shows a distinct change to 10^7–10^8 ohm cm with n-type conduction. The region As/Ga > 0.475 is called the semi-

* The system can be a closed system because it is covered with the melt of B_2O_3. In the closed system, the activity of As or P_{As} is automatically fixed by fixed temperature and melt composition.

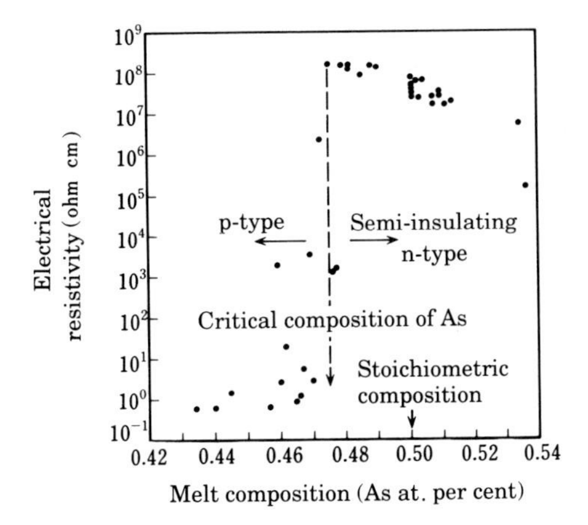

FIG. 3.50 Dependence of electrical resistivity ρ on the melt composition for GaAs crystals grown by the Czochralski method.[44]

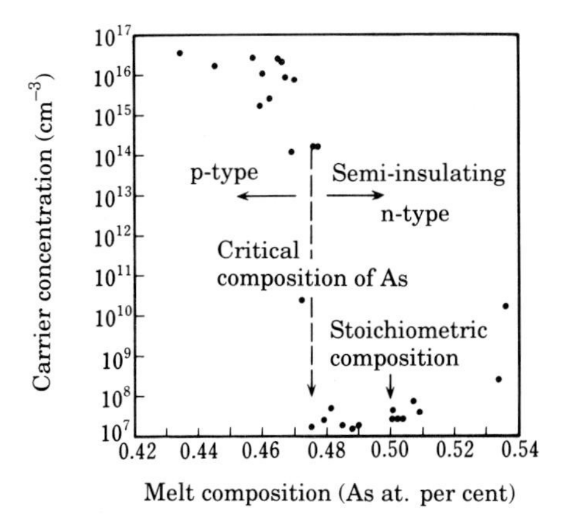

FIG. 3.51 Dependence of carrier concentration on the melt composition for GaAs crystals grown by the Czochralski method.[44]

insulating region. (Crystals having a semi-insulating character can also be grown by the horizontal Bridgman method. In that case, Cr metal is doped as acceptor to compensate for Si donors.) Figure 3.51 shows the relation between the carrier concentration and the melt composition.[44] The carrier concentration shows a sharp jump from 10^{16}–10^{17} cm^{-3} (p-character) to 10^{7}–10^{8} cm^{-3} (n-character) at the critical composition of As. This figure indicates that the change in ρ corresponds to a change in carrier concentration. Hall mobility does not show a clear jump at the critical concentration of As, as shown in Fig. 3.52.[44] Thus, GaAs crystals with semiconductive

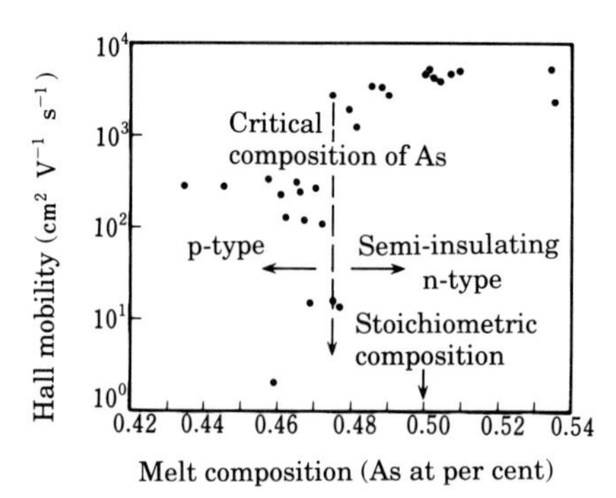

FIG. 3.52 Dependence of Hall mobility on the melt composition for GaAs crystals grown by the Czochralski method.[44]

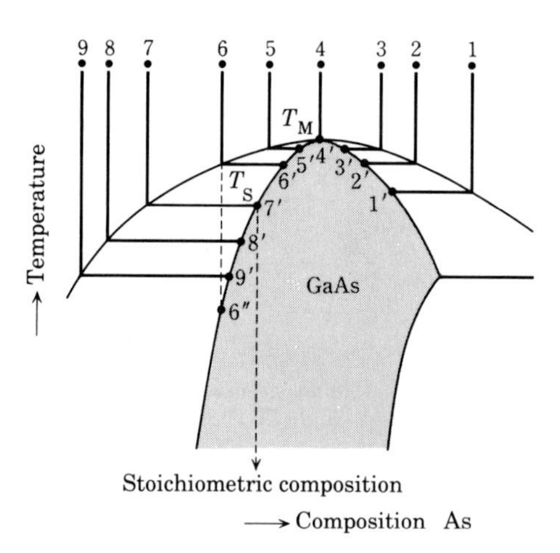

FIG. 3.53 Relation between the composition of grown crystal and the melt composition, referred to as the phase diagram of the Ga–As system.

p-character and with semi-insulating n-character were grown by controlling the melt composition. Semi-insulating GaAs is a promising material for the base of IC.

Let us consider the relation between the composition of the crystal and its electrical properties based on the phase diagram. Figure 3.53 shows a schematic drawing of the phase diagram of the Ga–As system around the GaAs phase, assuming that the incongruent melting temperature T_S for stoichiometric GaAs is lower than the congruent melting temperature T_M and the congruent melting composition is in the As-rich region. In the figure, numbers 1, 2, . . . , 9 denote the composition of the melt and 1′, 2′, . . . , 9′ the corresponding composition of grown crystal. As can be seen, the composition of the melt does not coincide with the composition of crystal except for the congruent melting composition 4. Under ideal conditions, that is, the amount of melt is infinite and the temperature of the liquid–solid interface (liquidus temperature) is strictly controlled,* the composition of grown crystals must be constant from the top to the end of the crystal. Then it follows that stoichiometric GaAs 7′ is grown from melt 7. We will assume that melt composition 7 is As/Ga = 0.475, taking Holmers' result into consideration.[44] Accordingly, crystals grown from the melt having a composition richer in Ga than 7 should show p-type character and those from the melt having a composition richer in As than 7 should show n-type character. This was proved to some extent by Holmers' experiment.[44]

The above assumptions that the amount of melt is infinite and that the temperature of the liquid–solid interface is strictly controlled are not realistic, and the following phenomena are observed experimentally. Crystals grown from the melt having a composition richer in As than 4 (congruent melting composition) show a continuous increase in As content from the top to the end of the crystal and n-type conduction. Crystals grown from the melt having a composition richer in Ga than 7, on the other hand, show a continuous increase in Ga content from the top to the end of the crystal and p-type conduction. Let us consider a crystal grown from melt 6. The composition of the crystal must show the change as $6' \rightarrow 7' \rightarrow 8' \rightarrow 9' \rightarrow 6''$ from the top to the end, accompanied by a change in the sign of carrier from n to p. This was experimentally also observed by Holmers *et al.*[44] They concluded that the reason why crystals with n-type and p-type character can be grown by choosing the composition of the melt is that the non-stoichiometry of the crystal is exactly controlled by this method, not that the change in impurity concentration is controlled.

The fact that n-type crystals thus grown are semi-insulating cannot be explained from the viewpoint of the phase diagram. The semi-insulating phase is regarded as a pseudo-intrinsic semiconductor, i.e. the concentration of free carriers is very low, due to the carrier compensation in some sense. Holmers *et al.*[44] have concluded from their data that the concentration of free carrier called 'EL2', N_D, is compensated for by that of acceptors derived from impurity carbon, N_A. Ta *et al.*[45] carried out a similar investigation independently and reached the same conclusion.

The Bridgman and Czochralski (LEC) methods as mentioned above, are useful for growing GaAs single crystals, from the viewpoint of practical applications. The following methods tried or proposed by many workers are

* It is also assumed that the composition of the grown crystal does not change in the cooling process. As is well known for the case of finite quantity of melt, the composition of melt (liquid) and solid change along the liquidus and solidus curves, respectively.

also of interest:

(1) zone melting method;

(2) chemical transport method (epitaxial);

(3) liquid phase electro-epitaxy method.

References

1. S. Sekido, *Chemistry*, 1983, **38**, 406 (in Japanese).
2. S. Sekido, *Modern Chemistry*, 1981, **12**, 28 (in Japanese).
3. J. Smit and H. P. Wijn, *Ferrites*, J. Wiley and Sons, New York, 1959.
4. P. I. Slick, *Proc. Int. Conf. Ferrites—1*, 1971, 81.
5. Y. Shichijo, *Trans. JIM*, 1961, **2**, 204.
6. P. Beyer, *Phys. Status Solidi*, 1971, **A6**, K55.
7. T. Akashi, I. Sugano, T. Okuda, and T. Tsuji, *Proc. Int. Conf. Ferrites—1*, 1971, 337.
8. Y. Shichijo and E. Takama, *Proc. Int. Conf. Ferrites—1*, 1971, 210.
9. T. Yamada, Y. Shimizu, and T. Ito, *IEEE Trans. Mag. Mag.*, 1975, **Mag-11**, 227.
10. T. Tanaka, Study on high-dense Mn–Zn Ferrite prepared by controlling the oxygen non-stoichiometry, Thesis, Keio University, 1981.
11. T. Tanaka, *Jap. J. Appl. Phys.*, 1974, **13**, 1235.
12. T. Tanaka, *Jap. J. Appl. Phys.*, 1975, **14**, 153.
13. T. Tanaka, *Jap. J. Appl. Phys.*, 1975, **14**, 1169.
14. T. Tanaka, *Jap. J. Appl. Phys.*, 1978, **17**, 349.
15. T. Tanaka, *J. Jap. Soc. Powder and Powder Metallurgy*, 1978, **25**, 26 (in Japanese).
16. T. Tanaka, *J. Am. Ceram. Soc.*, 1981, **64**, 419.
17. H. Bruning and A. Sieverts, *Z. Physik. Chem.*, 1933, **163**, 409.
18. J. R. Lacher, *Proc. R. Soc., London*, 1937, **A161**, 525.
19. J. S. Anderson, *Proc. R. Soc., London*, 1946, **A185**, 69.
20. G. G. Libowitz, *J. Appl. Phys.*, 1962, **33**, 399.
21. J. W. Simons and T. B. Flanagan, *Can. J. Chem.*, 1965, **43**, 1665.
22. T. B. Flanagan and W. A. Oates, *Ber. Bunsenges. Phys. Chem.*, 1972, **76**, 706.
23. P. Fisher, A. Furrer, G. Busch, and L. Schlapbach, *Helv. Phys. Acta*, 1977, **50**, 421.
24. J. F. Lakner, F. S. Uribe, and S. A. Steward, *J. Less-Common Met.*, 1980, **72**, 87.
25. G. D. Sandrock, J. J. Murray, M. L. Post, and J. M. Taylor, *Mat. Res. Bull.*, 1982, **17**, 887.
26. G. G. Libowitz, *The solid state chemistry of binary metal hydrides*, Benjamin, New York, 1965.
27. J. H. N. van Vucht, F. A. Kuijpers, and H. C. A. M. Bruning, *Philips Res. Rept.*, 1970, **24**, 133.
28. F. A. Kuijpers, *Philips Res. Rep. Suppl.*, 1972, No. 2, 20, 28, 48.
29. M. Kitada, *J. Jap. Inst. Metals*, 1977, **41**, 412 (in Japanese).
30. J. J. Reilly, *The 7th Int. Energy Conv. Eng. Conf.*, San Diego, 1972, BNL-16889.
31. R. H. Wismal, *Inorg. Chem.*, 1972, **11**, 1691.

32. M. Kitada, *Bull. Jpn Inst. Met.*, 1978, **17**, 345 (in Japanese).
33. S. M. Sze and J. C. Irrin, *Solid State Electron.*, 1968, **11**, 599.
34. W. Koster and B. Thoma, *Z. Metallk.*, 1955, **46**, 291.
35. M. E. Straumanis and C. D. Kim, *Acta Crystallogr.*, 1965, **19**, 256.
36. J. S. Harris, Y. Nannichi, and G. L. Pearson, *J. Appl. Phys.*, 1969, **40**, 2257.
37. J. R. Arthur, *J. Phys. Chem. Solids*, 1967, **28**, 2257.
38. J. van der Boomgard and K. Schol, *Philips Res. Rep.*, 1957, **12**, 127.
39. H. Otsuka, K. Ishida, and J. Nishizawa, *Jap. J. Appl. Phys.*, 1969, **8**, 632.
40. E. Munoz, W. L. Snyder, and J. L. Moll, *Appl. Phys. Lett.*, 1970, **16**, 262.
41. J. Nishizawa, H. Otsuka, S. Yamakoshi, and K. Ishida, *Jap. J. Appl. Phys.*, 1974, **13**, 46.
42. J. Nishizawa and K. Sudo, *J. Crystallogr. Soc. Jpn*, 1976, **3**, 15.
43. J. M. Parsey, Y. Nanishi, J. Lagouski, and H. C. Gatos, *J. Electrochem. Soc.*, 1982, **129**, 388.
44. D. E. Holmers, R. T. Chen, K. R. Elliot, C. G. Kirkpatrick, and P. W. Yu, *IEEE Trans. Electron Device*, 1982, **ED-29**, 1045.
45. L. B. Ta, H. M. Hobgood, A. Rohatgi, and R. N. Thomas, *J. Appl. Phys.*, 1982, **53**, 5771.

SUBJECT INDEX

FORMULA INDEX